TECHNIQUES AND INSTRUMENTATION IN ANALYTICAL CHEMISTRY – VOLUME 1

Evaluation and Optimization of Laboratory Methods and Analytical Procedures

A Survey of Statistical and Mathematical Techniques

TECHNIQUES AND INSTRUMENTATION IN ANALYTICAL CHEMISTRY

Volume 1 **Evaluation and Optimization of Laboratory Methods and Analytical Procedures. A Survey of Statistical and Mathematical Techniques**
by D.L. Massart, A. Dijkstra and L. Kaufman

TECHNIQUES AND INSTRUMENTATION IN ANALYTICAL CHEMISTRY – VOLUME 1

Evaluation and Optimization of Laboratory Methods and Analytical Procedures

A Survey of Statistical and Mathematical Techniques

Desiré L. Massart
Farmaceutisch Instituut, Vrije Universiteit Brussel

Auke Dijkstra
Laboratorium voor Analytische Chemie, Rijksuniversiteit Utrecht

Leonard Kaufman
Centrum voor Statistiek en Operationeel Onderzoek, Vrije Universiteit Brussel

with contributions by **Svante Wold, Bernard Vandeginste** and **Yvette Michotte**

ELSEVIER SCIENTIFIC PUBLISHING COMPANY
Amsterdam – Oxford – New York 1978

ELSEVIER SCIENTIFIC PUBLISHING COMPANY
335 Jan van Galenstraat
P.O. Box 211, 1000 AE Amsterdam, The Netherlands

Distributors for the United States and Canada:

ELSEVIER/NORTH-HOLLAND INC.
52, Vanderbilt Avenue
New York, N.Y. 10017

First edition 1978
Second impression 1980

Library of Congress Cataloging in Publication Data

Massart, D L
Evaluation and optimization of laboratory methods and analytical procedures.

(Techniques and instrumentation in analytical chemistry ; v. 1)
Bibliography: p.
Includes index.
1. Chemistry, Analytic--Statistical methods.
I. Dijkstra, Auke, joint author. II. Kaufman, Leonard, joint author. III. Title. IV. Series.
QD75.4.S8M37 543 78-12667
ISBN 0-444-41743-5

ISBN 0-444-41743-5 (Vol. 1)
ISBN 0-444-41744-3 (Series)

Printed in The Netherlands

CONTENTS

Errata

Page 99, 2nd equation, 3426^2 should read 3436^2
Page 536, 2nd line, $2\pi\omega_0$ should read $\omega_0/2\pi$

3rd line, $\sum_{n=1}^{\omega_m/\omega_0}$ should read $\sum_{n=-\omega_m/\omega_0}^{\omega_m/\omega_0}$

Page 545, Fig. 27.1, e^{t/τ_x} (ordinate) should read e^{-t/τ_x}
Page 549, 4th line, 3% should read 1%

INTRODUCTION

One of the major problems with which the analytical chemist is confronted is how to make the best possible use of the large amount of information on analytical principles, methods and procedures that is available in the analytical literature. As stated by Laitinen and Harris (1975), an analytical chemist can be judged in part by his skill in the critical selection of methods. Almost daily the analytical chemist is confronted with problems of optimizing analytical procedures or related problems such as the selection of the best procedure for solving a given problem. This situation is represented by the most recent definitions of analytical chemistry, stating that analytical chemists have to produce qualified, relevant information on materials and processes in an optimal way (Gottschalk, 1972 ; Kaiser, 1974). It is therefore surprising to note that analytical chemists in general do not seem to have taken pains to develop strategies for optimization. Until recently it was common for the choice of the best procedure for a given analytical problem to be made largely intuitively and based upon experience.

Recently, a number of papers have appeared that express the concern of an increasing number of analytical chemists with this situation. It has been our intention to discuss in this book the formal methods that are available at present for the optimization and selection of analytical methods.

Before it is possible to make a selection or to carry out an optimization, one must have criteria according to which this may be done. Consequently, the performance of analytical procedures has to be evaluated by determining one or more performance characteristics of the procedure that is to be optimized or of the procedures from which the best one is to be selected. The set of criteria has to be defined for each problem and will include quantities such as precision, accuracy, limits of detection and interferences. Up to now, most of these criteria have been used in quantitative analysis, and it is probable that another set of characteristics will be required for qualitative analysis. Measures of information may become important in this respect and, therefore, a large section is devoted to information theory. Performance characteristics are discussed in Part I.

The next two parts of the book are devoted to the optimization of procedures. One can discern different levels of optimization (Beveridge and Schechter, 1970), and for the purpose of this book we consider three such levels :

(1) Selection of one existing procedure from several alternatives. This is the simplest possible stage. There are several alternatives ; each of these is evaluated and the one which corresponds best to the exigencies of the application is selected. The main problem here is the evaluation of the procedures. This is described in Part I.

(2) Optimization of a procedure for which the outline is given. For example, given that the determination will be carried out colorimetrically with dithizone, select the optimal wavelength, the best concentration of dithizone, the pH, etc. This kind of problem usually consists in the selection of the optimal value of one or more continuously adjustable parameters. Occasionally, one may also have to include discrete parameters such as the kind of detector to be used. These problems are discussed in Part II.

(3) Optimization of combinations of analytical procedures or attributes of methods. There are many instances in which analytical procedures are combined, for instance

the combination of tests in a clinical laboratory to yield the optimal diagnostic or discriminatory power ;

the combination of GLC stationary phases to form a preferred set ;

the combination of elementary steps in a separation procedure to yield an optimal multicomponent separation scheme.

Such combinatorial problems are discussed in Part III. The analytical laboratory as a whole may be considered as a combination of methods, apparatus, etc. Some optimization problems concerning the functioning of an analytical laboratory are therefore also included in Part III.

Laitinen (1973), in an Editorial in *Analytical Chemistry*, stated that an analytical method is a means to an end, and not an end in itself. Analytical chemists tend to overlook this and sometimes develop methods that are more precise, faster, etc., while forgetting the intended applications. In such instances, one can hardly speak of optimization. The analytical procedure must be chosen in relation to the goal and questions such as "how useful is it to increase the precision of a procedure used for a particular purpose" then arise. Some problems of this kind are discussed in Part IV.

Problems of optimization in analytical chemistry are often related to other optimization problems. However, such analogies will only be recognized if problems are formulated in a more or less formal and generalized way. The analytical procedure and the analytical laboratory should be considered from the point of view of systems theory. Some aspects of such a generalizing approach are discussed in Part V. In fact, one observes that analytical chemists concerned with optimization problems intuitively follow a systems approach. It would therefore have been more logical to start this book with a discussion of systems theory, but as yet it is not possible to construct a complete systems theoretical picture of the analytical procedure and the analytical laboratory, and to some extent the topic is of academical interest. Therefore, we have started the discussion with those points which clearly are of direct value in analytical practice.

The trend towards a more formal approach of the selection of analytical methods is not really new, but it has definitely grown stronger in the last few years. It is one of the principal concerns of the very recent field of chemometrics (Kowalski, 1975). It is not only felt among those who make this their research field in general analytical chemistry, but also by analytical chemists who are concerned more directly with analytical practice, such as clinical chemists and by those who need the results of analytical determinations, such as physicians using laboratory tests for medical diagnosis. At about the same time that concepts such as information were introduced into general analytical chemistry, clinical chemists began to use multivariate data analysis techniques to investigate which analytical methods yield the most diagnostic information. If one looks at the literature cited by the "generalists" and the clinical chemists, one finds that they cite different literature and that, in general, there seems to be very little communication between the two groups.

In some applications, formal methods for the investigation of the performance of analytical methods were introduced many years ago. This is the case, for example, with official analytical chemists, who have developed methods for the evaluation of errors likely to occur in analytical procedures. Many analytical chemists from other specialities, but who are also concerned with the evaluation of analytical methods, seem, however, to ignore the existence of such methods.

We have tried to combine the knowledge stored in these (and other) different specialities in the hope of stimulating a more systematic application of formal selection methods in analytical chemistry. Our first idea was to limit this book to newer methods or concepts, such as information theory and operational research, but it was soon clear that it would be meaningless to try and make a synthesis and not include classical statistical concepts. Therefore, a number of chapters on classical statistical methods were added. Because, on the other hand, we did not want to duplicate the material already available in several books on statistics in chemical analysis, we have tried to eliminate statistical methods designed for the evaluation of results rather than of procedures, and we have not tried to discuss the subject exhaustively. This is also true for all other chapters. More specialized knowledge should be sought in the original literature or specialized books and chapters to which we refer. Since this book was written to introduce a number of formal optimization techniques to analytical chemists and not for specialists, we have not tried to cover the existing literature exhaustively. Instead, we usually have given some references to books, review articles and a few illustrative articles.

In writing this book, we have started from the belief that some of the newer mathematical methods or theories, such as pattern recognition, information theory, operational research, etc., are relevant to some of the basic aims of analytical chemistry, such as the evaluation, optimization, selection, classification, combination and assignment of procedures or sub-procedures - in short all those processes that intervene in determining exactly which analytical procedure or programme should be used. Unfortunately, most chemists are daunted by the task of learning how these mathematical methods function and this is not helped by the difficulty of establishing a link between the formal mathematics found in most books on this subject and analytical problems.

We have tried to treat the mathematical topics as lucidly as possible and to illustrate the text with examples, in some instances abandoning a rigorous mathematical treatment. However, we have tried to compensate for this by including a series of mathematical sections. The level of the mathematics is higher in Chapters such as 2, 3 and 4, where the subject treated is not completely unfamiliar to most analytical chemists. In those chapters where the subject matter is probably new to most analytical chemists, only the most elementary explanations are given, often in words, because we think it more important to emphasize the underlying philosophy than to explain the mathematics. In doing so, we hope we have removed the barriers of applying formal methods to optimization problems in analytical chemistry. One major difficulty encountered when writing this book was the mathematical symbolism. We have tried to present a coherent set of symbols throughout the book but, because of the diversity of the methods described, we have not been entirely succesful in this respect. Nevertheless, we think that the symbols used should be sufficiently clear.

Parts of the draft of this book were read by E. Defrise-Gussenhoven (Vrije Universiteit Brussel ; Chapters 2 and 3), H.C. Smit (University of Amsterdam ; Chapter 10), D. Coomans (Vrije Universiteit Brussel ; Chapter 20), G. Kateman (Catholic University of Nijmegen ; Chapters 26 and 27) and P.M.E.M. van der Grinten (DSM, Heerlen ; Chapters 26 and 27). Their suggestions were very valuable. The authors are also indebted to S. Peeters, who typed the final text, A. Langlet-De Schrijver and L. Maes, who prepared the figures, and C. Uytterhaegen-Hendrickx, G. De Boeck, A. Van Gend, F. Gheys and M. Segers-Geeroms who typed the manuscript. The research leading to the idea of writing this book was helped financially by organizations such as the Fonds voor Geneeskundig Wetenschappelijk Onderzoek and the Fonds voor Kollektief en Fundamenteel Onderzoek and was stimulated by collaboration with members from the Studiegroep voor

Laboratoriumoptimalisering and the Centrum voor Statistiek en Operationeel Onderzoek of the Vrije Universiteit Brussel, and with the following students, who obtained degrees based on research on subjects covered in this book : H. De Clercq, M. Detaevernier, J. Smeyers-Verbeke, J.H.W. Bruins Slot, P.F. Dupuis, G. van Marlen, P. Cley, T. Koppen and A. Eskes. The proofs were read by A. Kaufman. The authors express their thanks to all of these persons and organizations.

REFERENCES

G.S.G. Beveridge and R.S. Schechter, Optimization : Theory and Practice, McGraw-Hill, New York, 1970.
G. Gottschalk, Z. anal. Chem., 258 (1972) 1.
R. Kaiser, Z. anal. Chem., 272 (1974) 186.
B.R. Kowalski, J. Chem. Inf. Computer Sci., 15 (1975) 201.
H.A. Laitinen, Anal. Chem., 45 (1973) 1585.
H.A. Laitinen and W.E. Harris, Chemical Analysis, McGraw-Hill, New York, 1975.

PART I

EVALUATION OF THE PERFORMANCE OF ANALYTICAL PROCEDURES

Chapter 1

PERFORMANCE CHARACTERISTICS OF ANALYTICAL PROCEDURES

The purpose of this book is to survey methods for the optimal selection of an analytical procedure or a combination of such procedures. The first step that has to be accomplished in order to make any selection or optimization possible is to choose the criteria according to which a procedure will be chosen or optimized. In other words, procedures must be evaluated in order to make a selection possible. Garton et al. (1956) called these criteria the "performance characteristics", a term later adopted by Wilson (1970) and other workers. Kaiser (1973) called them "figures of merit" but, although this term is suitable for the description of most criteria, it is not generally useful as some important properties (such as safety) cannot be easily quantified.

The most important object in Part I is to discuss some of the technical criteria according to which the performance can be evaluated. Some topics, such as accuracy and precision and the use of t-tests, are no doubt familiar to analytical chemists and are discussed in many books on statistics, including those written especially for chemists, for example the excellent "The Handling of Chemical Data" by Lark, Craven and Bosworth (1969) and "Statistical Methods for Chemists" (Youden, 1951), or for analytical chemists, such as those by Gottschalk (1962) and Doerffel (1966).

The formal treatment of other criteria, such as noise, drift, reliability and information, is probably not so well known to most analytical chemists. Because, as far as we know, no book has appeared on the subject of performance characteristics, it seemed useful to discuss all of these characteristics, even those which should be familiar to every analytical chemist. For the latter type, we have given only a brief account, stressing any ambiguities that exist and any limitations in particular applications.

Several workers have suggested sets of performance characteristics (for example,

Morrison and Skogerboe, 1965 ; Gottschalk, 1962 ; Kaiser and Specker, 1956) and, according to Wilson (1970), they can be summarized under three headings : errors in the analytical results, the calibration graph and the time of analysis. If we follow this classification, Chapters 2, 3, 4, 5 and 7 can be considered to fall into the first category, Chapter 6 into the second and part of Chapter 9 into the third. Wilson's proposal is excellent for most types of analyses. We feel, however, that two fields of analytical chemistry, namely qualitative analysis and continuous analysis, require different or additional performance characteristics and Chapters 8 and 10 are devoted to these subjects.

In general, performance characteristics can be divided into two categories, economic and technical. The most obvious economic criterion is cost. The cost of a method is extremely important, particularly in routine laboratories where often it is necessary to make a profit or at least to be self-sustaining. Other characteristics, such as time, are often also used on an economic basis. These economic and a few other factors that are of importance for the selection of analytical procedures for use in actual applications are discussed in Chapter 9.

An important question, related to the choice of the optimization or selection criterion, concerns the relevance of the solution obtained. This question is part of the systems analysis approach of optimization, which is discussed in the last part of this book. However, we should state here that the optimal solution obtained according to a certain optimization criterion is not always of practical value, for three main reasons :

(a) As stated by Laitinen (1973), one may consider that the nearest approach to the ideal method is that which handles the problem in the most convenient way and therefore takes into account the equipment, personnel and reagents available. This is another way of saying that in general optimization systems are not without constraints. In fact, the restrictions mentioned by Laitinen (1973) are not the only ones possible and, as remarked by Beveridge and Schechter in their book "Optimization : Theory and Practice" (1970), it is uncommon to find unrestricted optimization problems.

(b) The optimization criterion chosen may appear not to be relevant in relation to the problem that has to be solved. For example, as discussed in Chapter 2 and Part IV, it may be of no practical importance to increase the precision of a method, because although a better method in the technical sense will be obtained it does not have a significant effect on the discriminatory power of the method.

(c) Multiple criteria. In general, criteria are interrelated. For example, a higher precision will usually be achieved at the cost of a slower speed of analysis. The optimal procedure from the point of view of one performance characteristic may be undesirable from the point of view of another. This topic is discussed further in section 9.4,and in Part IV.

REFERENCES

G.S. Beveridge and R.S. Schechter, Optimization : Theory and Practice, McGraw-Hill, New York, 1970.
K. Doerffel, Statistik in der Analytischen Chemie, VEB Deutscher Verlag für Grundstoffindustrie, Leipzig, 1966.
F.W.J. Garton, W. Ramsden, R. Taylor and R.J. Webb, Spectrochim. Acta, 8 (1956) 94.
G. Gottschalk, Statistik in der Quantitativen Chemische Analyse, Enke, Stuttgart, 1962.
H. Kaiser, Methodicum Chimicum, Band I Analytik, Teil I, Georg Thieme Verlag, Stuttgart and Academic Press, New York, London, 1973, p. 1.
H. Kaiser and H. Specker, Z. anal. Chem., 149 (1956) 46.
H. Laitinen, Anal. Chem., 45 (1973) 1585.
P.D. Lark, B.R. Craven and R.C.L. Bosworth, The Handling of Chemical Data, 2nd ed., Pergamon, Oxford, 1969.
G.H. Morrison and R.K. Skogerboe, in G.H. Morrison (Editor), Trace Analysis : Physical Methods, Interscience, New York, 1965, p. 2.
A.L. Wilson, Talanta, 17 (1970) 21.
W.J. Youden, Statistical Methods for Chemists, Wiley, New York, and Chapman and Hall, London, 1951.

Chapter 2

PRECISION AND ACCURACY *

2.1. GENERAL DISCUSSION OF CONCEPTS

2.1.1. Introduction

The purpose of carrying out a determination is to obtain a valid estimate of "true" values. When one considers the criteria according to which an analytical procedure is selected, precision and accuracy are therefore usually the first to be selected, and most text books concerned with analytical chemistry discuss and define these terms. One would therefore expect that there are universally accepted definitions of and methods for determining these quantities. However, a brief study of the literature shows that the ISO definition of precision is not the same as that used by *Analytical Chemistry*. This is only one example of the confusion that seems to exist and therefore a more thorough investigation of the meanings of precision and accuracy is necessary. The purpose is not to propose a new definition of these concepts, but to establish the factors that govern precision and accuracy and how they should be determined.

Of the many definitions proposed (only some of which are discussed here), we prefer the definitions given in *Analytical Chemistry* (1975), because these seem the most appropriate from both the analytical and statistical points of view.

2.1.2. Categories of errors in analytical chemistry

In analytical chemistry several types of error are encountered. Roughly, the following categories may be considered : random (indeterminate) errors cause

---

* This chapter has been written with the collaboration of Y. Michotte, Pharmaceutical Institute, Vrije Universiteit Brussel, Belgium.

imprecise measurements and are therefore assessed by means of the precision (or imprecision, as preferred by some workers), while systematic errors cause inaccurate (incorrect) results and are referred to in terms of accuracy. Usually the precision is studied first, because systematic errors can be determined only when random errors are sufficiently small.

When an analyst carries out a number of replicate determinations on the same sample with the same procedure, apparatus, reagents, etc., results that are subject to random and normally distributed errors are obtained. The results of replicate determinations are considered to be a random sample from a normal population of results obtained in the same way. The standard deviation of this distribution is generally called the precision of a procedure. It is usually obtained under favourable conditions and it is usually not what could be called the "real-life" precision. When the procedure is to be applied as a routine method, other sources of error will be introduced and the precision will decrease. For example, it is often observed that the precision calculated for samples analysed in several batches or on several days is worse than that for samples analysed in one batch or on the same day. The latter is sometimes called the day-to-day or between-days precision, while the former is the within-day precision.

These additional sources of variation are not necessarily random. When they are caused by unstable reagents or by ageing of parts of the apparatus (for example, ageing of the pump tubes in a continuous automatic analyser), they are systematic. Such a time-dependent error is sometimes known as drift and is discussed in more detail in Chapter 5.

In the same manner, when a procedure is carried out by several laboratories, each with their own personnel, apparatus, reagents, etc., on the same sample, one usually observes a normal distribution of errors broader than that obtained when a single analyst carries out the determinations. This effect results from the fact that each laboratory makes some systematic errors or bias owing to, for example, impurity of the reagents or incomplete directions for carrying out the procedure. Laboratory biases themselves are normally distributed (Youden and Steiner, 1975). Thus, the distribution obtained when the sample is analysed with the same method by several laboratories is also normal. The dispersion around the mean can again be considered to be a measure of precision and, in other definitions, this is called the reproducibility. Chapters 3 and 4 discuss how to assess this measure of precision by inter-laboratory comparison.

Procedures are also subject to inherent systematic (and therefore not normally distributed) errors. Systematic errors are generally said to influence the accuracy, although there is some divergence of opinion and terminology on this point (see section 2.1.5).

Systematic errors may be constant (absolute) or proportional (relative). A constant error refers to a systematic error independent of the true concentration of the substance to be determined and is expressed in concentration units. A proportional error is a systematic error that depends on the concentration of the analyte and is expressed in relative units, such as a percentage.

The main sources of constant error are : (a) insufficient selectivity, which is caused by another component that also reacts so that falsely high values are obtained ; measures of selectivity are discussed in Chapter 7 ; (b) interferences ; according to the terminology of Büttner et al. (1976), this source of error is due to the presence of a component, which does not by itself produce a reading but which inhibits or enhances the measurement (these interferences also cause an insufficient selectivity) ; (c) inadequate blank corrections.

Proportional errors are caused by errors in the calibration and more particularly by (a) the incorrect assumption of linearity over the range of analysis and (b) different slopes of the calibration lines for the sample and standard.

Systematic errors can be studied by a variety of methods. Some of these methods (standard addition or recovery experiments, linearity checks) detect only proportional errors while other methods (t-test) should not be used when a proportional error is present. These methods are discussed in Chapter 3.

There are other sources of error which cannot be classified easily in one of these categories. An example in automatic continuous analysis concerns the contamination caused by previous samples and is called the carry-over error, which occurs when successive samples take a common path in an automated system. Because of its dependence on the parameters of the method, it can be considered as a systematic error. On the other hand, it is neither constant nor relative as it also depends upon the concentration of the previous sample. In the analysis of a large series in a random sequence, this error may be considered part of the random error. Further discussion on this aspect can be found in an article by Broughton et al. (1969).

2.1.3. Precision and accuracy as criteria

Precision and accuracy together determine the error of an individual determination and their magnitude is one of the most important criteria for judging analytical procedures by their results. Many workers consider that these quantities describe the state of the art and the improvement of these criteria is regarded as the only possible aim of optimization studies. However, analysts proposing a method for a particular procedure should ask themselves whether an increase in the precision and accuracy of the determination is really important or even useful. All sources of variation must then be taken into account. For example, if sampling is to be regarded as part of the analysis, then sampling errors must also be considered. In some instances, these errors are very large and can dominate the total error. An example is a potassium determination carried out routinely in an agricultural laboratory (Vermeulen, private communication, see also 1957). It was found that 87.8% of the error was due to sampling errors (84% for sampling in the field and 3.8% because of laboratory sampling due to inhomogeneity of the laboratory sample), 9.4% to between-laboratory error, 1.4% to the sample preparation and only 1.4% to the precision of the measurement. It is clear that in this instance an increase in the precision of measurement is of little interest. A comparable situation is found in clinical chemistry, where the purpose of the analysis is to investigate whether the values fall in the normal range or not. Because of biological variability, this range can be very large. There have been interesting studies of the effect of analytical error on normal values (the values considered to be normal for a healthy population) and clinical usefulness. Whatever the results of these studies, it seems evident that there is no sense in trying to obtain a method with 0.01% precision and accuracy when the normal range is of the order of 20%. These aspects are discussed in more detail in Part IV, in which the relationship between analytical chemistry and its environment is considered. Therefore, this and the two following chapters are essentially descriptive in the sense that the assessment of precision and accuracy (or their components) is discussed without considering requirements for their magnitude.

2.1.4. Definition and measurement of precision (repeatability, reproducibility)

Different definitions of the above three terms have been proposed, and we shall restrict ourselves to two of them. The first was given by *Analytical Chemistry* (1975) : "Precision refers to the reproducibility of measurement within a set, that is, to the scatter or dispersion of a set about its central value". The term "set" is defined itself as referring to a number (n) of independent replicate measurements of some property. Readers are urged to use this definition with an understanding of its limitations, such as the fact that the values obtained are usually based on a small number of observations and should therefore be regarded as an estimate of the parameter. By adding this comment, the definition of *Analytical Chemistry* conforms with statistical usage. Statisticians make a careful distinction between a true quantity (for example a true concentration) and its estimate (for example the mean of a number of determinations of the concentration). This distinction is not always found in definitions of precision. As remarked by Wilson (1970), this was the case for a definition of bias by IUPAC (see also section 2.1.5).

The definition of *Analytical Chemistry* is (perhaps intentionally and to conform with current usage) somewhat ambiguous, as it is not specified whether or not the set of measurements is carried out by a single operator. As will be seen later, there is a large practical difference between the two possibilities.

On the other hand, the International Organization for Standardization (1966) prefers the following definitions. *Reproducibility* : closeness of agreement between individual results obtained with the same method or identical test material but under different conditions (different operator, different apparatus, different laboratory and/or different time). *Repeatability* : closeness of agreement between successive results obtained with the same method or identical test material and under the same conditions (same operator, same apparatus, same laboratory and same time).

According to the "Statistical Manual of the Association of Official Analytical Chemists" (Youden and Steiner, 1975), the precision is composed of random within-laboratory errors and unidentified systematic errors in individual laboratories (laboratory bias). These errors are also normally distributed. In this instance, precision is considered to be identical with the reproducibility as defined above, with repeatability as a component. Other terms such as scatter and analytical variability are also used occasionally. Some workers prefer the term imprecision to precision (Büttner et al., 1976) in order to avoid the linquistic difficulty that a procedure becomes more precise when its measure, the precision, decreases. In our view, collaboration between statisticians and analytical chemists is so important that no semantic difficulties should be created between them. Terms such as reproducibility, repeatability and imprecision, which are not used by statisticians, should not be used except in a colloquial sense, i.e., when there is no need to attach a precise meaning to them.

The following measures of precision within a set (as defined above) are proposed by *Analytical Chemistry*.

"Standard deviation is the square root of the quantity (sum of squares of deviations of individual results from the mean, divided by one less than the number of results in the set). The standard deviation, s, is given by

$$s = \sqrt{\frac{1}{n-1} \sum_{i=1}^{n} (x_i - \bar{x})^2} \qquad (2.1)$$

Standard deviation has the same units as the property being measured. It becomes a more reliable expression of precision as n becomes large. When the measurements are independent and normally distributed, the most useful statistics are the mean for the central value and the standard deviation for the dispersion". One observes that the symbol s is used for the estimate of the true standard deviation, σ. This is correct statistical practice. Recent rules approved by IUPAC (1976), state that, when the number of replicates is smaller than 10, s should be used instead of σ. In our view, it is preferable always to use s for an estimate, even a good one, and to reserve σ for the "true" value. It should be noted here that statisticians make a distinction between a biased and an unbiased estimate. The standard deviation as defined above is an unbiased estimate and should therefore be represented by $\hat{\sigma}$, where the "hat" on σ indicates that it is unbiased (see section 2.2), and we would prefer to use this symbolism throughout this book. As we do not want to introduce or create symbolism and terminology that would be unfamiliar to analytical chemists, we shall refrain from doing so, except occasionally when some distinction is important.

"Variance, s^2, is the square of the standard deviation".

"Relative Standard Deviation is the standard deviation expressed as a fraction of the mean : $s/\bar{x}$. It is sometimes multiplied by 100 and expressed as a percentage. Relative standard deviation is preferred over coefficient of variation".

Two other quantities are defined, although they are not to be recommended as measures of precision except when the set consists of only a few measurements. These quantities are the mean (or average) deviation, given by

$\sum_{i=1}^{n} |x_i-\bar{x}|/n$, and the range, given by the difference in magnitude between the largest and smallest results in a set.

One should also observe that the precision of the mean (called standard error in this instance) and equal to $s/\sqrt{n}$ is of no interest in evaluating the precision of a procedure, but only as a measure of the confidence one can have in a result stated as a mean. The measure of the precision of a procedure should clearly not depend on n, the number of replicate measurements. This is not the case for the standard deviation, but it is so for the standard error. However, many measurements are advantageous because s is only an estimate of σ and the more replicates one carries out, the better estimate s gives of the "true" precision, σ.

2.1.5. Definitions of bias and accuracy

When analytical determinations are carried out they yield (hopefully slightly) different results, x_i. A result can differ from the true value, μ_0, which is unknown and in statistical terminology this difference is referred to as the error :

$$e_i = x_i - \mu_0$$

If enough measurements are made, a stable mean $\bar{x}$ is obtained, where $\bar{x}$ is an estimate of the mean, μ, of an unlimited number of determinations. The absolute difference between μ as represented by $\bar{x}$ and the true value, μ_0, is called the bias or systematic error.

It should be noted that the bias obtained by experimentation is an estimate of the true bias, as it is calculated by using $\bar{x}$, which is itself an estimate. As observed by Wilson (1970), the IUPAC (1969) definition of bias (the difference between the mean of the results and the true values) is therefore valid in practice but not exact from the statistical point of view, as the difference between true values and estimates obtained by experimentation is not made.

It is necessary to consider at this stage the terms "laboratory bias" and "method bias". The former, as seen in the preceding section, contributes to the precision of a method (inter-laboratory precision), while the latter constitutes the systematic error. When defining the accuracy for an inter-laboratory trial, the accuracy is identical with the method bias. From the point of view of an individual laboratory, however, the systematic error is the sum of the method bias (common to all laboratories using the method) and the laboratory bias (for the laboratory in question). This, too, is called the accuracy and the meaning of the term accuracy is therefore not always clear. To illustrate this confusion, it is instructive to read the section on accuracy in the "Statistical Manual of the Association of Official Analytical Chemists" (Youden and Steiner, 1975). Steiner (p. 69) states that accuracy must be distinguished from precision. It "measures bias, that is to say the difference between the mean result given by the method and the true result (often unknown)". Clearly, Steiner considers method bias and accuracy to be identical. The other author, Youden, is more prudent. He writes (p. 25) that he has shunned the word accuracy because there are different interpretations, some of them englobing precision and bias, while others restrict it to bias. We should make it clear, however, that in citing this discrepancy, we are not criticizing this interesting and well written guide for the application of statistics to inter-laboratory trials, but are rather underlining a semantic difficulty.

Let us turn now to the definition of *Analytical Chemistry* (1975). It states that :

"Accuracy normally refers to the difference (error or bias) between the mean, $\overline{x}$, of the set of results and the value, $\hat{x}$, which is accepted as the true or correct value for the quantity measured. It is also used as the difference between an individual value x_i and $\hat{x}$. The absolute accuracy of the mean is given by $\overline{x} - \hat{x}$ and of an individual value by $x_i - \hat{x}$". $\hat{x}$ in this definition has the same meaning as the symbol μ used by us and in most statistical books. The definition is given with the same limitations as for the precision and its measures (see the preceding section). The definition by *Analytical Chemistry* is ambiguous because it consists, in fact, of two different sub-definitions. The first relates to the mean obtained with a particular method and is a synonym of systematic error, while the other relates to individual results and is therefore made up of a combination of the systematic error and the random error.

This introduces a new difficulty as this definition allows the use of the word accuracy for the sum of the errors due to systematic and random causes. The combination of both has been called total error by some workers (McFarren et al., 1970, and Westgard et al., 1974).

From the definition given above, it is clear that there is great confusion with the term accuracy, which has led us to three conclusions :

(1) Authors publishing numerical values for precision and accuracy should state how the calculation was carried out and the circumstances under which the results were obtained. This recommendation, made by Youden and Steiner (1975), is the best way of dispelling the confusion.

(2) In this book, we shall use the word accuracy in the general sense, i.e., in a colloquial way. When distinctions are important, we shall use the following terms : (i) *laboratory bias* (see section on precision), being the systematic error introduced by a laboratory. This bias is considered to be part of the inter-laboratory precision. (ii) *method bias*, being the systematic error introduced by the use of a particular method. It is the same for all laboratories. (iii) *total error*, for combinations of errors due to method bias and random errors (inter- and intra-laboratory precision).

(3) Owing to the confusion that already exists with terms connected with accuracy and precision, authors should not be encouraged to create new terms, which could increase these difficulties. In the same way, organizations should refrain from publishing their own definitions.

2.1.6. A demonstration of the importance of laboratory bias

Analytical chemists developing new methods should realize that these methods will be used by chemists in other analytical laboratories who may not have the same fundamental knowledge of the method and may therefore simply follow the procedure proposed with their own apparatus, reagents, etc. Very often a new method, when it is used under actual working conditions, gives poor results. To describe this frequently observed phenomenon in terms of precision, we have stated that the overall precision of a method, s, is composed of two terms, namely an intra-laboratory precision (s_r) and an inter-laboratory precision or laboratory bias (s_b). It is known that s_b is usually larger than s_r.

A typical example was given by Wernimont (1951) (see Fig. 2.1) in an article concerned with a study of sources of variation (laboratories, analysts within laboratories, different days for the same analyst, replicate determinations).

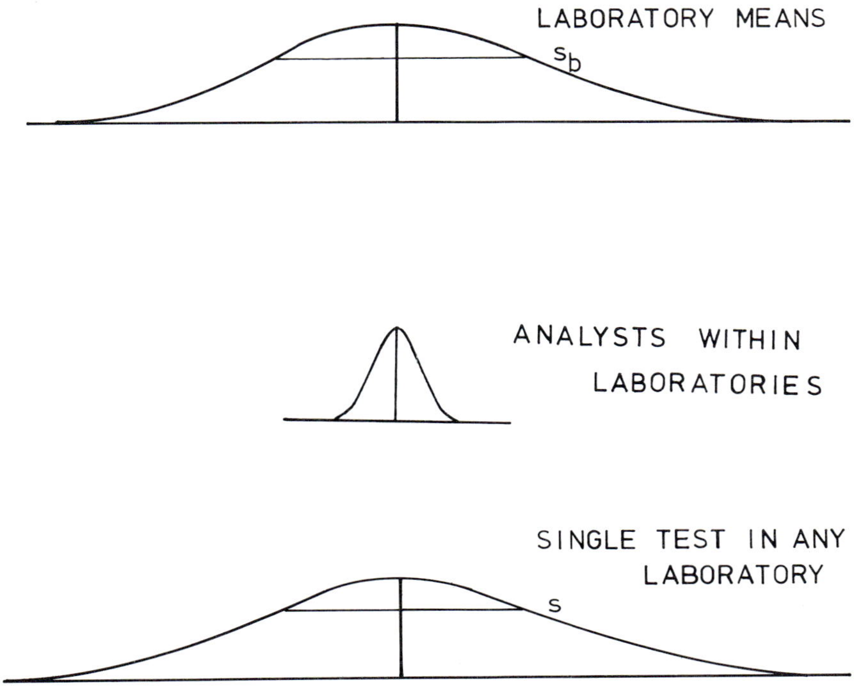

Fig. 2.1. A comparison of sources of variation in the determination of acetyl (adapted from Wernimont, 1951). Reprinted with permission. Copyright American Chemical Society.

It can be seen that the total variation for single tests carried out in any laboratory, on any day and by any analyst (s = 0.27) can be explained by the variation among laboratories (s_b = 0.25). The other sources of error are much less important.

Many analytical chemists consider that this effect is due to imperfect or incomplete descriptions of procedures and that, provided that procedures are described in sufficient detail, every laboratory should obtain results with the same precision and accuracy. In fact, this is not true. Very interesting work in this respect has been carried out under the auspicies of the Association of Official Analytical Chemists and their conclusions were given in their statistical manual (Youden and Steiner, 1975). This will be used as the basis for a discussion of inter- and intra-laboratory errors.

The only way of reaching a conclusion about the precision of an analytical method under actual working conditions is to homogenize a sample and to distribute it to a number of laboratories for analysis, i.e., to carry out an "intercomparison". Intercomparisons are carried out in two situations :

(a) when a method has been tested by one or a few laboratories, shown to be free of method bias and proved sufficiently precise in the laboratory of the promotors to warrant an examination of its general usefulness.

(b) when several methods are in use for a certain determination and one wants to know whether they yield the same or significantly different results.

Situation (b) is discussed in Chapters 3 and 4. In this section, only situation (a) is considered, in particular when one wants to demonstrate that laboratory biases are present.

Many intercomparisons or collaborative studies have been carried out to date and it has been shown that in most instances the overall precision, s , is much greater than the intra-laboratory precision, because not only are random errors present but also each laboratory obtains biased results. These laboratory biases have been shown to be normally distributed in most instances.

The general occurrence of systematic errors in user laboratories may appear surprising. However, it can be demonstrated easily by using a two-sample chart. If the collaborators in collaborative studies are asked to analyse two samples of more or less analogous constitution and there are no systematic laboratory (or method) biases, the chance of finding a high result (+) should be equal to the chance of obtaining a low result (-) for each participating laboratory. This means also that the combination of two high results (++), two low results (--) and both possibilities of obtaining one low and one high result (+- or -+) are equal. By plotting the result for sample No. 1 against the result for sample No. 2 for each laboratory, one should obtain a diagram similar to Fig. 2.2a. In fact, one nearly always obtains a result such as that shown in Fig. 2.2b, i.e., a significantly high prevalence of ++ and -- results, showing that more laboratories than expected deliver either two high or two low results.

Fig. 2.3 gives an example of the result of a collaborative experiment for an atomic-absorption spectrophotometric method for the determination of manganese in soy meat blends (Formo, 1974). An application of this method in clinical chemistry was given by Tonks (1963) concerning the accuracy and precision of 170 Canadian laboratories.

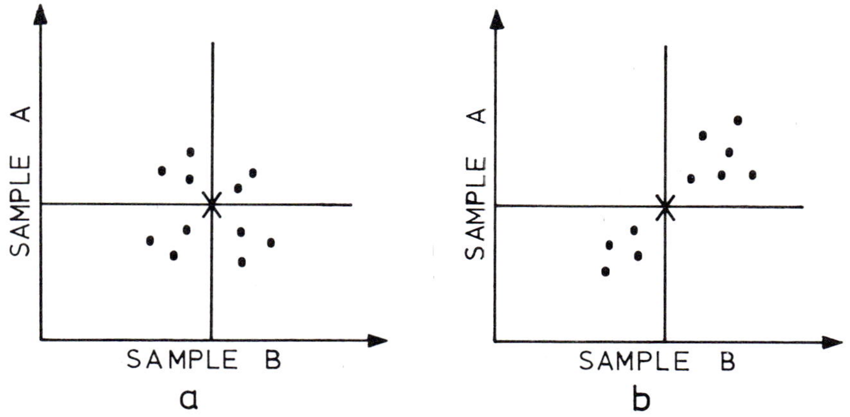

Fig. 2.2. The two-sample method for detection of laboratory bias : (a) No laboratory bias ; (b) laboratory bias. • represents individual results and x represents the mean.

The two-sample chart procedure has been applied also to intercomparisons of situation (b) to show whether systematic errors are predominant. This was done, for example, by Ekedahl et al. (1975) for many types of nutrient determinations in water. In this instance different methods were applied to the same type of determination so that method biases also exist. It is therefore predictable that in all instances systematic errors (laboratory and method biases) will predominate in view of the fact that this is usually also true when only laboratory biases have to be taken into account.

Further, as Youden and Steiner's two-sample test does not enable one to distinguish between method and laboratory bias, there is little reason to carry out intercomparisons of different methods in this way.

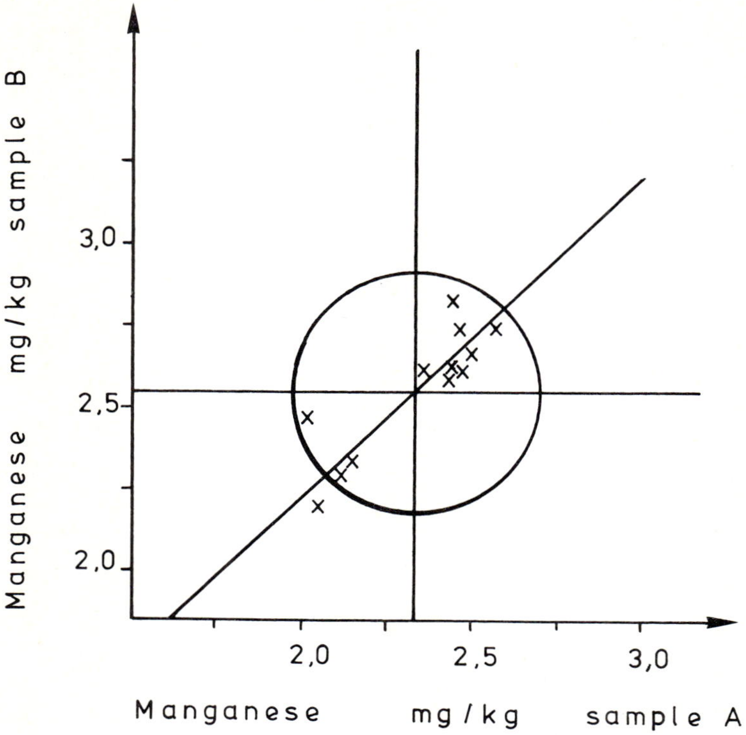

Fig. 2.3. Example of the occurrence of laboratory bias : determination of Mn in soy meat blends (Formo, 1974).

2.1.7. Some studies on precision

Several workers have investigated the factors that govern the magnitude of the precision obtained with a particular procedure or apparatus. As an example, one can cite a study on the precision characteristics of simple spectrophotometers by Ingle (1977). In this instance the conclusion was that the precision of the measurement is limited by irreproducibility of positioning of the cell and noise which is independent of the photon signal.

An interesting approach to this kind of problem was made by Aronsson et al. (1974), who studied the effect of many factors such as the number and precision of calibration samples using a computer simulation procedure.

Püschel (1968) showed that there is a linear correlation with a correlation coefficient of nearly -0.9 in the range from 1 ppm to 100% between the precision and the content to be determined.

2.1.8. The total error

For analytical chemists developing methods, it is necessary to separate the total observed error into its components, particularly when one wants to optimize a method as it permits a better understanding of the factors that contribute to the error. On the other hand, the user of an analytical method is often not interested in the types of errors present but rather in the total effect of these errors on the result. This fact has led several workers to define a total error which combines both random and systematic errors. As an example, McFarren et al. (1970) developed a criterion for judging the acceptability of analytical methods. They consider that the total error is due to the accuracy (which they seem to define in the sense of the first part of the definition of *Analytical Chemistry*) and random errors and they define the term total error as the relative error (i.e., mean error divided by the mean) plus twice the relative standard deviation. Recently, Midgley (1977) proposed a slightly different definition of the total error.

A much more elaborate study was carried out by Westgard et al. (1974), who developed criteria for judging precision and accuracy in method development and evaluation with special reference to clinical chemistry. The total error is defined for a concentration called x_c, at which critical medical decisions are made. It is equal to the systematic error (laboratory bias + method bias) at this concentration as estimated from regression methods (see next chapter) plus a term containing the standard deviation obtained from replicate determinations (intra-laboratory precision) and the confidence of the estimation made by regression analysis. This paper contains a lucid analysis of errors in analytical chemistry and is also important for those who have a practical interest in this topic as it contains some worked examples. Not everyone agrees with the use of terms combining precision and bias. In particular, Currie et al. (1972) noted in their review on statistical and mathematical methods in analytical chemistry that such a combination can be very misleading when the bias has been estimated with great imprecision.

2.2. MATHEMATICAL

The basic concepts of statistics have been discussed in many books and in many different ways. A complete discussion of these is beyond the scope of this book. However, for the purpose of reference, the books by Dunn (1964), Cooper (1975) and Morrison (1969) can be mentioned as good introductions to the subject. For readers particularly interested in statistics, the three volumes by Kendall and Stuart (1969) must be mentioned as giving a complete synthesis of modern statistics.

2.2.1. Frequency distributions

When a large number of measurements must be presented, it is often useful to group the data into classes and to calculate the number of measurements belonging to each class. This number is called the frequency of a class. Such a method of describing the data results in a frequency distribution.

The relative frequency of a class or of the value of a measurement is its frequency divided by the total number of measurements. It can also be expressed as a percentage. An example is given in Table 2.I.

Table 2.I

Frequency distribution of the concentration in a population of 100 samples

Concentration (%)	Number of samples	Relative frequency
0 - 3	4	0.04
4 - 7	15	0.15
8 - 11	24	0.24
12 - 15	30	0.30
16 - 19	18	0.18
20 - 23	6	0.06
24 - 27	3	0.03
	Total : 100	1

The total number of measurements less than or equal to the upper boundary of a class is called the cumulative frequency for that class. For example, in Table 2.I the cumulative frequency for the class 4 - 7 is equal to 4 + 15 = 19, which means that 19 samples gave a concentration of at most 7.5%. The other cumulative frequencies can be calculated in the same way and are given in Table 2.II.

Table 2.II

Cumulative frequency distribution of the concentration in a population of 100 samples

Concentration (%)	Cumulative frequency	Relative cumulative frequency
$\leqslant$ 3.5	4	0.04
$\leqslant$ 7.5	19	0.19
$\leqslant$ 11.5	43	0.43
$\leqslant$ 15.5	73	0.73
$\leqslant$ 19.5	91	0.91
$\leqslant$ 23.5	97	0.97
$\leqslant$ 27.5	100	1.00

It can be seen from Table 2.II that it is also possible to calculate relative cumulative frequencies by dividing the cumulative frequencies by the number of observations, n. This makes it possible to compare cumulative frequency distributions.

The frequency distribution can also be described by a diagram in which the measurements are represented by rectangles, the heights of which are proportional to the (relative) frequencies and the widths of which represent the class width. Such a representation is also called a histogram and examples of histograms are given in Fig. 2.4.

2.2.2. The mean of a frequency distribution

The arithmetic mean of a set of n values, x_1, x_2, x_3, ... x_n, is defined as

$$\overline{x} = \frac{1}{n} \sum_{i=1}^{n} x_i \qquad (2.2)$$

If the value x_1 occurs f_1 times, if x_2 occurs f_2 times and in general if x_i occurs f_i times, then the total number of measurements is given by

$$n = \sum_{i=1}^{k} f_i \tag{2.3}$$

The arithmetic mean is given by

$$\bar{x} = \frac{x_1 f_1 + x_2 f_2 + \ldots + x_k f_k}{n} = \frac{1}{n} \sum_{i=1}^{k} x_i f_i \tag{2.4}$$

When the data are grouped into classes, all measurements of a given class are considered to be equal to the middle of the class. By defining x_i as the middle of class i and if n is the total number of measurements, eqn. 2.4 gives the mean of the frequency distribution.

2.2.3. The variance and standard deviation of a frequency distribution

The mean of a frequency distribution gives a central value of the measurements that is representative of the data. In addition to this central value, it is also interesting to know the extent to which the different measurements are concentrated (or spread) around the mean.

In Fig. 2.4, two frequency distributions (histograms) are shown with the same mean value but with different concentrations around the mean value. The distribution in Fig. 2.4a is much less concentrated around its mean than that in Fig. 2.4b. This can be described by saying that the distance between the values of the measurements x_1, x_2 ... is, on average, larger for the distribution in Fig. 2.4a. A mathematical measure of concentration of the measurements is given by

$$s^2 = \frac{1}{n} \sum_{i=1}^{k} f_i \, (x_i - \bar{x})^2 \tag{2.5}$$

The squares of the differences are taken because it is necessary to prevent differences with opposite signs from being eliminated. Another measure of the concentration is given by the variance defined as

$$\hat{\sigma}^2 = \frac{1}{n-1} \sum_{i=1}^{k} f_i \, (x_i - \bar{x})^2 \qquad (2.6)$$

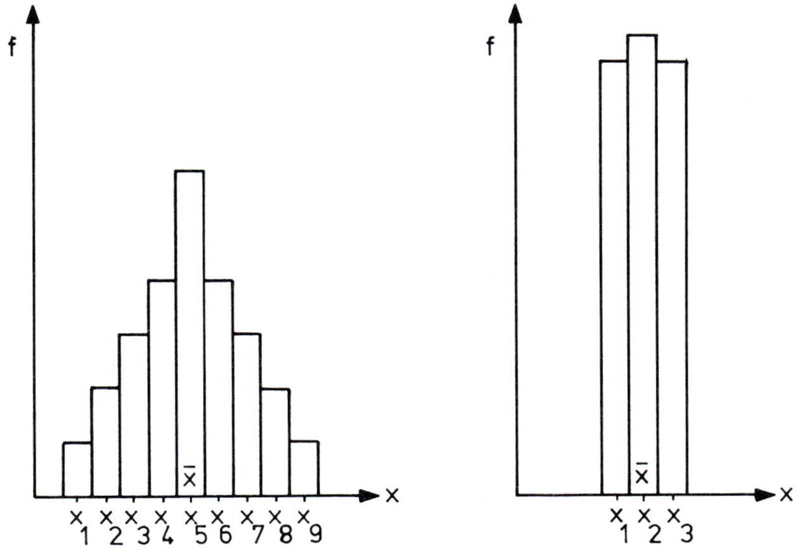

Fig. 2.4. Two histograms with different concentrations around the mean value.

The reason for dividing by n-1 instead of n in eqn. 2.6 is that the resulting value $\hat{\sigma}^2$ represents a better estimate of the variance of the entire population from which the sample is taken. For large values of n (> 25) there is virtually no difference between $\hat{\sigma}^2$ and S^2. The standard deviation, $\hat{\sigma}$, is defined as the square root of the variance :

$$\hat{\sigma} = \sqrt{\frac{1}{n-1} \sum_{i=1}^{k} f_i \, (x_i - \bar{x})^2} \qquad (2.7)$$

2.2.4. Discrete and continuous random variables

2.2.4.1. Discrete and continuous variables

A real function associates a real number with each element of a set or of a population. When considering an unspecified element of the population the real function is called a variable.

If the set or population contains a finite number of elements, or if the elements can be counted (in the same way as integers) the variable is called discrete. If this is not the case then the variable is called continuous.

2.2.4.2. Random variables

When there is an element of chance or probability associated with a variable it is generally called a random variable.

Very often the number of elements in the set or population is very large and then the best way of describing the random variable is by using a continuous function. For example the concentration of glucose in blood for the inhabitants of a country can be considered as a continuous random variable.

In other cases, when the number of elements is small or when the values taken by the random variable are well differentiated a discrete random variable is more appropriate. An example is the set of results from an experiment, where the probabilities form a discrete random variable. Another example is when a sample is drawn from a population. The set of values of the random variable for the elements of the sample is also a discrete random variable.

An intermediate case occurs when the sample drawn from the population is very large. In this case a widely used technique consists in grouping the values of the random variable into subsets and computing the number of elements in each subgroup. An example of this technique was given in section 2.2.1.

2.2.4.3. Characterization of continuous random variables

A continuous random variable, x, is characterized by its cumulative probability function, $F(x)$. This function gives, for each value of x that the variable can take, the probability that its value will be less than or equal to x. This function corresponds to the relative cumulative frequency distribution function defined in section 2.2.1 and it is also called the distribution function.

The changes or variations in the cumulative probability function F(x) correspond to the probabilities that the random variable will be equal to x. They are given by

$$f(x) = \frac{dF(x)}{dx} \tag{2.8}$$

This function is called the probability distribution function, or sometimes the (probability) density function.

When the variable is discrete these f(x) are the probabilities or frequencies of the different elements. When the elements are grouped as in section 2.1.1, they are called the relative frequencies.

2.2.4.4. Examples

(a) When a variable x can take all of the values of an interval (a, b) with equal probabilities, the following probability function is obtained :

$f(x) = k \qquad a \leqslant x \leqslant b$

$f(x) = 0 \qquad \text{elsewhere}$

where k is a constant. Such a function is called a rectangular probability

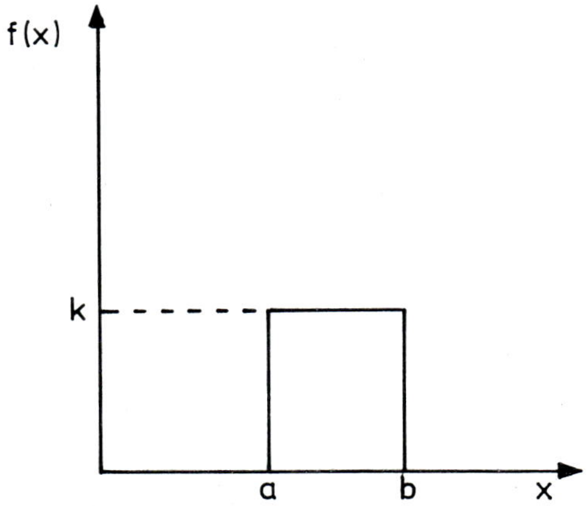

Fig. 2.5. A rectangular probability distribution function.

distribution function. A graphical representation of this function is shown in Fig. 2.5.

To calculate the value of k, it can be observed that the total probability of x having a value between a and b is equal to unity. As the variable x is continuous, this probability is calculated by taking the integral of f(x), which gives

$$\int_a^b f(x)dx = \int_a^b kdx = k \int_a^b dx = k\,[\,x\,]_a^b = k(b-a) = 1$$

Therefore $k = \frac{1}{b-a}$.

(b) An example of a similar discrete random variable is given by the values obtained on tossing a die. These are given by the following probability function :

$$p_1 = p_2 = p_3 = p_4 = p_5 = p_6 = \frac{1}{6}$$

The continuous function f(x) is now replaced by probabilities p_1, p_2, ... p_6.

The probabilities p_i have the property that their sum must be equal to unity, which corresponds to the property of the probability distribution function that its integral is equal to unity.

2.2.5. Parameters of a continuous random variable and their estimation

The mean value of a discrete random variable is given by

$$\overline{x} = \sum_{i=1}^{k} x_i f_i / n \qquad (2.9)$$

where the x_i values are the values taken by the variable. When considering a continuous random variable, the sum is replaced with an integral over the range of the variable. The mean value of a continuous random variable x is denoted by E(x) where, in general, the symbol E( ) stands for expectancy of ( ) :

$$E(x) = \int_{-\infty}^{+\infty} x\, f(x)\, dx \qquad (2.10)$$

This value is also called the expected value of x. The expected value of the jth power of x is called the jth moment of the variable :

$$E(x^j) = \int_{-\infty}^{+\infty} x^j\, f(x)\, dx \qquad (2.11)$$

It is also denoted by μ_j. The first moment, μ_1, is the mean value, which is usually denoted by μ. The expected value of the jth power of the difference between the variable and its mean value is called the jth moment around the mean :

$$E\left[(x - E(x))^j\right] = \int_{-\infty}^{+\infty} (x - E(x))^j\, f(x)\, dx$$

It is also denoted by μ_j' .

The variance of a random variable is defined as μ_2' :

$$\mathrm{Var}\,(x) = \mu_2' = E\left[(x - E(x))^2\right] \qquad (2.12)$$

It can be shown that

$$\mathrm{Var}\,(x) = E(x^2) - (E(x))^2 = \mu_2 - (\mu_1)^2 \qquad (2.13)$$

The variance is often written as σ^2. Its square root, σ, is called the standard deviation.

Example

Let us consider a random variable x with a rectangular probability distribution function

$$f(x) = \frac{1}{b-a} \qquad a \leqslant x \leqslant b$$

$$f(x) = 0 \qquad x < a \quad \text{or} \quad x > b \, ,$$

The mean value is given by

$$\mu_1 = E(x) = \int_{-\infty}^{+\infty} x.f(x)\,dx = \int_a^b x.\frac{1}{b-a}\,dx = \frac{a+b}{2}$$

and the variance is given by

$$\sigma^2 = \text{Var}(x) = E(x^2) - (E(x))^2$$

$$E(x^2) = \int_a^b x^2\,\frac{1}{b-a}\,dx = \frac{1}{3}\,\frac{1}{b-a}\,.\,(b^3-a^3) = \frac{1}{3}\,.\,(b^2+ ab + a^2)$$

$$\sigma^2 = \frac{1}{3}\,(b^2+ ab + a^2) - (\frac{a+b}{2})^2 = \frac{1}{12}\,(b - a)^2$$

It should be noted that, as shown in the examples, the terms standard deviation and variance are not confined to normal distributions, as is sometimes believed. At this point it is also necessary to make an important distinction, between population parameters and their estimators, i.e., the functions used to estimate these population parameters. To make this possible, one often adds "hat" to a parameter to denote the estimator. As the estimator of a population parameter is a function of measurements, it is itself a random variable possessing a probability distribution and its performance can be judged from the parameters of this distribution. If the mean value of the distribution of the estimator is equal to the parameter which it must estimate, it is called an unbiased estimator. For example, an unbiased estimator of the population mean μ is the sample mean $\overline{x}$ as defined in section 2.2.2. The estimator is unbiased as the expected value of $\overline{x}$ is the mean μ :

$$E(\overline{x}) = E(\frac{1}{n}\sum_{i=1}^{k} x_i f_i) = \frac{1}{n}\sum_{i=1}^{k} f_i\,E(x_i)$$

$$= \frac{1}{n}\sum_{i=1}^{k} f_i \mu = \frac{\mu}{n}\,n = \mu$$

This can be written as

$$\hat{\mu} = \overline{x} \tag{2.14}$$

The median (the middle value of a set of numbers arranged in order of magnitude) is also an unbiased estimator of the mean μ.

The distribution of $\overline{x}$ values has a variance of σ^2/n. The distribution of the median is not derived so easily but, in any case (except when the sample size is 2), the variance of the distribution for the mean is smaller than for the distribution of the median, which is why the former is preferred. An unbiased estimator of the standard deviation is given by

$$\hat{\sigma}^2 = \frac{1}{n-1} \sum_{i=1}^{n} (x_i - \overline{x})^2 \tag{2.15}$$

and a biased estimator is given by

$$S^2 = \frac{1}{n} \sum_{i=1}^{n} (x_i - \overline{x})^2 \tag{2.16}$$

The former expression is therefore used in most instances.

IMPORTANT NOTE :

There is a discrepancy between the notation used by analytical chemists and many statisticians, the former using the symbol s where the latter use $\hat{\sigma}$. In this text, we shall not use the biased estimator (equation 2.16) without expressly stating so. The symbol s is therefore equivalent to $\hat{\sigma}$ in this book. The symbol S will be used to relate the signal to a concentration (see also Chapter 6).

An extension of the notion of expectancy or mathematical expectation is obtained when considering functions of random variables. If x is a random variable with a probability distribution function f(x) and g(x) is a function of x, the mathematical expectation of g(x) is defined as the expected value of the function and is given by the equation

$$E(g(x)) = \int_{-\infty}^{+\infty} g(x)\ f(x)\ dx$$

An important case arises when considering the simple linear function

$$g(x) = ax$$

where a is constant. The mathematical expectation of g(x) is given by

$$E(ax) = \int_{-\infty}^{+\infty} ax\ f(x)\ dx = a\ E(x)$$

In the same way, it is possible to define the variance of a function g(x) as its second moment around its mathematical expectation :

$$\text{Var}\ (g(x)) = E\ [(g(x) - E(g(x))\ )^2\ ]$$

Taking the linear function g(x) = ax, we obtain

$$\begin{aligned} \text{Var}\ (ax) &= E\ ((ax - a\ E(x))^2\ ) = E\ (a^2(x - E(x))^2\ ) \\ &= a^2\ \text{Var}\ (x) \end{aligned}$$

This is of importance in analytical chemistry where a signal y is related to a concentration x through a constant S (see Chapter 6). If y = Sx, then it follows from the foregoing equation that $\mu(y) = S\mu(x)$ and $\sigma(y) = S\sigma(x)$.

2.2.6. Some special distributions

2.2.6.1. The normal distribution

The probability function of a normal distribution is given by

$$f(x) = \frac{1}{\sigma\sqrt{2\pi}}\ e^{-\frac{1}{2}(\frac{x-\mu}{\sigma})^2} \qquad (2.17)$$

where μ and σ are the mean value and standard deviation, respectively, of this probability function. By analogy with the relative cumulative frequency distribution, the cumulative frequency distribution function of the normal distribution is given by

$$F(x) = \int_{-\infty}^{x} \frac{1}{\sigma\sqrt{2\pi}} e^{-\frac{1}{2}\left(\frac{x-\mu}{\sigma}\right)^2} dx \qquad (2.18)$$

When a variable x has a normal distribution with a mean value μ and variance σ^2, this is written as

$$x \sim N(\mu,\sigma^2)$$

An important particular case arises when the mean value is zero and the variance is unity, which is called a standard or reduced normal variable

$$x \sim N(0,1)$$

In this instance the probability function is given by

$$f(x) = \frac{1}{\sqrt{2\pi}} e^{-\frac{x^2}{2}} \qquad (2.19)$$

and its cumulative frequency distribution function is given by

$$F(x) = \int_{-\infty}^{x} \frac{1}{\sqrt{2\pi}} e^{-\frac{x^2}{2}} dx \qquad (2.20)$$

The functions are illustrated in Fig. 2.6.

It can be shown that if a variable x has an $N(\mu,\sigma^2)$ distribution, the variable $z = (x-\mu)/\sigma$ has an $N(0,1)$ distribution ; z is called the reduced variable of x.

Values of the cumulative distribution function of z are given in the Appendix.

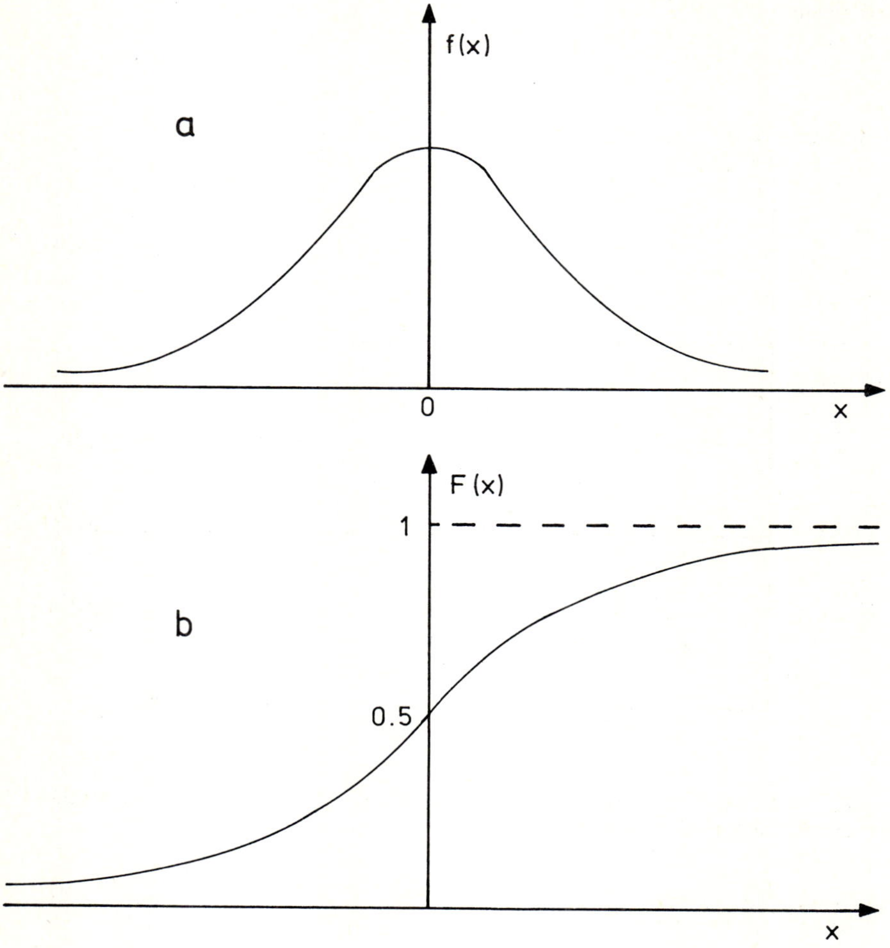

Fig. 2.6. The normal probability distribution function (a) and the resulting cumulative frequency distribution function (b).

In order to check whether a frequency distribution can be approximated by a normal distribution, one can use probability graph paper. Therefore, the given frequency distribution is converted into a cumulative frequency distribution. The cumulative relative frequencies are plotted against the upper class boundaries on probability graph paper. If a straight line is obtained, one can state that a normal distribution closely fits the data. Indeed, as $z = \frac{x - \mu}{\sigma}$, there is a linear relationship between z and x when x is normally distributed.

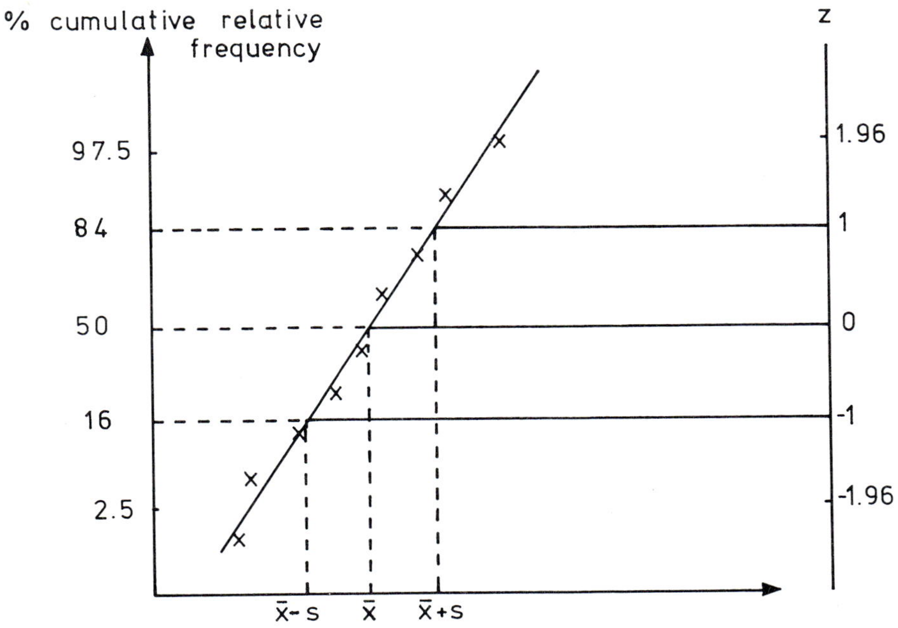

Fig. 2.7. Probability graph : % cumulative relative frequencies plotted against the upper class boundaries.

2.2.6.2. The chi-square distribution

If $x_1, x_2, \ldots, x_k$ are independent normal variables with a mean value of zero and a standard deviation of unity, then the variable

$$\chi_k^2 = x_1^2 + x_2^2 + \ldots + x_k^2 \tag{2.21}$$

is said to have a chi-square distribution with k degrees of freedom. Values for the chi-square distribution are given in Table II (Appendix). The χ^2 distribution is depicted in Fig. 2.8.

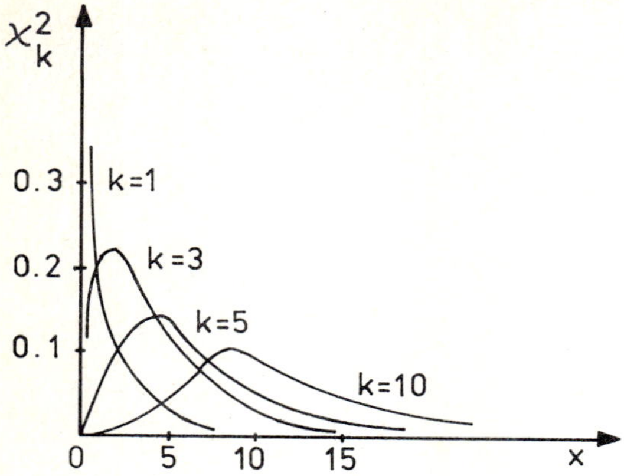

Fig. 2.8. χ^2 probability distribution function.

2.2.6.3. The t-distribution

Let us consider an N(0,1) normal variable z and a χ_k^2 variable independent of z. The variable t_k, given by

$$t_k = \frac{z}{\sqrt{\frac{\chi_k^2}{k}}} \tag{2.22}$$

is said to have a t-distribution with k degrees of freedom. It is also called a Student's distribution (see Fig. 2.9). Values of the cumulative distribution function of t_k are given in Table III (Appendix).

2.2.6.4. The F-distribution

The ratio

$$F_{k,m} = \frac{\chi_k^2 / k}{\chi_m^2 / m} \tag{2.23}$$

is said to have an F or Fisher-Snedecor distribution if the two chi-square distributions χ_k^2 and χ_m^2 are independent. The parameters of this distribution, which is depicted in Fig. 2.10, are k and m. Values of the F-distribution are tabulated in Table IV (Appendix). The normal distribution, the t, the χ^2 and the F distributions are used in many instances in the following chapters.

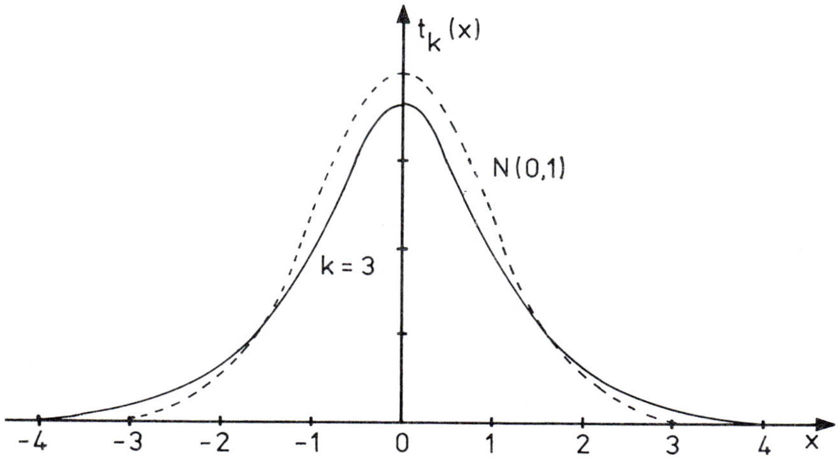

Fig. 2.9. The t-distribution for k = 3 compared with N(0,1).

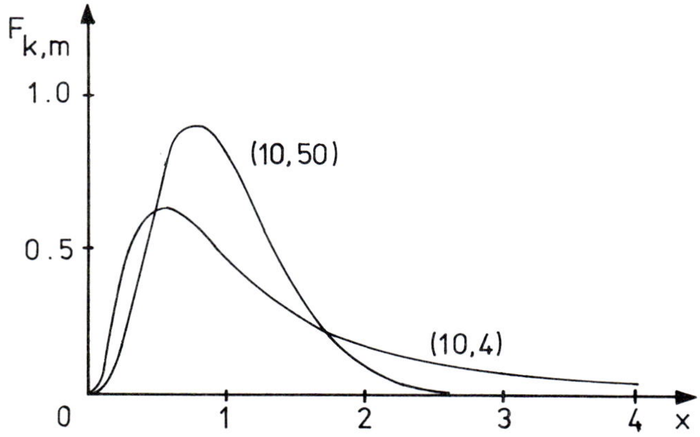

Fig. 2.10. The F-distribution.

REFERENCES

Analytical Chemistry, Guide for use of terms in reporting data, Anal. Chem.,47 (1975) 2527.
T. Aronsson, C.-H. de Verdier, T. Groth, Clin. Chem., 20 (1974) 738.
P.M.G. Broughton, M.A. Buttolph, A.H. Gowenlock, D.W. Neill and R.G. Skentelbury, J. Clin. Pathol., 22 (1969) 278.
J. Büttner, R. Borth, J.H. Boutwell, P.M.G. Broughton and R.C. Bowyer, Clin. Chim. Acta, 69 (1976) F1.
L.A. Currie, J.J. Filliben and J.R. DeVoe, Anal. Chem., 44 (1972) 497R.
B.E. Cooper, Statistics for experimentalists, Pergamon, Oxford, 1st ed., 1969, reprinted 1975.
O.J. Dunn, Basic statistics, a primer for the biomedical sciences, Wiley, New York, 1964.
G. Ekedahl, P. Junker and B. Röndell, J. Water Pollut. Contr. Fed., 47 (1975) 858.
M.W. Formo, G.R. Honold and D.B. McLean, J. Ass. Offic. Anal. Chem. 57 (1974) 841.
J.D. Ingle, Jr., Anal. Chim. Acta, 88 (1977) 131.
International Organization for Standardization, ISO/TC 69, 1966.
IUPAC, Analytical Chemistry Division, Commission on Analytical Nomenclature, report prepared by R.W. Fennel, T.S. West, Pure Appl. Chem., 18 (1969) 439.
IUPAC, Analytical Chemistry Division, Commission on spectrochemical and other Optical procedures for analysis, Pure Appl. Chem., 45 (1976) 99, reproduced in Anal. Chem., 48 (1976) 2295.
M.G. Kendall and A. Stuart, The advanced theory of statistics, Vol I, 3rd ed., 1969, Vol II, 2nd ed., 1967, Vol III, 2nd ed., 1968, Charles Griffin, London.
E.F. McFarren, R.J. Lishka and J.H. Parker, Anal. Chem., 42 (1970) 358.
D. Midgley, Anal. Chem., 49 (1977) 510.
D.F. Morrison, Multivariate statistical methods, McGraw-Hill, New York, 1969.
R. Püschel, Mikrochimica Acta (Wien), (1968) 783.
D.B. Tonks, Clin. Chem., 9 (1963) 217.
F.H.B. Vermeulen, Neth. Journal of Agric. Science, 5 (1957) 221.
G. Wernimont, Anal. Chem., 23 (1951) 1572.
J.O. Westgard, R.N. Carey and S. Wold, Clin. Chem., 20 (1974) 285.
A.L. Wilson, Talanta, 17 (1970) 31.
W.J. Youden and E.H. Steiner, Statistical Manual of the Association of Official Analytical Chemists, The Association of Official Analytical Chemists, Washington, D.C., 1975.

Chapter 3

EVALUATION OF PRECISION AND ACCURACY - COMPARISON OF TWO PROCEDURES *

3.1. GENERAL DISCUSSION OF METHODS AND CONCEPTS

3.1.1. Introduction

One of the simpler ways of optimizing a particular analytical problem is to compare several methods according to their performance characteristics and to select the best one. Often one procedure is already being used and one can consider replacing it with a cheaper or faster procedure or, in general, with a procedure with more desirable characteristics. A prerequisite for doing this is that the new method should be accurate, i.e., free from method bias, and this aspect is the main concern of this chapter.

The simplest means of obtaining some idea of the accuracy of a method is to use it to analyse a standard or reference material for which the concentration of the analyte is known with high accuracy and precision. The difference between the known true value and the mean of replicate determinations with the "test" method is due to the sum of method bias and random errors. It is therefore necessary to estimate the proportion of each type of error, and the strategy used for this purpose is to investigate first whether the deviation can be explained by random errors alone. This is done with a t-test and, when the answer is that the deviation can indeed be assigned to random errors, the method is considered to be accurate. If not, the deviation is considered to be a measure of the bias. Often, it is simply stated that the deviation is equal to the bias, whereas it is, of course, only an estimate of the method bias.

---

* This chapter has been written with the collaboration of Y. Michotte, Pharmaceutical Institute, Vrije Universiteit Brussel, Belgium.

This procedure of investigating accuracy has the disadvantage that the result is valid only for the particular reference material used. Often no standard material of known concentration is available. In this instance, one often compares the method being investigated or "test method" with an existing method called the "reference method", for which it is usually assumed that there is no method or laboratory bias. There may be satisfactory reasons for this assumption but often, however, the person developing a new method for a particular determination takes an existing method from the literature as a reference method. In view of our discussion of the components of inter-laboratory precision in the previous chapter, this is a hazardous assumption and, for this reason, the final evaluation of a method should preferably be carried out in an inter-laboratory study.

When one assumes that one possesses an accurate reference method, the reference and test methods are used to carry out a number of determinations. Sometimes one analyses replicates from the same sample but in this instance one will learn only whether the method is accurate for the particular material being analysed. It is therefore preferable to analyse a range of samples with both methods. The results obtained can be used in several ways :

(1) Ideally the results sbuld be completely correlated, i.e., the correlation coefficient (r) should be equal to unity. The correlation coefficient, however, cannot be interpreted directly in terms of accuracy. For example, does r = 0.95 mean that the method should be considered accurate or not ? Therefore, a calculation of the correlation coefficient will serve only as a preliminary indication and it will not be discussed further.

(2) Tests can be applied to investigate whether the differences obtained are significant or not. According to whether one assumes a normal distribution of errors or does not make any assumptions, a t-test (section 3.1.2) or a non-parametric test (section 3.1.3) will be carried out.

(3) If one plots the results from one method against those from the other, the regression line should ideally pass through the origin and have a slope of

unity. The intercept on the ordinate is therefore a measure of method bias, while the slope is a measure of proportional error. The standard deviation can also be calculated and is a measure of the precision. The application of regression analysis to method comparison is discussed in section 3.1.4. In section 3.1.5, the application of this technique to recovery experiments (standard addition techniques), used to detect proportional errors, is also considered.

(4) The standard deviations for the replicate analysis of one sample by two methods can be compared using the F-test (section 3.1.6).

So far, we have considered a comparison of two methods. More than two methods can be investigated by using the analysis of variance technique or the much less used principal components method (Carey et al., 1975). The latter method is discussed in Chapter 19 and the former in Chapter 4.

The literature abounds with examples of the evaluation of precision and accuracy. Extensive schemes have been proposed by some workers, including those written for clinical chemists by Barnett and Youden (1970) and for official analytical chemists by Youden and Steiner (1975). Both of these schemes include methods for the evaluation of precision and accuracy in a very simple way but with little detail about the underlying mathematics, as they are intended for users with little statistical knowledge. The procedures proposed, however, are correct and often very efficient. An interesting paper on method comparison studies was published by Westgard and Hunt (1973), and contains a simulation study of the errors that have an effect on the precision and accuracy of methods. This enables them to show clearly the limitations of the different statistical procedures used for method comparison purposes. A thorough, but unfortunately for most users too complex series of papers on method evaluation was published by Gottschalk (1976). They should be read, however, by every worker who has a more fundamental interest in this topic.

3.1.2. Evaluation of method bias using tests on the mean (t-test)

When one analyses a standard or reference sample (such as those proposed by organizations such as ASTM, NBS and IAEA) with a new method, one will have to decide whether the result obtained differs significantly or not from the stated concentration. The stated concentration is the mean obtained with a large number of careful determinations by the organization issuing the sample, while the result obtained with the new ("test") method is the mean of a number of replicate determinations. Statistically, one therefore compares the means of two populations. In practice, it is often impossible to carry out a meaningful statistical test as the only population parameter given for the reference material is the mean. Often no standard deviation is given.

Reference samples of this type have real value only when they have been certified with sufficient care. An example of how this should be done is the certification procedure used by the National Bureau of Standards (NBS) (Cali, 1976). The NBS uses three principal modes of measurement of reference samples : measurement with a method of known accuracy, by at least two analysts working independently ; measurement with at least two independent methods, the estimated accuracies of which are good compared with the accuracy required for certification ; and measurement according to a collaborative scheme, incorporating qualified laboratories.

Therefore, although there is an obvious need for standard materials, we would recommend individual laboratories and organizations not to issue their own standard materials except when unavoidable, but to leave it to the few organizations that have long and established experience in this field. The reservations made in the preceding paragraphs do not mean that individual laboratories should not test their methods by comparison with a reference method or a standard material with incomplete statistical information. It is obviously better to make a study of the accuracy of a method with the reference materials that are available, however imperfect the statistical data may be rather than make no study at all of the accuracy of the proposed method. However, it is necessary that the limitations in the conclusions that can be drawn from such comparisons should be borne in mind.

After these introductory cautionary remarks, we can turn to the statistical methodology. One has to decide whether or not there is a significant difference between the stated value which is accepted as the true value, μ_0, and its experimental estimate, $\bar{x}$, obtained with the test method. Let us investigate first the case where μ_0 has been determined with high precision so that the standard deviation can be considered to approximate to zero. The use of the symbol μ_0, which we defined as the "true" value of a sample in Chapter 2, can be criticized, as one has no way of being completely sure about this true value. However, we state here that the value given for the reference sample is by definition equal to the true value. It is intuitively clear that one has to take into account the difference between $\bar{x}$ and μ_0 and the precision on the determinations of $\bar{x}$. The smaller the ratio between $|\bar{x} - \mu_0|$ and s, the less probable it becomes that there is a method bias. Student has shown that one should calculate the value

$$t = \frac{\bar{x} - \mu_0}{s/\sqrt{n}} = \frac{\bar{x} - \mu_0}{s} \cdot \sqrt{n} \qquad (3.1)$$

where n is the number of determinations with the test method. The incorporation of n in the equation originates from the fact that the standard deviation that must be used is the standard deviation of the population of averages which is equal to $s/\sqrt{n}$. The larger t, the higher is the probability that the difference is not due to random errors and is therefore significant. This probability can be found in statistical tables. It is a function of the number of degrees of freedom, which in this instance is n - 1. In the tables for 20 degrees of freedom (i.e. for an experimental set-up with n = 21) and a probability level of 99%, the value 2.84 is found. If an experimental value equal to or greater than 2.84 is obtained, this means that the probability, that the observed difference is due to chance is 1% or less.

When the standard deviation on the reference sample is not negligible, it must be taken into account as an additional source of variation. The mean value given for the reference sample is an estimate of its true value and we shall represent it here by $\hat{\mu}$. The term t now becomes

$$t = \frac{\bar{x} - \hat{\mu}}{(s_1^2/n_1 + s_2^2/n_2)^{1/2}} \quad (3.2)$$

where n_1 and n_2 are the number of replicates on which the estimates s_1 (test method) and s_2 (reference sample) are based. The same equation can also be used when the test procedure is compared with a reference procedure by analysing the sample with both procedures. Then it is preferable to re-write eqn. 3.2 as

$$t = \frac{\bar{x}_1 - \bar{x}_2}{s\sqrt{1/n_1 + 1/n_2}} \quad (3.3)$$

where s is a pooled estimate of the standard deviation. It can be calculated in the following way

$$s^2 = \frac{\sum_{i=1}^{n_1} (x_{1i} - \bar{x}_1)^2 + \sum_{j=1}^{n_2} (x_{2j} - \bar{x}_2)^2}{n_1 + n_2 - 2} \quad (3.4)$$

where x_{1i}, $\bar{x}_1$ and n_1 refer to the test method and x_{2j}, $\bar{x}_2$ and n_2 to the reference method. The use of a pooled variance assumes that the variances (or the precisions) of both methods are identical (or do not differ too much). When the precisions cannot be considered to be identical, a more complicated calculation is necessary (see, for example, Lark et al., 1969).

This evaluation procedure enables one to conclude only that the method is accurate (or not) for the analysis of a sample of that particular concentration. In order to obtain more general conclusions, one can carry out one determination with each method on n different samples, which should preferably include a sufficient variety of matrices and a range of concentrations. The question to be asked now is whether the differences, d_i, between the results of the two methods are significantly different from zero. If this is not so, the methods will be considered to give the same result. Another consequence is that the differences between d_i and zero should then be due to random errors. One could say that $\overline{d}$, the mean value of d_i, is compared with the reference value zero. In statistical terminology, one says that the null hypothesis is that the true mean of the d_i values is zero. Mathematically, this is analogous to the first reference sample case investigated in this section. The t-test is therefore applied with s_d, the standard deviation on d

$$t = \frac{\overline{d} - 0}{s_d} \cdot \sqrt{n} \qquad (3.5)$$

Of all the evaluation procedures described in this section, the last one is the most desirable. One should be aware, however, of its limitations. In particular, the t-test will yield erroneous results in the following cases :

(a) if a systematic error is caused in only one or a few of the samples by an interferent present only in those samples, the random error in the samples can mask the systematic error, or else the systematic error in one sample may lead to such a high t-value that it is concluded that the method is generally inaccurate ;

(b) the t-test is valid for a constant systematic error or proportional errors in a very restricted concentration range but not for proportional errors over a wider range, as the research hypothesis (see section 3.2) is that the difference between both procedures (populations in statistical terminology) is independent of the concentration. Proportional errors depend on the concentrations so that the t-test is not valid. This was shown very elegantly by a simulation of method comparison studies by Westgard and Hunt (1973).

As Part I of this book is devoted to criteria, it should be stressed that the t-test enables one to investigate only whether a procedure is accurate or not or, more precisely, how large the probability is that it is accurate. The t-value should not be used, however, as a numerical criterion. Westgard and Hunt (1973) gave one reason for this in their study : t is a ratio of constant and random errors, whereas the quantity of importance to the user is the total error. A large difference term and a large standard deviation may yield a low t-value, indicating that the method is apparently acceptable when in fact it is not. Another reason is that the t-value depends on the number of observations. If one wants to use the result of a t-test as a criterion, one should employ the probability that the test is accurate obtained from the t-table.

In section 3.1.1, it was argued that method comparison studies should preferably be carried out on an inter-laboratory basis, so as to take into consideration the effect of laboratory biases on test and reference methods. This is not true when laboratory biases are considered to be of little importance.

One situation of this nature sometimes occurs in clinical laboratories. Clinical chemists are often less concerned with intercomparisons of their data with those of their colleagues from other laboratories than with the internal consistency of their data. A clinical laboratory which carries out statistical control will determine its own normal values for a particular test, thereby eliminating the importance of laboratory bias, or else adjust the values by the analysis of control sera. Therefore, a concept such as the inter-laboratory precision or laboratory bias is of less importance than it is to official analysts. To be fair, it should be noted that there is a trend towards more inter-laboratory quality control - proficiency testing - in the clinical laboratory.

On the other hand, it is vital for the validity of the statistical evaluation of clinical chemistry data to take into account the day-to-day precision, as clinical laboratories carry out the same tests daily for long periods. Therefore, a t-test between a standard method and a newly evaluated method can be carried out in the following way (Barnett and Youden, 1970). Samples from five patients or less are collected and analysed with both methods on successive days until a total of 40 samples has been analysed. In this way, the standard deviation used in the t-test will be representative for day-to-day precision, which is more meaningful than within-day precision, and it will also incorporate the effect of interfering substances (such as drugs) which affect patient values.

3.1.3. Non-parametric tests for the comparison of methods

In the previous sections, the comparison of methods by using the t-test was discussed. When using such tests, one implicitly accepts that the results for each method are normally distributed, but often this is not so or at least it cannot be proved conclusively. In fact, according to some studies, it seems that normal distributions are obtained in few instances. Clancey (1947) examined approximately 250 distributions for a total of 50,000 determinations of samples such as metals and alloys. According to his results, only 10-15% of the distributions were normal, 15% were truncated normal curves, 10% were symmetrical but high-peaked (leptokurtic) compared with normal, 20-25% were skewed, 20-25% were J-shaped and a few were bimodal. A number of reasons why non-gaussian distributions are obtained were given by Thompson and Howarth (1976). These include, for example, heterogeneity of samples, rounding off (producing a discontinuous distribution) and measurements near the detection limit (with sub-zero readings set to zero).

The use of tests based on a normal distribution can then lead to erroneous conclusions. Sometimes a transformation of variables makes it possible to obtain the normal distribution. A detailed discussion of such transformations for use in clinical chemistry was given by Martin et al. (1975), the most common being the log-normal distribution. When no gaussian distribution can be obtained, one can use methods that are not based on a particular distribution (so-called distribution-free methods). These methods do not require calculations of the usual parameters such as the mean or standard deviation and are therefore also called non-parametric. They can also be used for so-called ordinal scales and are discussed under this heading in the mathematical section. These methods have the advantage of being always valid and they require only very simple calculations. Therefore, once one knows these methods, one is tempted to use them on every occasion. However, it should be stressed that they are less efficient and require more replicate measurements than the "normal" methods. A typical and very simple non-parametric method is the so-called sign-test.

Suppose that two methods are compared and that measurements are carried out on n samples with both. If in all n cases method A yields a higher result than method B, it is probable that A and B differ significantly. If on the contrary, n/2 results obtained are higher when A is used and lower for the other n/2 samples, then it is probable that A and B do not yield significantly different results. To put this in probabilistic terms, if the two methods are equivalent, then the chances of obtaining higher results from method A (which we will call positive differences D) and higher results from method B (negative differences) are the same. In other words, the probabilities for $D > 0$ and for $D < 0$ are both 1/2. Let us suppose there are eight measurements and that only one is negative. The probability that at most one negative value would occur by chance is

$$p = \text{probability for 0 negatives} + \text{probability for 1 negative}$$
$$= (1/2)^8 + 8(1/2)^8 = 9(1/2)^8 = 0.035$$

One is then able to reject the null-hypothesis that both methods are equivalent with a 3.5% probability of error.

Wilcoxon's matched-pair test takes into account the values of the differences observed by carrying out two methods on the same n samples. A generalized version for k methods, the Kruskal-Wallis test, also exists. In Wilcoxon's test, one calculates the differences obtained for each sample by subtraction of the result obtained with method B from that obtained with method A. If, for example, method A yields significantly higher results than method B, then there will be more positive differences than negative and the positive differences will be larger. The differences are therefore ranked according to absolute value, with rank 1 for the smallest difference. Suppose the following results are obtained

Sample	A	B	C	Rank
1	11.2	10.9	+ 0.3	2
2	13.7	11.2	+ 2.5	8
3	14.8	12.1	+ 2.7	9
4	11.1	12.4	- 1.3	6
5	15.0	15.6	- 0.5	5
6	16.1	14.6	+ 1.5	7
7	17.3	13.5	+ 3.8	10
8	10.9	10.8	+ 0.1	1
9	10.8	11.2	- 0.4	3
10	11.7	11.2	+ 0.5	4

One then obtains the sum of the ranks of positive (T^+) and negative (T^-) differences.

$$T^+ = 1 + 2 + 4 + 7 + 8 + 9 + 10 = 41$$

$$T^- = 3 + 5 + 6 = 14$$

One compares the value of T^+ (or T^-) with the values in tables to conclude whether or not there is a real difference. This method is discussed in more detail in section 3.2.3, together with another commonly used non-parametric test, the Kolmogoroff-Smirnoff test.

Non-parametric tests have not been used very often in analytical chemistry. Gindler (1975) discussed non-parametric tests applied in the clinical laboratory.

3.1.4. Comparison of two methods by least-squares fitting

3.1.4.1. Philosophy

When the results obtained for a number of samples with the test procedure are plotted against those obtained with the reference procedure, a straight regression line should be obtained. In the absence of error, this line should have a slope, b, of exactly unity and an intercept on the ordinate, a, of zero, and all points should fall on the line. In this book, we have adopted the convention that x values relate to concentrations of a sample and y values to signals used to derive these concentrations. In comparing two procedures by least-squares techniques, we should therefore use the symbols x_1 and x_2. For ease of notation, in this section and in section 3.2.8 we shall represent the concentration obtained with one of these procedures by y and the other by x.

Let us now consider the effects of different kinds of error (see Fig. 3.1). The presence of random errors leads to a scatter of the points around the least-squares line and a slight deviation of the calculated slope and intercept from unity and zero, respectively. The random error can be estimated from the calculation of the standard deviation in the y-direction, s_y (also called the standard deviation of the estimate of y on x).

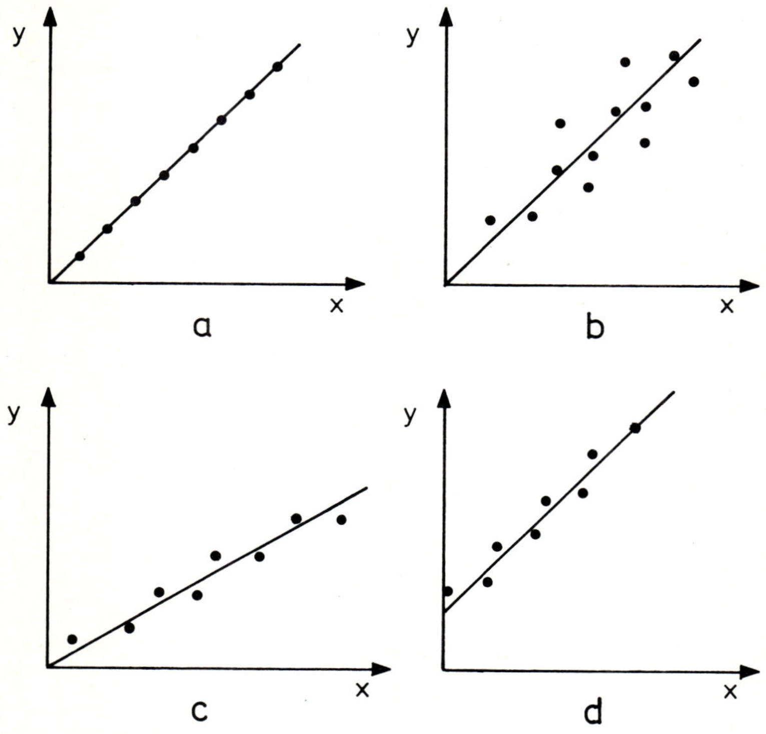

Fig. 3.1. Use of the regression method in the determination of systematic errors. (a) Ideal behaviour ; (b) accurate method with low precision ; (c) effect of proportional error ; (d) effect of constant error.

A proportional systematic error leads to a change in b so that the difference between b and unity gives an estimate of the proportional error. A constant systematic error shows up in a value of the intercept different from zero.

The study of the regression line leads therefore to estimates of the three types of error (random, proportional and constant), which enables one to conclude (see also Westgard and Hunt, 1973) that least-squares analysis is potentially the most useful statistical technique for the comparison of two methods. The least-squares method is a very general technique which enables one to fit data to a theoretical function. One can investigate whether this function does really describe the experimental observations by carrying out a goodness-of-fit test. In the present case, the equation is

$$y = x \tag{3.6}$$

where y = result of the test method and x = result of the reference method. Eqn. 3.6 is a particular case of

$$y = \alpha + \beta x \tag{3.7}$$

If the experimental estimate a for α is close enough to zero and the estimate b for β is close enough to unity, it will be concluded that eqn. 3.6 is true and that there are no systematic errors. This calculation requires two steps :

(1) the determination of a and b from the experimental data ; according to the statistical practice for symbolizing an unbiased estimate, these should, in fact, be called $\hat{\alpha}$ and $\hat{\beta}$, but following the practice in analytical chemistry we shall use a and b ;

(2) a test to investigate whether a and b differ significantly from zero and unity, respectively. These two steps will be described in the next section and in section 3.2.8. These sections closely follow Cooper's (1969) treatment of curve fitting. It should be noted that the regression model as applied here is somewhat arbitrary. The assumption is made that y depends on x, when in fact y and x are both independent methods. One could also make regression calculations assuming that x depends on y. This question and an alternative model are discussed in more detail at the end of section 3.1.4.2.

3.1.4.2. The fitting of a straight line

The available data consist of n pairs of values (x_i, y_i) where the x_i values are obtained with the test method and the y_i values with the reference method. The general situation, where the true relationship between x and y is given by $y = \alpha + \beta x$, is considered first. In the presence of random error this leads to the statistical model

$$y_i = \alpha + \beta x_i + e_i \tag{3.8}$$

where e_i is the random error (distributed normally around zero and with variance σ^2). Eqn. 3.8 is a particular case of the general linear model described in the mathematical section of Chapter 4. From the (x_i, y_i) data, one obtains a and b (estimates of α and β). In the mathematical section it is shown that

$$b = \frac{\sum_{i=1}^{n} (x_i-\bar{x})(y_i-\bar{y})}{\sum_{i=1}^{n} (x_i-\bar{x})^2} \tag{3.9}$$

This equation can be re-written in the more practical form

$$b = \frac{n \sum_{i=1}^{n} x_i y_i - \sum_{i=1}^{n} x_i \sum_{i=1}^{n} y_i}{n \sum_{i=1}^{n} x_i^2 - (\sum_{i=1}^{n} x_i)^2} \tag{3.10}$$

a can be obtained from

$$a = \bar{y} - b\bar{x} \tag{3.11}$$

or directly from

$$a = \frac{\sum_{i=1}^{n} y_i \sum_{i=1}^{n} x_i^2 - \sum_{i=1}^{n} x_i \sum_{i=1}^{n} x_i y_i}{n \sum_{i=1}^{n} x_i^2 - (\sum_{i=1}^{n} x_i)^2} \tag{3.12}$$

As we have introduced at this point the use of regression analysis for estimating the parameters of a straight line, we should comment here on the use of these methods and, in particular, we wish to point out some of the pitfalls that may be encountered in the indiscriminate use of least-squares straight lines. Four errors are commonly made :

(a) the true relationship is not linear ; when in doubt one should check this, for example by extending the range over which the values are obtained. Nonlinear calibration curves are discussed in a recent article by Schwartz (1976) ;

(b) the range of values chosen is so small that the least-squares estimates become unreliable ;

(c) the general equation is used, although it is known that the line must pass through the origin (for example, the calibration line of a colorimetric method). One should then use the particular equation $y = \beta x$;

(d) the estimated relationship is distorted by a few diverging points, which usually happens with points at one of the extremes of the measurement range. Often this reflects the fact that the measurement is less precise at that concentration. Consider, for example, the case of neutron-activation analysis. In this technique, the concentration is derived from a γ-counting measurement, the precision of which is proportional to the square root of the number of counts. As the latter is directly related to the concentration, the measurement is less precise at low than at high concentrations. One can cope with this situation by weighting the observations. The fitting of a straight line to weighted variables was described by Cooper (1969).

A further remark that should be made here concerns the model used in regression analysis. Eqns. 3.10 and 3.12 are obtained by minimizing the squares of the differences, d_1, between experimental results and computed results in the y-direction (see figure 3.2). One could also minimize the differences, d_2, in the x-direction. In fact, some workers present results for both kinds of regression lines. The most logical procedure when errors occur in both y and

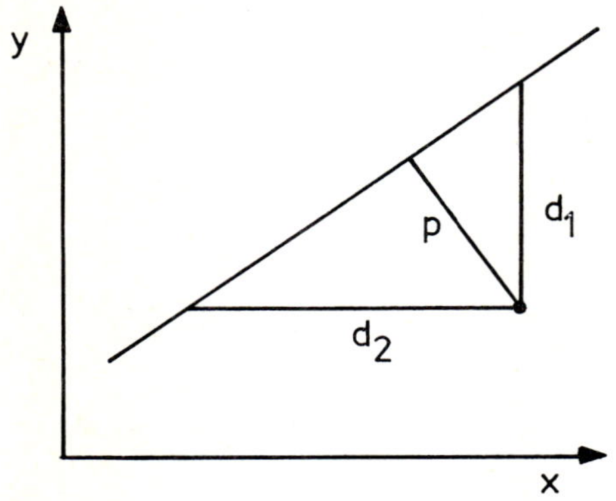

Figure 3.2. Models for regression analysis.

in x, however, is to minimize a distance p measured in a direction perpendicular to a line. Wakkers et al. (1975) proposed equations based on such a model, which they applied to a comparison of clinical analytical methods, and they also showed that this model is more reliable than the usual procedure.

Let us return now to the use of least-squares lines for a comparison of methods. When one has obtained a and b, it will be found that they differ from their ideal values, 0 and 1, even when the relationship y = x (eqn. 3.6) is true, owing to the occurrence of random errors. When a value of b = 0.95 is found, this is usually understood as a proportional error of 5%. It should not be forgotten that b is an estimate and therefore one should ask whether the observed difference is significant or, to put it another way, "does the line y = x fit the data ?". To do this, one must carry out an analysis of variance, as proposed for instance by Cooper (1969), or apply a t-test. The means of doing this is shown in section 3.2.8.

3.1.5. Recovery experiments

Proportional systematic errors are caused by the fact that the calibration line obtained with standards does not have the same slope as the functional relationship between the measurement result and the concentration in the sample or, to use the terminology introduced in Chapter 6, the sensitivity is different for standards and sample.

Consider, for example, neutron-activation analysis. In this technique, the concentration of element a in the unknown, $x_{a,u}$, is estimated by comparing the radioactivity $A_{a,u}$ with the activity $A_{a,s}$ of a standard with known concentration $x_{a,s}$ of a by using the relationship

$$\frac{A_{a,u}}{x_{a,u}} = \frac{A_{a,s}}{x_{a,s}}$$

These ratios are, in fact, the sensitivities in the samples and for the standards. The calculation procedure implies that they do not depend on the composition of the matrix, but analytical chemists know that often this is not so. In neutron-activation analysis, it is possible, for example, that a strongly neutron-absorbing isotope is present in the sample. The activity obtained per gram of substance u will then be smaller, i.e., the ratio $A_{a,u}/x_{a,u}$ is smaller than the ratio $A_{a,s}/x_{a,s}$, and a proportional systematic error is obtained. When such a difficulty is suspected, analytical chemists estimate the content of the unknown by the standard addition method, which requires the determination of a calibration line in the particular sample. Often such an experimental design is used simply to obtain the analytical result, which is outside the scope of this book. However, it can also serve to evaluate the occurrence of proportional systematic errors, and such an approach is then called a recovery experiment. In its simplest form it consists of the addition of a known amount of the analyte, the concentration of the analyte before and after the addition being determined. The difference $x_D = x_{before} - x_{after}$ should ideally be identical with the known amount added, Δx. Owing to the presence of random errors, in general this does not occur. If the standard deviation at both levels of concentration is known, one can test whether or not the difference between Δx and x_D is significant. One can state, for example, that it is considered to be significant when it exceeds twice the standard deviation on x_D. If the standard deviation, s, is assumed to be the same for both levels of concentration, the standard deviation on x_D is equal to $s\sqrt{2}$ and, therefore $\Delta x - x_D$ is considered to be significant when

$$\Delta x - x_D > 2\, s\sqrt{2}$$

One can also carry out several additions of known but different concentrations in such way as to arrive at the determination of the slope of a calibration line in the sample (Fig. 3.3).

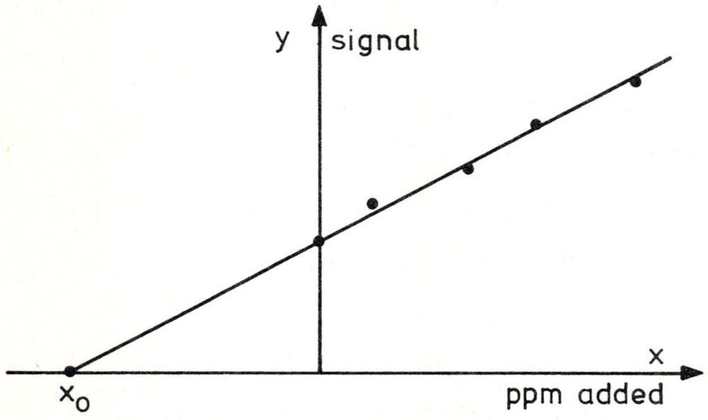

Fig. 3.3. A standard addition experiment.

This procedure can be exploited in several ways :

(a) One can compare the slopes of the regression line obtained in the recovery experiments and of the calibration line obtained with pure standards. These slopes are estimates of a true slope and one should therefore carry out a test to decide whether or not the slopes differ significantly.

(b) A second, but more indirect way, is to compare the results obtained from standard additions with those obtained by using the direct determination. The standard addition result is equal to the value measured without addition divided by the recovery slope. If one uses the measurement values determined from the regression lines $y = a + bx$ instead of the actual measurement results, this is given by a/b. As the intercept on the abscissa, x_0 (see Fig. 3.2) is equal to -a/b, one can determine the standard addition result graphically by measuring this intercept. A test of the significance of the difference between the concentration determined from the direct and the recovery experiments can be obtained in several ways. For example, if replicates of the determinations are carried out one can apply a t-test. One can also calculate the standard deviation on a and b (see Youden, 1951 and Doerffel, 1966) and therefore on the result obtained by the standard addition method. An equation for a confidence interval for the extrapolated line to the abscissa (x_0 in Fig. 3.2) was given by Larsen et al. (1973).

As linear regression lines are used, the remarks made in section 3.1.4.2 should be borne in mind. In particular, it is possible that the linear model does not correspond with reality. As an example, the work of Folsom et al. (1975) can be cited. They found that it is preferable to use an exponential equation of the type $y = A(1-e^{-\lambda x})$ for a standard additions procedure for the determination of sodium and potassium in fish blood.

3.1.6. Comparison of the precision of different methods (F-test)

It is common practice to compare the precision of two or more procedures by carrying out multi-replicate analyses with each of the procedures. This results in standard deviations, which are compared in order to select the most reproducible procedure. It is not always realized that, as standard deviations obtained from measurements are estimates, they are subject to sampling errors. Estimated standard deviations are subject to a distribution, the standard deviation of which is $\sigma_s = \sigma/\sqrt{2n}$, where n is the number of measurements. Therefore the fact that procedures 1 and 2 yield results such that $s_1 > s_2$ does not automatically mean that procedure 2 is more precise. The significance of the difference in standard deviations must be tested. Most analytical chemists know that in the analysis of variance, variances are compared by using the Fisher F-ratio. The same ratio can be used to compare variances in general, which is not always appreciated by analytical chemists comparing the reproducibility of methods.

Let us suppose that one carries out n_1 replicate measurements by using procedure 1 and n_2 replicate measurements by using procedure 2, all on the same sample. One asks whether $\sigma_1 = \sigma_2$. If the null hypothesis is true, then the estimates s_1 and s_2 do not differ very much and their ratio should not differ much from unity. In fact, one uses the ratio of the variances

$$F = \frac{s_1^2}{s_2^2}$$

This ratio is distributed around unity and its mathematical properties are discussed in section 3.2. As there is no *a priori* reason why s_1 should be smaller or larger than s_2, this means that the ratio can be both significantly smaller or larger than unity. If one sets a significance level of, for example, 5%, one has to compare F_{obs}, the observed F-value, with $F_{0.05,(n_1-1),(n_2-1)}$ - (2-sided) or $F_{0.025,(n_1-1),(n_2-1)}$ - (1-sided) from a double entry F table. If F_{obs} is smaller than the F value from the table, one concludes with 95% probability that the procedures are not significantly different in precision.

3.2. MATHEMATICAL SECTION

3.2.1. Theory of statistical tests ; statistical decisions

One of the most important aspects of applied science is the examination of the acceptability of hypotheses derived through theoretical considerations, and the rationalization of this aspect requires an objective technique for accepting or rejecting a hypothesis. Such a technique must be based on quantification of the available information ; it must take into account the risk a scientist is willing to take of making a wrong decision. This risk is the result of considering a sample instead of the entire population. The difference between characteristics of the sample and those of the population can lead to erroneous conclusions. The following procedure, which is a model for statistical decision making, will be used throughout this book. The procedure consists of several steps, which are considered in the following sub-sections.

3.2.1.1. The statement of the hypothesis

Two types of hypotheses will be encountered in statistics. The null hypothesis, H_0, is a hypothesis of no difference, and is the negation of an effect or a difference which has been measured by the scientist. The existence of this effect or difference is called the research hypothesis and is denoted by H_1.

3.2.1.2. The elaboration of the test

Choosing a statistical test for the examination of a hypothesis can present several difficulties. When several tests are available, the conditions for using each of them must be examined. The test is then selected for which these seem to give the best approach of the existing research conditions.

The different statistical models and scales used for constructing tests are considered below.

When a test is selected it must still be decided which level of significance will be given to it. This level, denoted by α, is defined as the probability of rejecting the null hypothesis, H_0, when it is true. This probability, which is defined as a risk, in fact corresponds to a small number of samples that yield extreme results. The error defined here is called the error of the first type and it is usually given an *a priori* maximum value of 1 or 5%.

In addition to this error, it is also possible that the null hypothesis should be accepted when it is false. This possibility is called the error of the second type and its probability is denoted by β. By giving a value to the error of the second type, it can be shown that the sample size is completely determined. In general, the sample size is given together with α and this determines β.

3.2.1.3. The statistical distribution

In general, a statistical test concerns the hypothesis one makes for the value or values of a parameter of a population. As the conclusions are based upon a sample, one must know how the set of samples behaves with regard to the parameter. This behaviour can usually be described by mathematical theorems and, in this way, a statistical test can be selected for the hypothesis. When several tests are available, one chooses that test which, for the same value of α and n, yields a smaller β.

3.2.1.4. The regions and decisions

The set of values of the parameter being studied can be divided into two sub-sets, the region of acceptance and the region of rejection. The region of rejection is defined in such a way that the probability of the parameter falling in it if the null hypothesis, H_0, is true is given by α. The region of acceptance is the set of points outside the region of rejection. Obviously, if the sample yields a value in the region of rejection the null hypothesis, H_0, is rejected and the research hypothesis is accepted.

3.2.1.5. Statistical scales

When selecting a test for solving a statistical problem, various factors must be taken into account, such as the nature of the population being studied, the way the sample was or will be drawn and the type of test to be used. The way in which the measurements are made forms the basis of the mathematical operations necessary for carrying out a test and therefore for testing a hypothesis. The types of measurements used are called statistical scales, and four types can be distinguished : the nominal, ordinal, interval and arithmetical scales. These scales, together with the mathematical operations associated with them, are discussed below.

The nominal scale

The nominal scale, which is mathematically the weakest scale, is used when the only information known about the elements of a sample is its classification into classes or groups. The symbols used for describing the groups or the names of the classes form the nominal scale. For example, when studying the results of a determination of glucose in blood, it is possible to classify the results into two groups : the values outside the normal range (abnormal values) and those within the normal range (normal values). This classification constitutes a nominal scale. As the names or symbols for the different groups only have a classification purpose, any arithmetical operation can be performed on a nominal scale provided that the new values obtained for the classes are differentiated in the same way.

The ordinal scale

It may be possible, in addition to the classification of the elements of a sample into classes, to compare the different classes and to define an order of these classes. If this order is complete, i.e., if each pair of classes can be compared, the scale of classification is called an ordinal scale.

If one considers again a series of glucose results, one can make a classification according to whether the results are below the normal range (low values), within the normal range (normal values) or above the normal range (high values), and an ordinal scale is defined.

One can observe that a classification does not imply a distance between the classes but only a sequence according to which "low values" are situated below "normal values" and normal values below "high values". The arithmetical operations carried out on an ordinal scale must preserve the order of the classification. This means that if an arithmetical operation is performed on two classes, A and B, such that A is smaller than B, the resulting values A' and B' must also satisfy this condition.

It should also be emphasized that statistical tests using parameters such as the arithmetic mean or standard deviation are not valid for data in an ordinal scale, as the distances between groups have no real meaning. Most statistical tests used in an ordinal scale are of the non-parametric type. Some of these tests will be described further in this chapter.

The interval scale

An interval scale has the same properties as the ordinal scale, but in addition the distance between any two classes of the scale can be measured. In an interval scale it is necessary to choose a zero point and a unit of measurement. An example is the Celsius temperature scale, which originally referred all temperatures to the melting point of ice. Each temperature measurement is then located a number of degrees above or below this level.

All arithmetic operations can be performed on an interval scale provided that the relative value of the difference between two measurements is maintained. The allowed operations may therefore change the zero point or the unit of measurement of the scale.

The arithmetical scale

An arithmetical scale has the same properties as the interval scale except that the zero point has an absolute value. Examples of arithmetical scaled variables are absolute temperature, weight and milligrams % glucose in blood. The values of the scale can be multiplied by a constant changing the unit of measurement.

This scale is the strongest statistical scale available. All tests can be carried out under this scale.

3.2.2. Parametric and non-parametric statistical tests

The fundamental assumption made to ensure the validity of a statistical test is that the observations used should be independent and drawn in a random way. Further, in most classical statistical tests assumptions are also made about the nature and shape of the populations being considered.
These assumptions usually imply that the variables involved must have been measured in at least an interval scale. Such tests are called parametric statistical tests. Later two important parametric tests, the t-test and the F-test, will be examined.

More recently, tests have been introduced that do not specify any conditions about the parameters or shape of the population being considered and are called non-parametric statistical tests. These tests are especially important when studying problems that involve variables measured in an ordinal or nominal scale for which no other tests are available or when the distribution is not normal. Two non-parametric tests will be examined : the Kolmogorof-Smirnof test and the Wilcoxon test ; another was introduced in section 3.1.3. It must be observed that the non-parametric tests are more general then the parametric tests as they can also be used for interval and arithmetical scales. On the other hand, when this is done the parametric tests yield more useful results.
A complete review of nonparametric methods in statistics can be found in the books of Siegel (1956) and Conover (1971).

3.2.3. Tests for ordinal scales

3.2.3.1. The one-sample case (Kolmogorov-Smirnov test)

An important problem that arises when studying a population is the distribution of the variable being studied. For an ordinal scale the distribution of the variable can be described only by the frequencies of the different groups or by the relative cumulative frequency distribution.

When an assumption is made about the shape of the relative cumulative frequency distribution of a population, the Kolmogorov-Smirnov test makes it possible, by drawing a sample from the population, to check whether the sample can reasonably be thought to have been extracted from a population with the given relative cumulative frequency distribution. This is achieved by measuring the differences of the observed and the theoretical relative cumulative frequency distributions for each group and calculating the largest of these differences.

Let us call $F_0(x)$ the theoretically assumed relative cumulative frequency distribution and $F(x)$ that measured by using a sample of size n. For each group x the difference of these values is given by

$$D(x) = |F_0(x) - F(x)|$$

and the largest of these differences, D by

$$D = \max D(x)$$

If the hypothetical relative cumulative frequency distribution is correct, it is reasonable that this value should be small. The distribution function of D is called the Kolmogorov-Smirnov function. Values of this function are given in Table V of the Appendix.

3.2.3.2. The two-sample case (Wilcoxon test)

In the Wilcoxon test, it is supposed that a random sample is drawn from the population and that its elements are matched into pairs. Subsequently, one element is chosen randomly from each pair to undergo the experiment and the other is used as a control element. The variables used for the Wilcoxon test must be measurable at least on an ordinal scale, i.e., it must be possible at least to compare the result of the two elements of a pair by saying which is greater or better. Further, it must also be possible to order the differences between the elements of the pairs, i.e., to compare any two pairs in terms of the importance of their difference. Let us call D_i the difference score for the two elements of pair i. All pairs must be ranked in ascending order according to the importance of D_i, ignoring the sign of D_i. After disregarding all pairs for which the two experiments gave the same results, i.e., those for which $D_i = 0$, rank 1 is assigned to the pair with the smallest difference D_i.

Occasionally, two or more pairs yield the same difference, D_i, and in this instance they are all given the same rank. This rank is taken as the average of the ranks they would have received had they all been slightly different. For example if the fourth and fifth differences are equal to 4 and -4, these pairs are both given the rank 4.5.

Under the null hypothesis, H_0, for which there is no difference between the two experiments, one would expect that within the larger ranks approximately as many positive as negative values would occur and that the same should happen with the smaller ranked pairs. Hence, it can be expected that the sum of all ranks for positive differences would be close to the sum for negative differences. Usually these ranks are used in two different ways, depending on the size of the sample. If n_0 is the number of pairs with non-zero difference D_i, we shall distinguish between n_0 values smaller or larger than 25.

Small samples ($n_0 < 25$) :

Let us call T^+ the sum of the ranks corresponding to positive D_i and T^- the sum of the ranks corresponding to negative D_i. Then T, the smallest of these values, is

$$T = \mathrm{Min}\ (T^+, T^-)$$

Values of the cumulative distribution function of T are given in Table VI of the Appendix.

Large samples ($n_0 > 25$) :

It can be shown in this instance that the sum of the ranks, V, given by

$$V = T^+ - T^- \tag{3.13}$$

is normally distributed with the following parameters :

$$\text{Mean} : \mu = \frac{n_0(n_0+1)}{4} \tag{3.14}$$

$$\text{Standard deviation} : \sigma = \frac{n_0(n_0+1)(2n_0+1)}{24} \tag{3.15}$$

The reduced variable z given by

$$z = \frac{V - \mu}{\sigma} \tag{3.16}$$

is normally distributed with mean zero and variance unity.

The above makes it possible to test the hypothesis that V is significantly different from zero. For this purpose, V, μ and σ are calculated with eqns. 3.13, 3.14 and 3.15, which makes it possible to calculate z witn eqn. 3.16.

To test the hypothesis, two values $\zeta_{1-\alpha/2}$ and $\zeta_{\alpha/2}$ are given in Table I of the Appendix. These are equal to the values of the N(0,1) variable for which the distribution function is equal to $1-\alpha/2$ and $\alpha/2$, respectively. If

$$\zeta_{\alpha/2} \leqslant z \leqslant \zeta_{1-\alpha/2}$$

the hypothesis is rejected.

3.2.4. Tests for interval or arithmetical scales

When studying variables in an interval or arithmetical scale, it is useful to examine the value of the parameters of a population obtained by arithmetical calculations. In this section, a random variable x will be examined which will be assumed to have a normal distribution. The tests described in this section will concern the values of the mean value, μ, and the standard deviation, σ.

3.2.4.1. The one-sample case

3.2.4.1.1. Test on the mean with variance known

Suppose it is known that a variable x has a normal distribution $N(\mu,\sigma)$, and further suppose the standard deviation, σ, is known but the mean value, μ, is not. The object of the following test is to establish whether μ is or is not equal to a hypothetical value μ_0. This can be stated as

$H_0 : \mu = \mu_0$: the null hypothesis

$H_1 : \mu \neq \mu_0$: the research hypothesis

To test this hypothesis, it is possible to measure the value of x for each member of the population, to calculate the mean value and to compare it with μ_0. Clearly, when the population is large this approach is impossible, and we therefore suppose that a random sample containing n elements is drawn from the population.
Let us call the mean value of the frequency distribution of the sample $\bar{x}$. If several random samples of size n are drawn from the population, $\bar{x}$ will take different values. It can be shown that $\bar{x}$, which is also a random variable, also has a normal distribution. The mean value of the distribution is μ and its standard deviation is $\sigma/\sqrt{n}$. This can be written as

$$\bar{x} \sim N(\mu,\frac{\sigma^2}{n})$$

The larger n becomes, the smaller is the standard deviation of $\bar{x}$ and the surer we are that $\bar{x}$ will be close to μ.

By reducing $\bar{x}$, we can now obtain an N(0,1) variable

$$z = \frac{\bar{x} - \mu}{\sigma/\sqrt{n}} \sim N(0,1)$$

If the null hypothesis is true ($\mu = \mu_0$), the variable $z = \dfrac{\bar{x} - \mu_0}{\sigma/\sqrt{n}}$ has an N(0,1) distribution. In this instance two points, $z_{\alpha/2}$ and $z_{1-\alpha/2}$, can be found such that the probability of being outside the interval $(z_{\alpha/2}, z_{1-\alpha/2})$ is equal to α. This is illustrated in Fig. 3.4. The values of F(z), the cumulative distribution function of z_α are given in Table I of the Appendix.

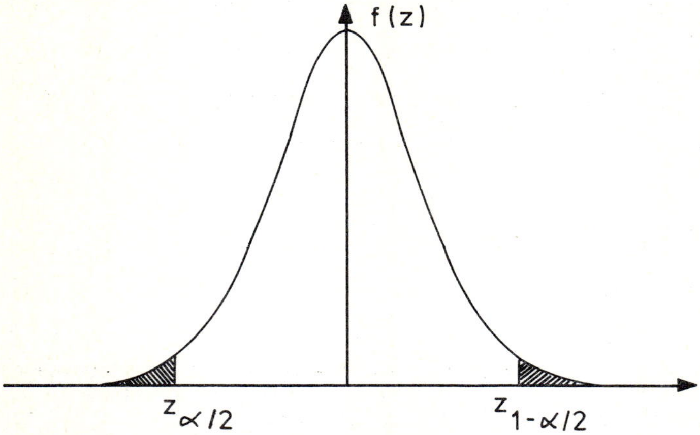

Fig. 3.4. An interval with probability 1 - α.

After choosing a small value of α (for example, α = 5%), z is calculated. If the value of z lies outside of the interval $(z_{\alpha/2}, z_{1-\alpha/2})$, the null hypothesis is rejected because if the null hypothesis had been true the probability of this event is very small (α). In this instance the research hypothesis can be accepted. If the value of z lies inside the interval, the null hypothesis is accepted as it is an acceptable value. The null hypothesis can therefore be accepted if

$$z_{\alpha/2} \leqslant \frac{\bar{x} - \mu_0}{\sigma/\sqrt{n}} \leqslant z_{1-\alpha/2}$$

or if

$$\mu_0 + \frac{\sigma}{\sqrt{n}} \cdot z_{\alpha/2} \leqslant \bar{x} \leqslant \mu_0 + \frac{\sigma}{\sqrt{n}} \cdot z_{1-\alpha/2} \qquad (3.17)$$

In practical situations, the standard deviation σ is unknown and therefore this case is of little importance for applications. In the next section, the case of an unknown standard deviation will be examined.

3.2.4.1.2. Test on the mean with variance unknown (t-test)

Again, one wishes to test the hypothesis that the mean value μ of a variable is equal to a hypothetical value μ_0

$H_0 : \mu = \mu_0$: null hypothesis

$H_1 : \mu \neq \mu_0$: research hypothesis

However, as is frequent in experimental situations, in this instance the standard deviation, σ, of the population is unknown and it is not possible to use the variable z defined in the previous section. In this instance an estimation of σ must be made.

It can be shown by estimation theory that the "best" estimation of σ is given by the standard deviation, s, of the frequency distribution of a random sample. As we have seen in section 2.2.5, s is given by

$$s = \sqrt{\frac{1}{n-1} \sum_{i=1}^{n} (x_i - \overline{x})^2 f_i}$$

This makes it possible to define a new statistical value given by

$$t = \frac{\overline{x} - \mu_0}{s/\sqrt{n}}$$

It can be shown that if the null hypothesis H_0, is true the variable t has the student distribution with n - 1 degrees of freedom. This makes it possible to find two points, $t_{\alpha/2,n-1}$ and $t_{1-\alpha/2,n-1}$, such that the probability of t being outside the interval $(t_{\alpha/2,n-1}, t_{1-\alpha/2,n-1})$ is equal to α if the null hypothesis is true. The values of $t_{\alpha,k}$ for several values of α and k are given in Table III of the Appendix. The null hypothesis can therefore be accepted if

$$t_{\alpha/2,n-1} \leqslant \frac{\overline{x} - \mu_0}{s/\sqrt{n}} \leqslant t_{1-\alpha/2,n-1}$$

or if

$$\mu_0 + \frac{s}{\sqrt{n}} \cdot t_{\alpha/2,n-1} \leqslant \overline{x} \leqslant \mu_0 + \frac{s}{\sqrt{n}} \cdot t_{1-\alpha/2,n-1} \qquad (3.17)$$

A particular case arises when two variables, x_1 and x_2 are measured for each element of the sample. To compare the means of the two variables the differences, d_i, are calculated and it is tested whether the mean difference does or does not differ significantly from zero, by using the equation

$$\frac{\overline{d} - 0}{s_d} \cdot \sqrt{n}$$

where s_d is the standard deviation of the differences.

3.2.4.2. The two-sample case

In the next two sections, the means and standard deviations of two random variables are compared.

3.2.4.2.1. Comparison of two means

Consider two populations, A and B. Suppose the first population has an $N(\mu_1, \sigma_1^2)$ distribution and the second population an $N(\mu_2, \sigma_2^2)$ distribution and that both variances σ_1^2 and σ_2^2 are known. The hypothesis one wishes to test is the equality of the two mean values, μ_1 and μ_2

$H_0 : \mu_1 = \mu_2$: null hypothesis
$H_1 : \mu_1 \neq \mu_2$: research hypothesis

Two samples of sizes n_1 and n_2 are drawn from the populations. The mean values $\overline{x}_1$ and $\overline{x}_2$ of the two samples have the following distributions

$$\overline{x}_1 \sim N(\mu_1, \frac{\sigma_1^2}{n_1})$$

$$\overline{x}_2 \sim N(\mu_2, \frac{\sigma_2^2}{n_2})$$

The difference $\overline{x}_1 - \overline{x}_2$ has the following distribution

$$\overline{x}_1 - \overline{x}_2 \sim N\ (\mu_1 - \mu_2, \frac{\sigma_1^2}{n_1} + \frac{\sigma_2^2}{n_2})$$

If the two means are equal, this difference becomes

$$\overline{x}_1 - \overline{x}_2 \sim N\ (0, \frac{\sigma_1^2}{n_1} + \frac{\sigma_2^2}{n_2})$$

If the variances are known, the following relationship is used for testing the equality of the means

$$\frac{\overline{x}_1 - \overline{x}_2}{\sqrt{\frac{\sigma_1^2}{n_1} + \frac{\sigma_2^2}{n_2}}} \sim N\ (0,1) \qquad (3.18)$$

In practice, the variances are unknown and two estimates, s_1^2 and s_2^2, are calculated. It can then be shown that an estimate for the equal variances of the population is given by the pooled variance

$$s^2 = \frac{(n_1-1)\ s_1^2 + (n_2-1)\ s_2^2}{n_1+n_2-2}$$

In this instance, the expression

$$\frac{\overline{x}_1 - \overline{x}_2}{\sqrt{\frac{s^2}{n_1} + \frac{s^2}{n_2}}}$$

has a $t_{n_1+n_2-2}$ distribution. This last expression can also be written as

$$\frac{\overline{x}_1 - \overline{x}_2}{s\sqrt{\frac{1}{n_1} + \frac{1}{n_2}}} \sim t_{n_1+n_2} - 2$$

The following two-sided test then makes it possible to verify the hypothesis : the means are considered non-significantly different (and the research hypothesis is rejected) if

$$t_{\alpha/2,n_1+n_2-2}\sqrt{\left(\frac{1}{n_1}+\frac{1}{n_2}\right)s^2} \leqslant \bar{x}_1 - \bar{x}_2 \leqslant t_{1-\alpha/2,n_1+n_2-2}\sqrt{\left(\frac{1}{n_1}+\frac{1}{n_2}\right)s^2} \qquad (3.19)$$

3.2.4.2.2. Comparison of two variances

Often the parameters of two normally distributed populations must be compared. In this section, it is tested whether the variances σ_1^2 and σ_2^2 of two normally distributed populations are equal

$H_0 : \sigma_1^2 = \sigma_2^2$: null hypothesis

$H_1 : \sigma_1^2 \neq \sigma_2^2$: research hypothesis

To compare the variances, random samples are drawn from the two populations, s_1^2 and s_2^2 are calculated and a new variable is defined

$$F = \frac{s_1^2}{s_2^2}$$

It can easily be shown that if the variances σ_1^2 and σ_2^2 are equal, F has a Fisher-Snedecor distribution with parameters n_1-1 and n_2-1. Two points $F_{1-\alpha/2,n_1-1,n_2-1}$ and $1/F_{1-\alpha/2,n_2-1,n_1-1}$ can be found such that the probability of F being outside the interval

$$\left(\frac{1}{F_{1-\alpha/2,n_2-1,n_1-1}} , F_{1-\alpha/2,n_1-1,n_2-1}\right)$$

is α if the null hypothesis is true. Values of $F_{\alpha,k,m}$ for several values of α, k and m are given in Table IV of the Appendix. The null hypothesis will be accepted if

$$\frac{1}{F_{1-\alpha/2,n_2-1,n_1-1}} \leqslant \frac{s_1^2}{s_2^2} \leqslant F_{1-\alpha/2,n_1-1,n_2-1} \qquad (3.20)$$

For a one-sided test, the hypotheses are formulated as

$H_0 = \sigma_1^2 = \sigma_2^2$: null hypothesis

$H_1 = \sigma_1^2 > \sigma_2^2$: research hypothesis

The null hypothesis will then be accepted if

$$s_1^2 < s_2^2 \; F_{1-\alpha,n_1-1,n_2-1} \qquad (3.21)$$

3.2.5. Two-dimensional frequency distribution

3.2.5.1. Definition

A one-dimensional frequency distribution was obtained by grouping the measurements of a variable into groups and by calculating the number of values in each group. Often, when studying the elements of a population, two or more variables are of interest. To represent such a population, the number of measurements (also called the frequency) of each combination of possible values for the two variables must be calculated, and these values can then be given in a table.

An example giving the concentrations of two materials, A and B, for 100 samples is shown in Table 3.I.

Table 3.I.

A two-dimensional frequency distribution

Concentration of B	Concentration of A			
	0-3	4-7	8-11	12-15
0-3	10	3	0	1
4-7	4	20	4	1
8-11	1	6	23	8
12-15	0	2	7	10

To obtain a better comparison of populations of different sizes, an alternative table gives the relative frequencies of each combination. For the example above, Table 3.II is obtained.

Table 3.II

A two-dimensional relative frequency distribution

Concentration of B	Concentration of A			
	0-3	4-7	8-11	12-15
0-3	0.10	0.03	0	0.01
4-7	0.04	0.20	0.04	0.01
8-11	0.01	0.06	0.23	0.08
12-15	0	0.02	0.07	0.10

3.2.5.2. Marginal distributions

A two-dimensional frequency distribution can be described by using the parameters of the two variables separately. If the variables are called x_1 and x_2, the marginal x_1 and x_2 distributions are defined as the frequency distributions of the separate variables. The frequency of these two distributions will be denoted by f_{x_1} and f_{x_2}. In Table 3.III the marginal distributions for the example in section 3.2.5.1 are given.

Table 3.III

Marginal distributions

x_2	x_1				f_{x_2}
	0-3	4-7	8-11	12-15	
0-3	10	3	0	1	14
4-7	4	20	4	1	29
8-11	1	6	23	8	38
12-15	0	2	7	10	19
f_{x_1}	15	31	34	20	100

The values $\overline{x}$, S^2, s^2 can then be calculated for both marginal distributions, and they will be called $\overline{x}_1$, $S^2_{x_1}$, $s^2_{x_1}$ and $\overline{x}_2$, $S^2_{x_2}$, $s^2_{x_2}$.

3.2.5.3. Covariance and correlation

The disadvantage of the marginal distributions is that they do not describe the connection between the two variables. In this section, two parameters will be introduced for this purpose.

A two-dimensional frequency distribution with variables x_1 and x_2 is considered. Variable x_1 takes the values $x_{11}, x_{12}, \ldots, x_{1i}, \ldots, x_{1n_1}$ and variable x_2 the values $x_{21}, x_{22}, \ldots, x_{2j}, \ldots, x_{2n_2}$. The number of elements for which the values of the variables are x_{1i} and x_{2j} is called f_{ij}.

The covariance between the variables x_1 and x_2 is defined as

$$C = \frac{1}{n} \sum_{i=1}^{n_1} \sum_{j=1}^{n_2} f_{ij} (x_{1i} - \bar{x}_1)(x_{2j} - \bar{x}_2) \tag{3.22}$$

where n is the total number of measurements ($n = \sum_{i=1}^{n_1} \sum_{j=1}^{n_2} f_{ij}$). It can be shown that the covariance is also given by

$$C = \frac{1}{n} \sum_{i=1}^{n_1} \sum_{j=1}^{n_2} f_{ij}\, x_{1i}\, x_{2j} - \bar{x}_1 \cdot \bar{x}_2 \tag{3.23}$$

The covariance between variables is sometimes also written $\mathrm{Cov}(x_1, x_2)$ or $C(x_1, x_2)$. Another much used measure of the association between two variables is the correlation coefficient, r, which is given by

$$r = \frac{C}{S_{x_1} \cdot S_{x_2}} \tag{3.24}$$

S_{x_1} and S_{x_2} are the biased estimators of the standard deviation of the marginal distribution (eqn. 2.5).

It can be shown that the correlation coefficient always takes a value between -1 and 1. A further discussion of the correlation coefficient is given in section 3.2.6.2.

3.2.6. Two-dimensional random variables

3.2.6.1. Definitions

As in the one-dimensional case, two-dimensional random variables are characterized by a cumulative probability distribution function, $F(x_1, x_2)$. This function is defined for each pair of values (x_1, x_2) that the variables can take. It is the probability that the first variable be less than or equal to x_1 and the second less than or equal to x_2. The probability distribution function or density function is then given by

$$f(x_1, x_2) = \frac{\delta^2 F(x_1, x_2)}{\delta x_1 \, \delta x_2} \tag{3.25}$$

It must satisfy the condition that its integral is equal to unity

$$\int_{-\infty}^{+\infty} \int_{-\infty}^{+\infty} f(x_1, x_2) \, dx_1 \, dx_2 = 1$$

When the probability distribution function is discrete, its values are written as p_{x_1, x_2} and are called probabilities. These must satisfy the conditions

$$\sum_{x_1} \sum_{x_2} p_{x_1, x_2} = 1$$

The marginal probability distribution functions are obtained by considering only one variable and taking the integral or sum over the others

$$g_1(x_1) = \int_{-\infty}^{+\infty} f(x_1, x_2) \, dx_2$$

$$g_2(x_2) = \int_{-\infty}^{+\infty} f(x_1, x_2) \, dx_1$$

3.2.6.2. Parameters of a two-dimensional random variable

When considering a two-dimensional random variable, parameters can be calculated for each of the two random variables. Considering, for example, x_1

its mean is defined as

$$\mu_{x_1} = E(x_1) = \int_{-\infty}^{+\infty} x_1 g_1(x_1)\, dx_1 = \int_{-\infty}^{+\infty} \int_{-\infty}^{+\infty} x_1\, f(x_1, x_2)\, dx_1\, dx_2$$

and its variance as

$$\sigma_{x_1}^2 = \mathrm{Var}\,(x_1) = E((x_1 - E(x_1)^2) = \int_{-\infty}^{+\infty} (x_1 - E(x_1 - E(x_1))^2\, g_1\,(x_1)\, dx_1$$

In the same way, the mean and variance of x_2 can be calculated. A parameter more adapted to the two-dimensional nature of the random variable is the covariance, which is defined as

$$\gamma\,(x_1,x_2) = E((x_1 - E(x_1))\,(x_2 - E(x_2)))$$

$$= \int_{-\infty}^{+\infty} \int_{-\infty}^{+\infty} (x_1 - E(x_1))\,(x_2 - E(x_2))\, f(x_1,x_2)\, dx_1\, dx_2$$

It can easily be shown that

$$\gamma\,(x_1,x_2) = E(x_1.x_2) - E(x_1)\,E(x_2) \tag{3.26}$$

To obtain a parameter independent of the scales in which x_1 and x_2 are measured, the covariance is divided by the standard deviations of x_1 and x_2, which gives the correlation coefficient

$$\rho\,(x_1,x_2) = \frac{\gamma(x_1,x_2)}{\sigma_{x_1} \cdot \sigma_{x_2}} \tag{3.27}$$

The greek letter symbols Γ and ρ are population parameters and symbols C and r are estimates (in fact, C is a biased estimate, since one divides by n instead of n-1).

3.2.6.3. Independent and uncorrelated random variables

Two random variables x_1 and x_2 are considered to be independent if their joint probability distribution function, $f(x_1, x_2)$, can be obtained by the multiplication of the two marginal probability distribution functions, $g_1(x_1)$ and $g_2(x_2)$

$$f(x_1,x_2) = g_1(x_1) \cdot g_2(x_2) \tag{3.28}$$

An important property of independent random variables is that their covariance, and therefore also their correlation coefficient, is zero.

Proof

$$E(x_1 . x_2) = \int_{-\infty}^{+\infty} \int_{-\infty}^{+\infty} x_1\, x_2\, f(x_1,x_2)\, dx_1\, dx_2$$

$$= \int_{-\infty}^{+\infty} \int_{-\infty}^{+\infty} x_1\, x_2\, g_1(x_1)\, g_2(x_2)\, dx_1\, dx_2$$

$$= \int_{-\infty}^{+\infty} x_1\, g_1(x_1)\, dx_1 \int_{-\infty}^{+\infty} x_2\, g_2(x_2)\, dx_2$$

$$= E(x_1) \cdot E(x_2)$$

This implies that

$$\gamma\,(x_1,x_2) = E(x_1 . x_2) - E(x_1)\, E(x_2) = 0$$

and

$$\rho\,(x_1,x_2) = \frac{\gamma(x_1,x_2)}{\sigma_{x_1}\, \sigma_{x_2}} = 0$$

When the correlation coefficient $\rho(x_1,x_2)$ is zero, the variables are called uncorrelated : independent random variables are uncorrelated. It can be observed that the reciprocal does not necessarily hold.

The covariance is also used to calculate the variance of the sum of two random variables

$$\mathrm{Var}(x_1 + x_2) = E\,(((x_1 + x_2) - E(x_1 + x_2))^2)$$

$$= E\,((x_1 - E(x_1) + x_2 - E(x_1))^2)$$

$$= E\,((x_1 - E(x_1))^2 + E(x_2 - E(x_2))^2) + 2E\,((x_1 - Ex_1))\,(x_2 - E(x_2)))$$

$$= Var(x_1) + Var(x_2) + 2\gamma(x_1,x_2)$$

A corollary of this result is that the variance of the sum of two independent random variables is the sum of their variances.

3.2.7. Multi-dimensional random variables

3.2.7.1. Definitions *

The two-dimensional probability distribution function and cumulative probability distribution function can easily be extended to the multi-dimensional case : if the variables are called x_1, x_2, ..., x_n, the probability distribution function or density function is called $f(x_1,x_2, \ldots,x_n)$. It must satisfy

$$\int_{-\infty}^{+\infty} \cdots \int_{-\infty}^{+\infty} f(x_1,x_2, \ldots,x_n)\, dx_1\, dx_2 \ldots dx_n = 1$$

The marginal density functions for some of the variables are obtained by integrating over the others. For example, the marginal density function $g(x_1,x_2)$ is given by

$$g(x_1,x_2) = \int_{-\infty}^{+\infty} \cdots \int_{-\infty}^{+\infty} f(x_1,x_2, \ldots,x_n)\, dx_3\, dx_4 \ldots dx_n$$

3.2.7.2. Variance-covariance matrix

The covariance between variables x_i and x_j is defined by

$$\gamma_{ij} = \gamma(x_i,x_j) = \int_{-\infty}^{+\infty} \int_{-\infty}^{+\infty} (x_i - E(x_i))\,(x_j - E(x_j))\, g(x_i,x_j)\, dx_i\, dx_j \qquad (3.29)$$

* Matrices and vectors are discussed in Chapter 17.

In the same way, the variance of variable x_i is defined by

$$\sigma_i^2 = \gamma_{ii} = \gamma(x_i x_i) = \int_{-\infty}^{+\infty} (x_i - E(x_i))^2 \, g(x_i) \, dx_i \tag{3.30}$$

Sometimes the variables $x_1, x_2, \ldots, x_n$ are arranged in a vector $\vec{x}$

$$\vec{x} = \begin{bmatrix} x_1 \\ x_2 \\ \vdots \\ x_n \end{bmatrix}$$

By considering the values σ_i^2 and Γ_{ij} as the variances of and covariances of and between the elements of vector $\vec{x}$, the following notation will be used

$$\Gamma = \begin{bmatrix} \sigma_1^2 & \gamma_{12} & \cdots & \gamma_{1n} \\ \gamma_{21} & \sigma_2^2 & \cdots & \gamma_{2n} \\ \vdots & & & \\ \gamma_{n1} & \gamma_{n2} & \cdots & \sigma_n^2 \end{bmatrix} \tag{3.31}$$

Matrix Γ is called the variance-covariance matrix of the multi-dimensional random variable $\vec{x}$.

3.2.7.3. The multi-dimensional normal distribution

The density function of a multi-dimensional or multi-variate normal distribution is given by

$$f(x_1, x_2, \ldots, x_n) = \frac{e^{-1/2(\vec{x}-\vec{\mu})' \Gamma^{-1} (\vec{x}-\vec{\mu})}}{(2\pi)^{n/2} \, |\Gamma|^{1/2}} \tag{3.32}$$

where

$$\vec{\mu} = \begin{bmatrix} \mu_1 \\ \mu_2 \\ \vdots \\ \mu_n \end{bmatrix}$$

is the vector of means of the variables, Γ is the variance-covariance matrix and $|\Gamma|$ is its determinant.

3.2.8. The fitting of a straight line by the least-squares method

A pair of measurements made for each element of a population or of a sample can be considered as a two-dimensional frequency distribution. If the first measurement of each pair is called x and the second is called y, the two-dimensional frequency distribution can be represented by the following table

Element	First variable	Second variable
1	x_1	y_1
2	x_2	y_2
⋮	⋮	⋮
n	x_n	y_n

If one considers the second variable as a function of the first, the following general relationship can be considered

$$y = f(x)$$

In the specific and very restricted case when the relationship is linear, it can be written as

$$y = \beta x + \alpha$$

where β and α are unknown parameters.

As each measurement of the variable y is influenced by a measurement of x, the following model is used to describe the relationship

$$y_i = \beta x_i + \alpha + e_i \qquad i = 1, 2, \ldots, n \tag{3.33}$$

where e_i represents the error during measurement i. The object of the model is then to find estimates for the values of the unknown parameters β and α. These estimates, called a and b, will be chosen in such a way that the difference between the measured y_i and the ones given by the model, $bx_i + a$, will be as small as possible. This condition is obtained by considering the sum of the squares of these differences

$$\sum_{i=1}^{n} (y_i - \beta x_i - \alpha)^2$$

A minimum for this function of β and α is obtained by setting to zero the partial derivatives with respect to α and β.

$$\frac{\delta \sum_{i=1}^{n} (y_i - \beta x_i - \alpha)^2}{\delta\beta} = -2 \sum_{i=1}^{n} x_i (y_i - \beta x_i - \alpha) = 0$$

$$\frac{\delta \sum_{i=1}^{n} (y_i - \beta x_i - \alpha)^2}{\delta\alpha} = -2 \sum_{i=1}^{n} (y_i - \beta x_i - \alpha) = 0$$

This yields minimum square estimates of the parameters

$$\sum_{i=1}^{n} (x_i y_i - bx_i^2 - ax_i) = 0$$

$$\sum_{i=1}^{n} y_i - b \sum_{i=1}^{n} x_i - na = 0$$

This gives the following equations

$$b = \frac{\sum_{i=1}^{n} (x_i - \overline{x})(y_i - \overline{y})}{\sum_{i=1}^{n} (x_i - \overline{x})^2} \qquad (3.34)$$

$$a = \overline{y} - b\overline{x} \qquad (3.35)$$

The straight line $y = a + bx$ obtained by the least-squares method can now be compared with a given hypothetical straight line with $\alpha = 0$ and $\beta = 1$. The hypothetical straight line can be written in the form

$$y = \overline{x} + \beta\ (x - \overline{x})$$

and the fitted line can be written as

$$y = a + b\overline{x} + bx - b\overline{x}$$

or as

$$y = \overline{y} + b\ (x - \overline{x})$$

To examine the value of the hypothetical straight line, the total sum of the squares of the deviations from the observed values is calculated. One such deviation is given by

$$y_i - \overline{x} - (x_i - \overline{x})$$

This value can be broken down in the following way

$$\begin{aligned} y_i - \overline{x} - (x_i - \overline{x}) \\ &= (y_i - \overline{y}) - b\ (x_i - \overline{x})\ ) \\ &\quad + (\overline{y} - \overline{x}) \\ &\quad + (b - 1)\ (x_i - \overline{x}) \end{aligned}$$

These terms correspond to the deviation from the least-squares straight line and to the differences between $\overline{y}$ and $\overline{x}$ and between b and 1. The sums of the squares of these terms can be written in the following form

Source	Sum of squares	Degrees of freedom
Slope of line	$SS_b = (b-1)^2 \sum_{i=1}^{n} (x_i-\overline{x})^2$	1
Constant of line	$SS_a = n\ (\overline{y} - \overline{x})^2$	1
About line	$SS_1 = \sum_{i=1}^{n} (y_i-\overline{y}-b(x_i-\overline{x})\ )^2$	n-2
Total	$SS_T = \sum_{i=1}^{n} (y_i-\overline{x}-(x_i-\overline{x})\ )^2$	n

To test the hypothesis, the following result is used. The expressions $\frac{SS_b}{SS_1/n-2}$ and $\frac{SS_a}{SS_1/n-2}$ both have an F distribution with (1, n-2) degrees of freedom.

Another method of testing the hypotheses that $\beta = 1$ and $\alpha = 0$ consists in performing a Student's t-test.

(1) β has a specified value (e.g. $\beta = 1$).

The statistic $t = \frac{b - \beta}{\sqrt{1-r^2}} \sqrt{n - 2}$ has Student's distribution with n - 2 degrees of freedom (Spiegel, 1972).

(2) α has a specified value (e.g. $\alpha = 0$).

The statistic $t = \frac{a - \alpha}{Var(a)}$ has Student's distribution with n - 2 degrees of freedom (Grémy, 1969).

$$Var(a) = s^2 \left(\frac{1}{n} + \frac{\bar{x}^2}{\Sigma(x_i-\bar{x})^2} \right)$$

$$s^2 = \frac{\Sigma(y_i-y'_i)^2}{n-2}$$

where y'_i is the y value obtained with the estimated regression coefficient :

$y'_i = a + bx_i$.

REFERENCES

R.N. Barnett and W.J. Youden, Amer. J. Clin. Pathol., 54 (1970) 454.
J.P. Cali, Anal. Chem., 48 (1976) 202A.
R.N. Carey, S. Wold and J.O. Westgard, Anal. Chem., 47 (1975) 1824.
V.J. Clancey, Nature, 159 (1947) 339.
W.J. Conover, Practical Nonparametric Statistics, Wiley-Interscience, New York, 1971.
B.E. Cooper, Statistics for Experimentalists, Pergamon, Oxford, 1st ed., 1969, reprinted 1975.
T.R. Folsom, N. Hansen, W.E. Weitz, Jr and G.J. Parks, Jr, Appl. Spectros., 29 (1975) 404.
E.M. Gindler, Clin. Chem., 21 (1975) 309.
G. Gottschalk, Z. anal. Chem., 280 (1976) 205.
F. Grémy and D. Salmon, Bases Statistiques, Dunod, Paris, 1969.
P.D. Lark, B.R. Craven and R.C.L. Bosworth, The Handling of Chemical Data, Pergamon, Oxford, 2nd ed., 1969.
I.L. Larsen, N.A. Hartmann and J.J. Wagner, Anal. Chem., 45 (1973) 1511.

H.F. Martin, B.J. Gudzinowicz and H. Fanger, Normal Values in Clinical Chemistry : A Guide to Statistical Analysis of Laboratory Data, Marcel Dekker, New York, 1975.
L.M. Schwartz, Anal. Chem., 48 (1976) 2287.
S. Siegel, Nonparametric Statistics for the behavioral sciences, McGraw-Hill, New York, 1956.
M.R. Spiegel, Theory and problems of statistics, McGraw-Hill, New York, 1972.
M. Thompson and R.J. Howarth, Analyst, 101 (1976) 690.
P.J.M. Wakkers, H.B.A. Hellendoorn, G.J. Op de Weegh and W. Heerspink, Clin. Chim. Acta, 64 (1975) 173.
J.O. Westgard and M.R. Hunt, Clin. Chem., 19 (1973) 49.
W.J. Youden, Statistical Methods for Chemists, Wiley, New York, and Chapman and Hall, London, 1951.
W.J. Youden and E.H. Steiner, Statistical Manual of the Association of Official Analytical Chemists, The Association of Official Analytical Chemists, Washington, D.C., 1975.

Chapter 4

EVALUATION OF PRECISION AND ACCURACY - ANALYSIS OF VARIANCE *

4.1. GENERAL DISCUSSION

4.1.1. Introduction ; definitions

In section 3.1.2, the t-test was used to decide whether two methods yield significantly different results. The profusion of analytical methods is such that often more than two possible procedures have to be compared.

A two-by-two comparison of procedures using the t-test makes it possible to investigate whether some differ significantly from others. However, one may wish to investigate the whole body of data with a single statistical procedure. This is possible with the analysis of variance (ANOVA) technique. ANOVA is used in many areas of science and there are also several important applications within the scope of this book.

The basic problem to which the analysis of variance is applied is to determine which part of the variation in a population is due to systematic reasons (called factors) and which is due to chance (Jonckheere, 1966). Scheffé (1959), who is the author of an important book on ANOVA, defines it as a statistical technique for analysing measurements that depend on several kinds of effects operating simultaneously to decide which kinds of effects are important and to estimate the effects.

In the comparison of procedures discussed above, the procedures to be investigated may be subject to systematic error ; the choice of a procedure is called a (controlled) factor. Moreover, the results of the analytical determinations are subject to random errors. The analysis of variance compares

---

* This chapter has been written with the collaboration of Y. Michotte, Pharmaceutical Institute, Vrije Universiteit Brussel, Belgium.

both causes of error, with the purpose of deciding whether or not the controlled factor has a significant effect.

The importance of ANOVA can be further illustrated by the following two examples, the first of which is taken from Amenta (1968). Amenta's work was concerned with quality control in a clinical laboratory, and involved the analysis of 50 samples from the same pool, two a day at different places in the run during consecutive days. The questions asked were then as follows :

(1) is there a significant contribution to the total variance from day-to-day variations ? ;

(2) is there a significant effect due to the position in the run ?

The other example relates to the work of official or public analysts and concerns the statistical analysis of a collaborative test of a procedure considered for official adoption. In such a test, a number of laboratories are asked to analyse a number of samples with the same procedure using a pre-determined number of replicate determinations. The analysis serves to distinguish between sources of variance between laboratories, between samples and between replicates.

In the case studied by Amenta, there are two controlled factors (time and position). In the second example there are also two controlled factors (laboratories and samples), the variance between replicates being considered to be the effect of chance. An ANOVA with n factors is called an n-way layout, and therefore both examples consist of two-way layouts. It should be added that, in reality, the second application is more complex. It may be that one of the samples consists of a fine powder, while another consists of a more granular material that has to be ground in order to obtain smaller particles before the analysis. A laboratory that has efficient grinding equipment may obtain accurate results for both kinds of samples, while another laboratory with inadequate equipment may produce more or less correct results for the first sample but systematically low results for the coarser material. In other words, the effect of the laboratory is not the same for all samples ; this is called a sample-laboratory interaction and may have to be taken into account in the ANOVA. ANOVA can be used to detect such an interaction. This is a typical example of the use of ANOVA in optimization studies. If, in the example given, significant sample-laboratory interaction is detected, this should lead to a better specification of the procedure or to greater standardization of the equipment and therefore to smaller overall errors.

The introduction of the interaction concept leads to a difficulty in terminology. Statisticians often make a distinction between the terms "analysis of variance" and "factorial analysis". Some workers reserve the former term for cases when there is only one parameter being investigated (one controlled factor) or else for cases when there are several independent controlled factors (i.e., no interactions). Factorial analysis is then either an ANOVA for more than one controlled factor or for the interaction case. Other workers do not use the term factorial analysis at all. Here we shall adopt the policy of considering that factorial analysis is a sophisticated part of the more general ANOVA technique and we use the term to indicate the case when the possibility of interaction exists. The term ANOVA can be used in the restricted sense, i.e., the controlled factors are independent, or in its broader sense, when it also includes factorial analysis. This chapter considers ANOVA in the restricted sense, while Chapter 12 discusses factorial analysis.

ANOVA is such an important technique that we have decided to explain its mathematics in two forms. In this section, the equations for one-way ANOVA are derived. The mathematical development is not complete in the sense that a few assertions are made without proof, and it is specific for the one-way layout. After these equations have been obtained, more practical equations for manual calculations are derived. The equations for two-way ANOVA are given without mathematical proof. In the mathematical section, the same equations are derived in a much more formal and also a more general way starting from the general linear model.

4.1.2. A one-way layout

Let us consider the example given above, namely the comparison of p procedures. To avoid consideration of between-laboratory errors, the procedures are assumed to be carried out by one laboratory. If this were not the case, one would have to carry out an ANOVA with two controlled factors (and perhaps interaction), namely the laboratories and the procedures.

As stated above, the object of the ANOVA here is to compare systematic errors with the random error obtained for the replicates, i.e., the precision of the procedures. It is important to note that we have written "the precision of the procedures" and not "the precisions". This choice illustrates one of the important suppositions behind ANOVA : the random error is considered to be the same for the whole body of data, which means that it is stated that each of the procedures shows the same precision (or rather, as precision was defined as the estimate of a σ_r in Chapter 2, the same "true" precision, σ_r). It is well known that in practice this is improbable. However, ANOVA remains valid when the σs do not differ too greatly. Nevertheless, it should be remembered that ANOVA could lead to erroneous conclusions if methods with widely differing precision are compared.

Let the procedures be called i = 1, 2, ..., p and let there be j = 1, 2, ..., J determinations ; y_{ij} is the jth result with the ith procedure. The same reasoning can be applied throughout the following sections for the case when one wishes to determine the effect of days (between-day precision) by analysing the same sample J times on p days or the effect of the laboratory (between-laboratory precision) by carrying out J determinations in p laboratories. The number of determinations is considered here to be the same for all procedures. This is not necessary and in the mathematical section the more general case of different numbers, J_i, of replicates is considered.

The mean $\bar{y}_{i.} = \frac{1}{J} \sum_{j=1}^{J} y_{ij}$ estimates the mean μ_i that would be obtained for all determinations with procedure i. This leads us to the following model for the ith procedure :

(i) y_{ij} is normally distributed with μ_i and σ_r^2 as the parameters of the distribution ;

(ii) $\overline{y}_{i.}$ estimates μ_i ; this estimated value is distributed normally around μ_i with a standard deviation of $\sigma_r/\sqrt{J}$ (standard deviation on a mean, see section 3.2.4.1); $\overline{y}_{i.}$ is an unbiased estimate of μ_i ;

(iii) $s_r^2 = \frac{1}{J-1} \sum_{j=1}^{J} (y_{ij} - \overline{y}_{i.})^2$ is an unbiased estimate of σ_r^2 ; s_r is in fact the precision as defined in Chapter 2. If several laboratories or days were to be compared instead of procedures, s_r would be the within-laboratory or within-day precision.

To obtain a model for the p procedures (respectively laboratories, days), one must add to this supposition that the μ_i of the procedures (laboratories, days) are normally distributed around μ and with standard deviation σ_p^2. This means that we consider that a procedure, etc., is chosen randomly from an infinite number of procedures. While it has not been proven that the means from an infinite number of determinations for an infinite number of procedures are normally distributed, this assumption is generally accepted for days or laboratories. Hence between-days or between-laboratory errors are normally distributed around the true mean (in the absence of systematic error), a fact which has been accepted as true in section 2.1.2. The supposition that the inter-procedure error is normally distributed is therefore not as surprising as might appear at first. It illustrates one of the limitations of ANOVA : the variations due to the controlled factors are considered to be described by normal distributions. To avoid difficulties with the analytical and statistical terminology, we should remember that the error due to one laboratory or procedure is considered to be systematic but that the errors due to all laboratories or procedures are random.

Calling ε_{ij} the error on the measurement y_{ij} and μ_o the true mean (true value of analyte), one can write

$$\varepsilon_{ij} = y_{ij} - \mu_o$$

or

$$\varepsilon_{ij} = (y_{ij} - \mu_i) + (\mu_i - \mu) + (\mu - \mu_o)$$

with

$y_{ij} - \mu_i$ = random deviation within procedure i ;

$\mu_i - \mu$ = random deviation due to procedure i, i.e., due to the systematic errors of the procedure ;

$\mu - \mu_o$ = systematic deviation.

The difference between μ and μ_o is due to the fact that all procedures, etc., may have something in common, which can cause an error. This is easiest to understand when ANOVA is used for a comparison of laboratories using the same procedure. The grand mean of all results would estimate a value μ differing from μ_o by the systematic error of the procedure. When the object is to compare procedures for the same kind of determination carried out in one laboratory, then the deviation of μ from μ_o might, for example, be caused by the fact that the samples must always be dried before the actual determination begins and that a systematic error is made during this step. If it is assumed that $\mu = \mu_o$ (as is often done because the systematic error is not of interest to the investigator who has designed his experiment specifically to investigate certain sources of variation), the equation can be rewritten in the more usual ANOVA notation

$$\varepsilon_{ij} = e_{ij} + \alpha_i$$

with

$$e_{ij} = y_{ij} - \mu_i$$

$$\alpha_i = \mu_i - \mu$$

This leads to

$$y_{ij} = \mu + \alpha_i + e_{ij} \qquad (4.1)$$

In this way, a linear model has been developed according to which each measurement result is the sum of a constant (μ) and two random variables, the first of which, α_i, varies between the procedures and the other, e_{ij}, within the procedures. The latter is often called the error or the residual error. y_{ij} is normally distributed around μ with a standard deviation $\sigma_{t(otal)}$. α_i is the deviation from μ and, as the expected deviations are zero, they are distributed normally around zero with a standard deviation σ_p and e_{ij}, the deviation of the replicate measurements for a particular procedure from the mean for that procedure, is also distributed normally around zero with a standard deviation σ_r. From the foregoing equation it follows, as the variables e_{ij} and α_i are independent, that

$$\sigma_t^2 = \sigma_r^2 + \sigma_p^2$$

The ANOVA method now consists in the calculation of the following sums of squares

$$SS_1 = \sum_{i=1}^{p} \sum_{j=1}^{J} (y_{ij} - \bar{y}_{i.})^2$$

$$SS_2 = \sum_{i=1}^{p} J(\bar{y}_{i.} - \bar{y}_{..})^2$$

where $\bar{y}_{..}$ is the grand mean

$$\bar{y}_{..} = \frac{\sum_{i=1}^{p} \sum_{j=1}^{J} y_{ij}}{Jp}$$

These sums have to be divided by their degrees of freedom $\left[\text{p-1 for } SS_2 \text{ and } (J-1)p \text{ for } SS_1\right]$.

Clearly, SS_1 / (J-1)p estimates the variance due to the e_{ij} term, i.e., σ_r^2.
It can be shown that SS_2 / p-1 estimates $\sigma_r^2 + (Jp - \frac{J^2}{Jp})\ \sigma_p^2$ / p-1.
The null hypothesis is that no part of the variance σ_t^2 is due to the difference between the procedures, i.e., $\sigma_p^2 = 0$. If this is so SS_1 / (J-1)p and SS_2 / p-1 estimate the same variance σ_r^2, and should therefore not be significantly different. As SS_1 and SS_2 estimate variances, an F-test is made to test the null hypothesis. The total variance of the data can be estimated from

$$\frac{\sum_{i=1}^{p} \sum_{j=1}^{J} (y_{ij} - \bar{y}_{..})^2}{Jp - 1} = \frac{SS_3}{Jp - 1}$$

It can be shown that $SS_3 = SS_2 + SS_1$. Hence, one can also calculate SS_3 and SS_1 and obtain SS_2 by difference or SS_3 and SS_2 and obtain SS_1 by difference. In practice, the latter method is usually preferred. The calculation can be rendered more practical by the following modification

$$SS_3 = \sum_{i=1}^{p} \sum_{j=1}^{J} (y_{ij} - \bar{y}_{..})^2 = \sum_{i=1}^{p} \sum_{j=1}^{J} (y_{ij})^2 - \frac{(\sum_{i=1}^{p} \sum_{j=1}^{J} y_{ij})^2}{Jp}$$

$$= \sum_{i=1}^{p} \sum_{j=1}^{J} (y_{ij}^2) - C$$

where C is the correction term (or correction term for the mean).

$$SS_2 = \sum_{i=1}^{p} J(\bar{y}_{i.} - \bar{y}_{..})^2 = J \left[\sum_{i=1}^{p} (\bar{y}_{i.}^2) - \frac{(\sum_{i=1}^{p} \bar{y}_{i.})^2}{p} \right]$$

$$= J \left[\sum_{i=1}^{p} \bar{y}_{i.}^2 - \frac{(\sum_{i=1}^{p} \sum_{j=1}^{J} y_{ij})^2}{J^2 p} \right]$$

$$= J \sum_{i=1}^{p} \bar{y}_{i.}^{2} - \frac{(\sum_{i=1}^{p} \sum_{j=1}^{J} y_{ij})^2}{Jp}$$

$$= J \sum_{i=1}^{p} \bar{y}_{i.}^{2} - C$$

$$= \frac{\sum_{i=1}^{p} y_{i.}^{2}}{J} - C$$

where

$$y_{i.} = \sum_{j=1}^{J} y_{ij}$$

In practice, one makes the calculations using Table 4.I.

Table 4.I

One-way layout table

	Procedures		
Replicates	1	2 ...	p
1	y_{11}	y_{12}	y_{1p}
2	y_{21}	y_{22}	y_{2p}
3	y_{31}	y_{32}	y_{3p}
J	y_{J1}	y_{J2}	y_{Jp}
Sum for procedures	$y_{.1}$	$y_{.2}$	$y_{.p}$
Total sum	$y_{..} = y_{.1} + y_{.2} + \ldots y_{.p}$		

The following operations are then carried out :

(1) calculation of the correction factor :

$$C = (y_{..})^2 / Jp$$

(2) calculation of the sums of squares :

$$SS_{total} \; (= SS_3) = \sum_{i=1}^{p} \sum_{j=1}^{J} (y_{ij}^{2}) - C$$

$$SS_{procedures}\ (= SS_2) = \frac{1}{J} \sum_{i=1}^{p} (y_{i.})^2 - C$$

(3) calculation of the sum of squares of the residual errors (SS_1) :

$$SS_{residual} = SS_{total} - SS_{procedures}$$

(4) calculation of the estimates of the variances through division by the number of degrees of freedom :

$$s^2_{procedures} = SS_{procedures} / p-1$$
$$s^2_{residual} = SS_{residual} / p(J-1)$$

note that p-1 + p(J-1) = Jp-1, i.e., (total number of data - 1), which is also the number of degrees of freedom for the total variance ;

(5) calculation of the F ratio :

$$F = s^2_{procedures} / s^2_{residual}$$

When a critical value of F is not obtained (see section 4.2.2), one considers that the controlled factor induces no variation. In other words, all of the procedures yield the same result and, at least if one of the methods is a reference method, all of the procedures are accurate. If the null hypothesis is rejected, at least one of the procedures is at variance with the others. If one of the procedures is a reference procedure, a series of t-tests will indicate which one(s) give significantly different results.

4.1.3. Fixed- and random-effects models

Eqn. 4.1 is the result of the general linear model to be discussed in section 4.2, and is derived from eqn. 4.5. In this model the observations are represented as linear combinations of several unknown quantities $\beta_1, \ldots, \beta_q$ and of a random error, e_i (see also section 4.2.1). In ANOVA, the constant coefficients, v_{ji}, of the linear combinations are 0 to 1. The unknown quantity β represents an effect and the presence of a v_{ji} factor means that it is a real effect. When applying the general linear model to ANOVA, one usually includes all the effects that could be meaningful. One of the βs is usually an unknown constant such as the grand mean, μ or μ_o, and the others are random variables or unknown constants. The case when all of the β terms represent unknown constants is called a fixed-effects model and the case when all of the terms minus one (the mean) are random variables is called a random-effects model. The model of eqn. 4.2 is therefore a random-effects model as it is composed of a grand mean, an error term and one or more randomly distributed effects. The models in Chapter 12, on the contrary, are fixed-effects models. Intermediate cases when at least one Greek letter parameter is a constant and does not represent the grand mean are called mixed models.

The problem in which one considers p procedures can also be stated as a fixed-effects model. In this instance, one does not consider that the p procedures come from an infinite population of procedures. In the mathematical section, we give both formulations of the ANOVA model.

4.1.4. A two-way layout

Let us consider the problem discussed by Amenta (1968) and which was introduced earlier. As there are two controlled factors (the positions and the days-, the technique is called a two-way ANOVA. In this instance the linear model leads to

$$\sigma_t^2 = \sigma_d^2 + \sigma_p^2 + \sigma_r^2$$

where

$\sigma^2_{t(otal)}$ = the variance for the total set of observations ;

$\sigma^2_{d(ays)}$ = the variance due to variations between days ;

$\sigma^2_{p(osition)}$ = the variance due to variations between positions ; and

$\sigma^2_{r(esidual)}$ = the variance due to individual results.

When the days and the positions have no significant effect, one can conclude that $\sigma_t^2 = \sigma_r^2$ and σ_r^2 is therefore the experimental error that occurs in the absence of other effects. The analysis of variance is based here on the comparison of the terms σ_d and σ_p with σ_r. Table 4.III contains the 50 results obtained.

Table 4.II

Two-way layout table for the clinical laboratory problem

Day	1	2	3	4	...	25	Sum for positions
Position 1	y_{11}	y_{12}	y_{13}	y_{14}		$y_{1,25}$	$y_{1.}$
Position 2	y_{21}	y_{22}	y_{23}	y_{24}		$y_{2,25}$	$y_{2.}$
Sum for days	$y_{.1}$	$y_{.2}$	$y_{.3}$	$y_{.4}$		$y_{.25}$	Grand sum = $y_{..}$

$$y_{i.} = \sum_{j=1}^{25} y_{ij}$$

$$y_{.j} = y_{1j} + y_{2j}$$

$$y_{..} = y_{1.} + y_{2.} = \sum_{j=1}^{25} y_{.j}$$

The following operations are then carried out :

(1) Calculation of the correction factor for the mean :

$$C = \frac{y_{..}^2}{50}$$

(2) Calculation of the sums of squares :

$$SS_p = \frac{y_{1.}^2 + y_{2.}^2}{25} - C$$

$$SS_d = \frac{1}{2} \sum_{j=1}^{25} y_{.j}^2 - C$$

$$SS_t = \sum_{i=1}^{2} \sum_{j=1}^{25} y_{ij}^2 - C$$

(3) Calculation of the sum of squares of the residuals :

$$SS_r = SS_t - SS_d - SS_p$$

Table 4.III

Two-way layout table for the clinical laboratory problem

Days	1	2	3	4	5	6	7	8	9	10	11	12	13	14	15	16	17	18	19	20	21	22	23	24	25	Sum for positions $y_{i.}$
Position 1	138	137	137	136	137	136	140	139	137	135	132	136	138	141	137	136	137	138	138	138	136	140	139	140	138	3436
Position 2	140	137	136	139	138	137	139	138	139	135	136	137	142	139	137	139	135	138	138	138	139	140	141	141	139	3457
Sum for days $y_{.j}$	278	274	273	275	275	273	279	277	276	270	268	273	280	280	274	275	272	276	276	276	275	280	280	281	277	grand sum $y_{..} = 6893$

$$C = \frac{y_{..}^2}{50} = \frac{6893^2}{50} = 950268.98$$

$$SS_p = \frac{y_{1.}^2 + y_{2.}^2}{25} - C = \frac{3426^2 + 3457^2}{25} - C = 8.82$$

$$SS_d = \frac{1}{2} \sum_{j=1}^{25} y_{.j}^2 - C = \frac{1}{2} (278^2 + 274^2 + \ldots + 277^2) - C = 128.52$$

$$SS_t = \sum_{i=1}^{2} \sum_{j=1}^{25} y_{ij}^2 - C = 138^2 + 137^2 + \ldots + 139^2 - C = 172.02$$

$$SS_r = SS_t - SS_d - SS_p = 34.68$$

(4) Calculation of the variances by division through the number of degrees of freedom. These are equal to the number of levels at which the factors are considered minus one, except for the residual, which, by analogy with the one-way example, is equal to the total number of observations minus one (i.e., the total number of degrees of freedom) minus the degrees of freedom used up by the other factors

$$s_d^2 = SS_d \; / \; 24$$

$$s_p^2 = SS_p \; / \; 1$$

$$s_r^2 = SS_r \; / \; (50 - 1 - (24 + 1)) = SS_r \; / \; 24$$

(5) Calculation of the F ratios :

$$F_d = s_d^2 \; / \; s_r^2$$

$$F_p = s_p^2 \; / \; s_r^2$$

and testing of the null hypothesis.

How to do this in practice, is shown in a worked example, using the data from Table 4.III. These data are used to construct Table 4.IV.

Table 4.IV

ANOVA table

Source	SS	D.F.	M.S.	F
Total	172.02	49	3.51	
Days	128.52	24	5.36	3.72
Position	8.82	1	8.82	6.12
Residual	34.68	24	1.44	

H_0 : (1) s_d^2 is not significant ;

(2) s_p^2 is not significant.

If both hypotheses are exact, then $s_t^2 = s_r^2$.

At the 1% confidence level, the first null hypothesis is rejected (F = 2.66, degrees of freedom = 24,24) which means that there is a significant contribution to the total variance due to variations between days. The second hypothesis is accepted (F 1% = 7.82, degrees of freedom = 1,24), which means that there is no significant contribution to the total variance due to variations between positions. Therefore, $s_t^2 = s_d^2 + s_r^2$.

4.1.5. Applications

A recent review about the application of ANOVA in analytical chemistry is due to Hirsch (1977).

The application of the ANOVA technique that is encountered most frequently in analytical chemistry is the breakdown of a total precision into its components such as between-days and within-days, between-laboratories and within-laboratories, etc. (see also Doerffel, 1962). If these components are known, one can decide which components should be optimized first. ANOVA is also used to decide whether a certain factor has a meaningful effect on the results. When the factor is the choice of the procedure, this means that one determines whether the procedures give the same result. If they do, then one concludes that all of the procedures are accurate. A warning is necessary here : the procedures to be compared should not contain common steps, such as the same preliminary separation step, because if there is such a common step and it introduces a systematic error, one will not be able to detect this error, i.e., in the symbolism used in section 4.1.2, ANOVA does not permit one to observe whether $\mu = \mu_o$.

One of the more important applications of fixed-effects ANOVA in the context of this book is its use as a preliminary step in the experimental optimization of procedures. As will be explained in Part II, the selection of the factors that have an influence on the performance characteristic which is chosen as the optimization criterion is an important part of such an optimization. For this application, it is recommended that one should take into account the fact that the variables may depend on each other. Therefore, factorial analysis (Chapter 12) is indicated.

Let us now consider a few applications concerning the evaluation of precision and accuracy that have appeared in the literature. One-way ANOVA is rarely used. One example, due to Gooszen (1960), concerns quality control in the clinical laboratory. On each of k (typically 20) days, n (usually 2-5) determinations are carried out. ANOVA is used to divide the variance into between-days and within-days components. If a significant between-days variance is obtained, it may be due to systematic reasons (a trend) or not. Trend detection is also discussed in the next chapter.

Two-way ANOVA is described much more frequently. One example, also concerning the clinical chemistry laboratory (Amenta, 1968) has already been discussed in detail in section 4.1.4.

We have already stated that interaction between factors can occur and that this is considered in more detail in Chapter 12. To avoid the impression that this is only of interest in the context of experimental optimization, the work of Riddick et al. (1972) can be cited here. Riddick et al. considered the same problem as Amenta, i.e., they studied the sources of variation in the results from a clinical laboratory with two serum pools, one being "normal" and the other "abnormal". These two samples (usually with very different concentrations) are analysed as randomly placed samples in a run on each of 30 days. The variance can be split up into a residual error, a between-days component, a between-pools component and a pool by day interaction. The last interaction occurs when some factor has different influences on the two pools for different days. It indicates, for example, a change of slope of the standard curve or a depression of the high end of the standard curve when an inadequate amount of substrate is present in enzymatic methods.

The problem of how to determine the between-laboratory and within-laboratory components of the precision from collaborative experiments is important in analytical chemistry. Very often the between-laboratory component is the larger indicating that some of the parameters of the procedure should be controlled more strictly. An important text concerning this problem is due to Youden and Steiner (1975). A collaborative testing programme for the Association of Official Analytical Chemists typically involves distributing 3-5 sufficiently different samples to 8-15 sufficiently proficient laboratories. Each laboratory is asked to report two replicate determinations on each sample. The statistical analysis involves the following steps :

(1) Reduction of the results reported to equal numbers. Some laboratories report more than two replicates. The ANOVA calculations can be carried out on unequal numbers of replicates, but in this instance the procedure becomes very complex. Therefore, some of the results are eliminated, which must be done randomly, using a random numbers table, for example.

(2) Elimination of laboratories reporting systematically high or low results. This can be done using a simple non-parametric ranking test as described by Youden and Steiner.

(3) Elimination of outlying results using Dixons's test (Dixon, 1953).

(4) Examination of the homogeneity of the variance. It was assumed that the variances are the same in the laboratories. Indeed, one assumes that the laboratory biases are normally distributed, and they are therefore thought to come from the same population. If there is a statistically significant difference between the variances from the laboratories, this usually means that one laboratory has been working with much lower precision than the others, and this laboratory should be eliminated. In the same way, the residual errors should be normally distributed with constant mean. This follows from the discussion in section 4.1.2. Tests to check these homogeneities were given by Youden and Steiner.

(5) The ANOVA procedure, in this instance a two-way layout (samples ; laboratories) with interaction between samples and laboratories (see section 4.1.1).

(6) Calculation of what Youden and Steiner call reproducibility and repeatability (for definitions, see section 2). A general plan for the interlaboratory evaluation of a method has also been proposed by Gottschalk (1975). The experiment should include k laboratories, each carrying out m replicate determinations. The product k x m should be at least 24 and m should be at least 4. In this instance, simple equations permit an evaluation of the two precision components. Gottschalk discussed the same limitations and sources of error in this application of ANOVA as Youden and Steiner. As more replicate determinations are necessary in this plan, more data are available and other tests can be applied to make decisions concerning, for example, the homogeneity of the variances (Bartlett's test ; see section 4.1.6).

An analytical application of a special kind of layout for ANOVA, called a nested design, was proposed by Wernimont (1951). He studied a procedure for an acetyl determination and compared several sources of variation by having two analysts in each of eight laboratories perform two replicate tests on each of three days. The usual ANOVA procedure would have consisted of having the same two analysts perform two replicate tests on three days in the eight laboratories. Clearly, there is little sense in organizing an experiment that would have required two analytical chemists to run from one laboratory to the other, and therefore the nested design was preferred. Nested designs are incomplete layouts. One says that the levels of a factor (analysts) are nested within the levels of another factor (laboratories) if every level of the first factor appears with only one level of the second factor in the observations. Each analyst appears only in one of the eight laboratories. The nested design used by Wernimont is represented in Fig. 4.1.

Wernimont's conclusion was that the laboratories are responsible for the largest source of variation. The mathematics of nested designs are not discussed in the mathematical section of this book, but have been discussed, for example, by Scheffé (1959).

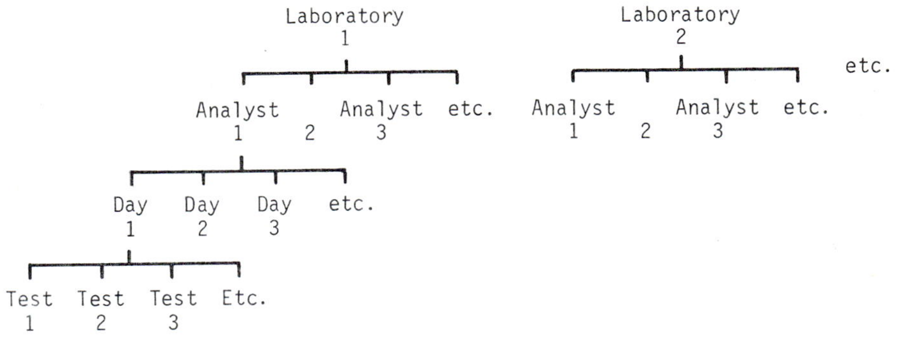

Fig. 4.1. Nested sampling design (Wernimont, 1951). Reprinted with permission. Copyright by the American Chemical Society.

One can use layouts of a higher order than the two-way layout. The equations in section 4.1.4 can be generalized without much difficulty. The theory was given comprehensively by Scheffé (1959). Practical calculation schemes can be found, for example, in Barrie Wetherill (1972), Scheffé (1959), Lindman (1974) and several other books such as those cited in Chapter 12 and 13. Additional applications in analytical chemistry can be found in Hirsch (1977), Maurice (1957) and Walker (1977).

4.1.6. Bartlett's test for the comparison of more than two variances

The largest section of this chapter is devoted to a comparison of the means of more than two procedures through ANOVA. One can also query whether the variances of these procedures can be considered to be equal. In some applications of ANOVA (see section 4.1.5), this is also necessary. The best known test in this respect is Bartlett's test.

The parameter proposed by Bartlett is

$$M = \sum_{i=1}^{p} (n_i-1)\log_e s^2 - \sum_{i=1}^{p} (n_i-1)\log_e s_i^2 \qquad (4.2)$$

where s_i is the variance of procedure i with n_i determinations and n_i-1 degrees of freedom, and

$$s^2 = \frac{\sum_{i=1}^{p} (n_i-1)\, s_i^2}{\sum_{i=1}^{p} (n_i-1)} \qquad (4.3)$$

Alternatively, we can write

$$M = \sum_{i=1}^{p} (n_i-1) \log_e \frac{s^2}{s_i^2} \qquad (4.4)$$

If the null hypothesis is true (the variances are equal), M follows a χ^2 distribution with p-1 degrees of freedom. An example is given below.

Eight replicate determinations were performed using four different procedures. The data are summarized in Table 4.V.

Table 4.V.

Data for comparison of more than two variances

Procedure i	Mean of 8 determinations $\bar{y}_i$	Variance s_i^2	Degrees of freedom n_i-1	$\log_e s_i^2$
1	124.8	143.89	7	4.9690
2	156.2	112.14	7	4.7197
3	139.5	84.45	7	4.4362
4	118.2	40.16	7	3.6929

Calculation of the first term in eqn. 4.2 :

$$s^2 = \frac{1}{28} (7 \times 143.89 + 7 \times 112.14 + 7 \times 84.45 + 7 \times 40.16) = 95.16$$

$$\log_e s^2 = 4.5556$$

$$\sum_{i=1}^{p} (n_i-1) \log_e s^2 = 28 \times 4.5556 = 127.4125$$

Calculation of the second term in eqn. 4.2

$$\sum_{i=1}^{p} (n_i-1) \log_e s_i^2 = 7 \times 4.9690 + 7 \times 4.7197 + 7 \times 4.4362 + 7 \times 3.6929$$

$$= 124.7246$$

M = 127.4125 - 124.7246 = 2.6879

The null hypothesis can be formulated as : H_0 = the variances s_i^2 are equal.
A significance level of 0.05 is chosen.

χ^2 (0.05) = 7.815 for the 3 (= p-1) degrees of freedom.

The null hypothesis is accepted.

4.2. MATHEMATICAL SECTION

4.2.1. The general linear model

4.2.1.1. Introduction

Let us consider values $y_1, y_2, \ldots, y_n$ taken by n random variables, and assume that each value y_i can be considered as a linear combination of q unknown quantities, $\beta_1, \beta_2, \ldots, \beta_q$, plus an error, e_i, which is also a random variable

$$y_i = v_{1i}\,\beta_1 + v_{2i}\,\beta_2 + \ldots + v_{qi}\,\beta_q + e_i \quad (i = 1, 2, \ldots, n) \qquad (4.5)$$

The v_{ji} are known constant coefficients. It is generally assumed that the random variables e_i have a zero mean, are uncorrelated and have equal variance. This can be expressed in the following way

$$\begin{aligned} &E(e_i) = 0 && i = 1, 2, \ldots, n \\ &E(e_i^2) = \sigma^2 && i = 1, 2, \ldots, n \qquad\qquad (4.6) \\ &E(e_i e_k) = 0 && i,k = 1, 2, \ldots, n \quad i \neq k \end{aligned}$$

The purpose of this general linear model is the examination of the unknown quantities β_j (j = 1, 2, ..., q). These β_j represent "factors of influence" which are felt to be of importance in a research problem.
Estimations of the values of the β_j and inferences for the research problem

will be discussed in the next section.

The notations just introduced can be considerably simplified by using matrix algebra. Matrix algebra is introduced in section 17.7. By using the following definitions

$$\vec{Y} = \begin{bmatrix} y_1 \\ y_2 \\ y_3 \\ \vdots \\ y_n \end{bmatrix} \qquad \vec{\beta} = \begin{bmatrix} \beta_1 \\ \beta_2 \\ \beta_3 \\ \vdots \\ \beta_q \end{bmatrix} \qquad \vec{E} = \begin{bmatrix} e_1 \\ e_2 \\ e_3 \\ \vdots \\ e_n \end{bmatrix}$$

$$V = \begin{bmatrix} v_{11} & v_{12} & \cdots & v_{1n} \\ v_{21} & v_{22} & & \\ \vdots & & & \\ v_{q1} & & & v_{qn} \end{bmatrix}$$

the model can then be written as

$$\vec{Y} = V'\vec{\beta} + \vec{E} \qquad (4.7)$$

where V' is defined as a matrix obtained by permuting the rows and columns of matrix V. V' is called the transpose of V (see Chapter 17).

According to the nature of the coefficients v_{ij}, three types of problems can be defined. If all coefficients v_{ji} are equal to 0 or 1, this defines an analysis of variance problem. For this type of problem, v_{ji} represents the belonging of a measurement y_i to a set j. When the measurement belongs to set j, v_{ji} is equal to unity and is equal to zero otherwise. If the v_{ji} are values taken by n observations of a set of q variables (then called independent variables) and the y_i are n values taken by a variable called the dependant variable, it is a problem of multiple regression analysis. Finally, if some of the v_{ji} are values taken by observations of variables and some are equal to zero or unity, it is called analysis of covariance.

The unknown quantities $\beta_1, \beta_2, \ldots, \beta_q$ can be defined either as unknown constants or parameters, or as unobservable random variables. In the first instance the model will be called a fixed-effects model and in the second a random-effects model. When both types of quantities β are used, the model is one of mixed effects.

4.2.1.2. Estimation

The model and hypotheses in section 4.2.1.1 will now be examined in detail.

Using matrix notation this model can be written as

$$(M_1) \begin{cases} \vec{Y} = V'\vec{\beta} + \vec{E} & (1) \\ E(\vec{E}) = 0 & (2) \\ E(\vec{E}\vec{E}') = \sigma^2 I & (3) \end{cases} \qquad (4.8)$$

Equality 2 expresses the fact that the mathematical expectation of the error vector E is a vector of zeros, i.e., all of the errors have a distribution with zero mean. Equality 3 indicates that the mathematical expectation of matrix EE' given by

$$\vec{E}\vec{E}' = \begin{bmatrix} e_1^2 & e_1e_2 & \cdots & e_1e_n \\ e_2e_1 & e_2^2 & \cdots & e_2e_n \\ \vdots & & & \\ e_ne_1 & & \cdots & e_n^2 \end{bmatrix}$$

is equal to the matrix $\sigma^2 I$

$$\sigma^2 I = \begin{bmatrix} \sigma^2 & 0 & \cdots & 0 \\ 0 & \sigma^2 & & 0 \\ \vdots & & \ddots & \vdots \\ 0 & & & \vdots \\ \vdots & & & \\ 0 & & & \sigma^2 \end{bmatrix}$$

When considering an element of $\vec{E}\vec{E}'$, which is not on the diagonal, for example, e_1e_2, equality 3 indicates that its expectation is zero or, in other words, that the two errors are uncorrelated. The expectations of the diagonal elements are all σ^2. The meaning of this is that the variances of all errors are equal to σ^2, indeed

$$\begin{aligned} \text{Var}\,(e_i) &= E(e_i^2 - E(e_i)\ ) \\ &= E(e_i^2) = \sigma^2 \end{aligned}$$

In future, eqns. 1-3 will be referred to as model (M_1). The expected value of $\vec{Y}$ is given by

$$\begin{aligned} E(\vec{Y}) &= E(V'\vec{\beta} + \vec{E}) \\ &= E(V'\vec{\beta}) + E(\vec{E}) = E(V'\vec{\beta}) = V'\vec{\beta} \end{aligned}$$

as the $V'\vec{\beta}$ are fixed values. The variance-covariance matrix of $\vec{Y}$ is given by

$$\begin{aligned} C(\vec{Y}) &= E(((\vec{Y} - E(\vec{Y}))\ '\ (\vec{Y} - E(\vec{Y}))) \\ &= E(\vec{E}'\vec{E}) = \sigma^2 I \end{aligned}$$

The problem after formulating (M_1) is to find a "good" estimation of the unknown parameters $\beta_1, \ldots, \beta_q$. Suppose that $b_1, \ldots, b_q$ denote estimates of $\beta_1, \ldots, \beta_q$ and call $\vec{B}$ the vector of these values

$$\vec{B} = \begin{bmatrix} b_1 \\ b_2 \\ \vdots \\ b_q \end{bmatrix}$$

A possible measure of the quality of an estimate vector $\vec{B}$ is given by the sum of squares of the differences of the measured values y_i and the values given by the linear function when replacing the β_j with b_j. This value is given by the vector $V'\vec{B}$ and the sum of squares of differences is given by the function $\psi(\vec{Y},\vec{B})$

$$\psi(\vec{Y},\vec{B}) = \sum_i (y_i - \sum_j v_{ji} b_j)^2$$

$$= (\vec{Y} - V'\vec{B})'(\vec{Y} - V'\vec{B})$$

A set of values $b_1, b_2, \ldots, b_p$ that minimize the function $\psi(\vec{Y},\vec{B})$ is called a least-squares estimate of the $\beta_1, \beta_2, \ldots, \beta_p$. These values are found by setting the partial derivatives of $\psi(Y, B)$ with respect to the b_j equal to zero. This gives the following equations

$$\frac{\partial\psi(\vec{Y},\vec{B})}{\partial b_j} = 0 \qquad j = 1, 2, \ldots, q$$

which give the following results

$$\frac{\partial\psi(\vec{Y},\vec{B})}{\partial b_j} = -2 \sum_i (y_i - \sum_k v_{ki} b_k)\, v_{ji} = 0 \qquad j = 1, 2, \ldots, q$$

or

$$\sum_i \sum_k v_{ki} v_{ji} b_k = \sum_i y_i v_{ji}$$

These equations can be written in matrix form

$$VV'\vec{B} = V\vec{Y}$$

and are called the normal equations.

A solution of these equations will be called $\vec{B}$, denoting a least-squares estimate of $\vec{\beta}$. When the rank of matrix V is equal to q (maximal rank) it can easily be shown that the matrix (VV') is non-singular and therefore $(VV')^{-1}$ exists. In this instance, the unique solution of the normal equations is given by

$$\vec{B} = (VV')^{-1} V\vec{Y} \qquad (4.9)$$

Often, as in the analysis of variance, the rank of V is smaller than q. In this instance, a unique solution to the normal equations can be found by imposing one or several conditions on the parameters β. A further discussion of this aspect was given by Kendall and Stuart (1967) and Scheffé (1959).

4.2.2. The analysis of variance

In this section, a particular case of the general linear model considered in section 4.2.1 will be examined in detail.

The values v_{ji} in eqns. 4.5 and 4.6 are all known constant parameters. In the models in this section, all of these parameters take the value 0 or 1. These models are referred to as analysis of variance models.

4.2.2.1. The one-way layout

The one-way layout problem is a comparison of the means of a variable measured in several populations. p populations are considered with unknown means $\mu_1, \mu_2, \ldots, \mu_p$. As these mean values are considered as fixed but unknown parameters the model defined in this way is called a fixed-effects model. It will be assumed that the mean of each population is composed of a general mean, μ, and a term α_i specific to population i

$$\mu_i = \mu + \alpha_i$$

Further, as the α_i represent deviations from the general mean, it will be assumed that

$$\sum_{i=1}^{p} \alpha_i = 0$$

The variables y are all assumed to have uncorrelated normal distributions with equal variance σ^2. Independent random samples of sizes J_1, J_2, ..., J_p are taken from the p populations. The values of the variable for the sample of the first population are given by

$$y_1, y_2, \ldots, y_{J_1}$$

Another subscript (1) is added to denote that these values concern the first population. The notation then becomes

$$y_{11}, y_{12}, y_{13}, \ldots, y_{1J_1}$$

In general, the values of the variable for the sample of the ith population are called

$$y_{i1}, y_{i2}, y_{i3}, \ldots, y_{iJ_i}$$

and the value of the variable for the jth element of the ith population is called y_{ij}. The model can then be written as

$$y_{ij} = \mu + \alpha_i + e_{ij} \qquad i = 1, \ldots, p \; ; \; j = 1, \ldots, J_i \qquad (4.10)$$

$$\sum_{i=1}^{p} \alpha_i = 0 \qquad (4.11)$$

$$e_{ij} \sim N(0, \sigma^2) \text{ and uncorrelated} \qquad (4.12)$$

This model can be identified with the general linear model. The sum of the sizes of the samples gives the number of observations

$$n = \sum_{i=1}^{p} J_i$$

The observations y_i are now defined as double-subscript observations, y_{ij}.

The following relationships hold

$$\vec{Y} = \begin{bmatrix} y_1 \\ y_2 \\ \cdot \\ \cdot \\ \cdot \\ \cdot \\ \cdot \\ y_n \end{bmatrix} = \begin{bmatrix} y_{11} \\ \vdots \\ y_{1J_1} \\ y_{21} \\ \vdots \\ y_{2J_2} \\ \vdots \\ y_{pJ_p} \end{bmatrix}$$

$$\vec{E} = \begin{bmatrix} e_1 \\ e_2 \\ \cdot \\ \cdot \\ \cdot \\ \cdot \\ e_n \end{bmatrix} = \begin{bmatrix} e_{11} \\ \vdots \\ e_{1J_1} \\ e_{21} \\ \vdots \\ e_{2J_2} \\ \vdots \\ e_{pJ_p} \end{bmatrix}$$

The matrix V' can be found by using equations 4.10

$$V' = \begin{bmatrix} 1 & 0 & 0 \dots & 0 & 0 \\ 1 & 0 & 0 \dots & 0 & 0 \\ \vdots & \vdots & \vdots & \vdots & \vdots \\ 1 & 0 & 0 \dots & 0 & 0 \\ 0 & 1 & 0 \dots & 0 & 0 \\ \vdots & \vdots & \vdots & \vdots & \vdots \\ 0 & 1 & 0 \dots & 0 & 0 \\ \vdots & \vdots & \vdots & \vdots & \vdots \\ 0 & 0 & 0 & 0 & 1 \\ \vdots & \vdots & \vdots & \vdots & \vdots \\ 0 & 0 & 0 & 0 & 1 \end{bmatrix}$$

The sum of squares to be minimized to obtain good estimates of the unknown means is given by

$$\psi(\vec{Y},u,\vec{A}) = \sum_{i=1}^{p} \sum_{j=1}^{J_i} (y_{ij} - u - a_i)^2 \tag{4.13}$$

where u is an estimate of the general mean μ and $\vec{A}$ is a vector of estimates a_i of the deviations α_i.

The normal equations are then obtained by setting the partial derivatives of ψ with respect to a_i and to u to zero

$$\frac{\partial\psi(\vec{Y},u,\vec{A})}{\partial u} = - \sum_{i=1}^{p} \sum_{j=1}^{J_i} 2(y_{ij} - u - a_i) = 0$$

$$\frac{\partial\psi(\vec{Y},u,\vec{A})}{\partial a_i} = - \sum_{j=1}^{J_i} 2(y_{ij} - u - a_i) = 0$$

Using the condition $\sum_{i=1}^{p} a_i = 0$, the following least-squares estimates $\hat{\mu}$ and $\hat{\alpha}_i$ are obtained

$$\hat{\mu} = \frac{1}{n} \sum_{i=1}^{p} \sum_{j=1}^{J_i} y_{ij} \tag{4.14}$$

and

$$\hat{\alpha}_i = \frac{1}{J_i} \sum_{j=1}^{J_i} y_{ij} - \hat{\mu} \tag{4.15}$$

Using the definitions

$$\overline{y}_{..} = \frac{1}{n} \sum_{i=1}^{p} \sum_{j=1}^{J_i} y_{ij}$$

and

$$\overline{y}_{i.} = \frac{1}{J_i} \sum_{j=1}^{J_i} y_{ij}$$

the estimates can then be written as

$$\hat{\mu} = \overline{y}_{..}$$

and

$\hat{\alpha}_i = \bar{y}_{i.} - \bar{y}_{..}$

The sum of the squares of the deviations of all measurements with respect to the general average is given by

$$\sum_{i=1}^{p} \sum_{j=1}^{J_i} (y_{ij} - \bar{y}_{..})^2$$

This sum can be broken down as follows

$$\sum_{i=1}^{p} \sum_{j=1}^{J_i} (y_{ij} - \bar{y}_{..})^2 = \sum_{i=1}^{p} \sum_{j=1}^{J_i} y_{ij}^2 - 2\bar{y}_{..} \sum_{i=1}^{p} \sum_{j=1}^{J_i} y_{ij} + \sum_{i=1}^{p} \sum_{j=1}^{J_i} \bar{y}_{..}^2$$

$$= \sum_{i=1}^{p} \sum_{j=1}^{J_i} y_{ij}^2 - 2n\bar{y}_{..}^2 + n\bar{y}_{..}^2$$

$$= \sum_{i=1}^{p} \sum_{j=1}^{J_i} y_{ij}^2 - n\bar{y}_{..}^2$$

This leads to the following breakdown

$$\sum_{i=1}^{p} \sum_{j=1}^{J_i} y_{ij}^2 - \sum_{i=1}^{p} J_i\bar{y}_{i.}^2 + \sum_{i=1}^{p} J_i\bar{y}_{i.}^2 - n\bar{y}_{..}^2 = \sum_{i=1}^{p} \sum_{j=1}^{J_i} (y_{ij} - \bar{y}_{i.})^2 + \sum_{i=1}^{p} J_i(y_{i.} - \bar{y}_{..})^2$$

The first of these two sums of squares is a measure of the distances between the observations of a group and the mean of the group. It will be called the sum of squares within groups, denoted by SS_r (or residual sum of squares). The second sum of squares is a measure of the distances between the groups, denoted by SS_p.

The total sum of squares will be denoted by SS_t, and gives the following equation

$$SS_t = SS_r + SS_p$$

It can be shown that, when the hypothesis that all deviations α_i are zero is true, i.e., when H_0 is true

$$H_0 : \alpha_1 = \alpha_2 = \ldots = \alpha_p = 0 \tag{4.16}$$

then the two sums of squares, SS_r and SS_p, are independent. Further, each term in the sum of squares has an N(0,1) distribution. As the number of independent terms in SS_r and SS_p are n-p and p-1, respectively, then $\frac{SS_r}{(n-p)^2 \sigma^2}$ and $\frac{SS_p}{(p-1)\sigma^2}$ are independent χ^2_{n-p} and χ^2_{p-1} variables. Therefore,

$$F = \frac{\dfrac{SS_p}{(p-1)\sigma^2}}{\dfrac{SS_r}{(n-p)\sigma^2}} = \frac{n-p}{p-1} \cdot \frac{SS_p}{SS_r} \tag{4.17}$$

has an F distribution with parameters (p-1) and (n-p).

The above information can be summarized in the analysis of variance shown in Table 4.V.

Table 4.V.

ANOVA table

	Sum of squares	Degrees of freedom	Mean square
Within groups	SS_r	p-1	$\frac{SS_r}{p-1}$
Between groups	SS_p	n-p	$\frac{SS_p}{n-p}$
Total	SS_t	n-1	$\frac{SS_t}{n-1}$

This leads to the following test for deciding whether the differences between the means are significant : if $F > F_{\alpha,p-1,n-p}$, we reject H_0 at level α. When the number of procedures (or, as called here, populations) is very large, a sample of procedures is drawn in a random way. The results obtained with each procedure (from each population) constitute values of a random variable. It is assumed that each of these random variables has a normal distribution with the same variance σ^2 ; the ith selected population (or variable) has an $N(\mu_i,\sigma)$

distribution. As the mean values μ_i depend upon the sample selected, they can also be considered as random variables. Considering the set of all possible mean values from which the sample was drawn, it can be assumed to have a normal distribution, the mean and variance of which will be called μ and σ_A^2. If all of the mean values of the populations are equal, the variance σ_A^2 is zero. Therefore, the null hypothesis of the problem becomes

$$H_0 : \sigma_A^2 = 0$$

If the hypothesis is true, all mean values, both those belonging to the sample and the others, can be assumed to be equal.

The linear model used to test this hypothesis can then be formulated as follows

$$\begin{aligned} y_{ij} &= \mu_i + e_{ij} \\ &= \mu + \alpha_i + e_{ij} \\ y_{ij} &\sim N(\mu_i, \sigma^2) = N(\mu + \alpha_i, \sigma^2) \\ e_{ij} &\sim N(0, \sigma^2) \\ \mu_i &\sim N(\mu, \sigma_A^2) \text{ or } \alpha_i \sim N(0, \sigma_A^2) \end{aligned} \tag{4.18}$$

The e_{ij} and α_i are uncorrelated. Eqns. 4.14 and 4.15 for estimating the values of μ and α_i and 4.17 for testing the hypothesis are the same as for the fixed-effects model.

4.2.2.2. The two-way layout without interaction

In this section, another particular case of the general linear model is examined. The two-way layout problem is that of examining the simultaneous influence of two different factors (A and B) on a given variable.

The components or levels of factor A will be denoted by h and the components of factor B by i. It will be assumed that h varies from 1 to p_1 and i from 1 to p_2 :

$h = 1, 2, \ldots, p_1$

$i = 1, 2, \ldots, p_2$

A combination of components (h,i) is called a cell. To simplify the notations, it will be assumed that for each cell combination (h,i) an identical number of observations, J, were made. The results obtained can be generalized to the case of unequal numbers of observations.

y_{hij} ($h = 1, 2, \ldots, p_1$; $i = 1, 2, \ldots, p_2$; $j = 1, 2, \ldots, J$)

is defined as the jth value taken by a random variable for which the factors A and B have components h and i. The unknown mean value of the random variable of cell (h,i) is called μ_{hi}. The μ_{hi} are unknown parameters. The model obtained in this way is a fixed-effects model and it can be generalized to the random-effects model for which the values μ_{hi} are random variables. All of the variables are assumed to have uncorrelated normal distributions with equal variance σ^2. The model can then be written

$$y_{hij} = \mu_{hi} + e_{hij} \qquad \begin{array}{l} h = 1, 2, \ldots, p_1 \\ i = 1, 2, \ldots, p_2 \\ j = 1, 2, \ldots, J \end{array} \qquad (4.19)$$

e_{hij} have uncorrelated $N(0,\sigma^2)$ distributions.

The following definitions are also needed

$$\text{(a)} \quad \mu = \frac{1}{p_1 p_2} \sum_{h=1}^{p_1} \sum_{i=1}^{p_2} \mu_{hi} \qquad (4.20)$$

where μ is the general mean for factors A and B ;

$$\text{(b)} \quad \alpha_h = \frac{1}{p_2} \sum_{i=1}^{p_2} \mu_{hi} - \mu \qquad h = 1, 2, \ldots, p_1 \qquad (4.21)$$

where α_h is the mean of factor B at the level h of factor A ;

(c) $\beta_i = \frac{1}{p_1} \sum_{h=1}^{p_1} \mu_{hi} - \mu \qquad i = 1, 2, \ldots, p_2$ (4.22)

where β_i is the mean of factor A at level i of factor B ; it can easily be seen that $\sum_{h=1}^{p_1} \alpha_h = 0$ and $\sum_{i=1}^{p_2} \beta_i = 0$.

(d) $\gamma_{hi} = \mu_{hi} - \frac{1}{p_2} \sum_{i=1}^{p_2} \mu_{hi} - \frac{1}{p_1} \sum_{h=1}^{p_1} \mu_{hi} + \mu$ (4.23)

$$= \mu_{hi} - \alpha_h - \beta_i - \mu$$

The last definition makes it possible to write the model in the following way

$$y_{hij} = \mu + \alpha_h + \beta_i + \gamma_{hi} + e_{hij} \qquad \begin{array}{l} h = 1, 2, \ldots, p_1 \\ i = 1, 2, \ldots, p_2 \\ j = 1, 2, \ldots, J \end{array} \qquad (4.24)$$

e_{hij} have uncorrelated $N(0,\sigma^2)$ distributions.

In this section, it will be assumed that all terms γ_{hi} are zero. This hypothesis is called the additivity hypothesis. It is fulfilled when there is no interaction between the two factors A and B. It follows from 4.23 that this is equivalent to

$$\mu_{hi} = \mu + \alpha_h + \beta_i$$

The average for cell (h,i) consists of a general average μ and two averages for the components h and i of factors A and B. The model then becomes

$$y_{hij} = \mu + \alpha_h + \beta_i + e_{hij} \qquad \begin{array}{l} h = 1, 2, \ldots, p_1 \\ i = 1, 2, \ldots, p_2 \\ j = 1, 2, \ldots, J \end{array}$$

$$\sum_{h=1}^{p_1} \alpha_h = \sum_{i=1}^{p_2} \beta_i = 0$$

e_{hij} have uncorrelated $N(0,\sigma^2)$ distributions.

The hypotheses that seem to be of most interest are

$$H_1 : \alpha_h = 0 \qquad h = 1, 2, \ldots, p_1$$

$$H_2 : \beta_i = 0 \qquad i = 1, 2, \ldots, p_2$$

Hypothesis H_1 states that all deviations from the general mean due to factor A are zero and therefore all values taken by factor A have the same effect ; hypothesis H_2 states the same for factor B. Again, this model can easily be identified with the general linear model. The number of observations n is given by the product of the number of cells and the number of elements per cell, J

$$n = p_1 \cdot p_2 \cdot J$$

The observations y_i are now defined with three subscripts, y_{hij}. The following relationships hold

$$\vec{Y} = \begin{bmatrix} y_1 \\ y_2 \\ \cdot \\ \cdot \\ \cdot \\ \cdot \\ \cdot \\ \cdot \\ \cdot \\ y_n \end{bmatrix} = \begin{bmatrix} y_{111} \\ y_{112} \\ \vdots \\ y_{11J} \\ y_{121} \\ \vdots \\ y_{p_1 p_2 J} \end{bmatrix}$$

$$\vec{E} = \begin{bmatrix} e_1 \\ e_2 \\ \cdot \\ \cdot \\ \cdot \\ \cdot \\ \cdot \\ \cdot \\ e_n \end{bmatrix} = \begin{bmatrix} e_{111} \\ e_{112} \\ \vdots \\ e_{11J} \\ e_{121} \\ \vdots \\ e_{p_1 p_2 J} \end{bmatrix}$$

Matrix V' can be found using equalities (4.24). If there is only one observation per cell (J = 1), it becomes

$$
V' = \left[\begin{array}{ccccccccc}
\mu & \alpha_1 & \alpha_2 & \cdots & \alpha_{p_1} & \beta_1 & \beta_2 & \cdots & \beta_{p_2} \\
1 & 1 & 0 & \cdots & 0 & 1 & 0 & \cdots & 0 \\
1 & 1 & 0 & \cdots & 0 & 0 & 1 & \cdots & 0 \\
\vdots & \vdots & \vdots & & \vdots & \vdots & \vdots & & \vdots \\
 & & & & & & & & 0 \\
1 & 1 & 0 & & 0 & 0 & 0 & & 1 \\
\hline
1 & 0 & 1 & \cdots & 0 & 1 & 0 & \cdots & 0 \\
\vdots & \vdots & \vdots & & \vdots & 0 & 1 & & \vdots \\
 & & & & & \vdots & \vdots & & \\
1 & 0 & 1 & \cdots & 0 & 0 & 0 & \cdots & 1 \\
\hline
\vdots & \vdots & \vdots & \vdots & \vdots & \vdots & \vdots & \vdots & \vdots \\
\hline
1 & 0 & 0 & \cdots & 1 & 1 & 0 & \cdots & 0 \\
\vdots & \vdots & \vdots & & \vdots & 0 & 1 & & \vdots \\
 & & & & & \vdots & \vdots & & \\
1 & 0 & 0 & \cdots & 1 & 0 & 0 & \cdots & 1
\end{array}\right]
$$

Each column of V' represents an unknown parameter μ, α_h or β_i, and each row represents an observation or a cell (h,i). Since there are J observations in each cell (h,i), each row of this matrix is replicated J times.

The sum of squares to be minimized in order to obtain good estimates of the unknown parameter μ, α_h and β_i is given by

$$\psi(\vec{Y},u,\vec{A},\vec{B}) = \sum_{h=1}^{p_1} \sum_{i=1}^{p_2} \sum_{j=1}^{J} (y_{hij} - u - a_h - b_i)^2$$

where $\vec{A}$ and $\vec{B}$ represent the vectors of estimates of the parameters α_h and β_i. The normal equations are found by setting the partial derivatives of ψ to zero

$$\frac{\partial\psi(\vec{Y},u,\vec{A},\vec{B})}{\partial u} = -2 \sum_{h=1}^{p_1} \sum_{i=1}^{p_2} \sum_{j=1}^{J} (y_{hij} - u - a_h - b_i) = 0$$

$$\frac{\partial\psi(\vec{Y},u,\vec{A},\vec{B})}{\partial a_h} = -2 \sum_{i=1}^{p_2} \sum_{j=1}^{J} (y_{hij} - u - a_h - b_i) = 0$$

$$\frac{\partial\psi(\vec{Y},u,\vec{A},\vec{B})}{\partial b_i} = -2 \sum_{h=1}^{p_1} \sum_{j=1}^{J} (y_{hij} - u - a_h - b_i) = 0$$

Using the conditions $\sum_{h=1}^{p_1} a_h = \sum_{i=1}^{p_2} b_i = 0$, this leads to the following least square estimates

$$\hat{\mu} = \frac{1}{n} \sum_{h=1}^{p_1} \sum_{i=1}^{p_2} \sum_{j=1}^{J} y_{hij}$$

$$\hat{\alpha}_h = \frac{p_2}{n} \sum_{i=1}^{p_2} \sum_{j=1}^{J} y_{hij} - \frac{1}{n} \sum_{h=1}^{p_1} \sum_{i=1}^{p_2} \sum_{j=1}^{J} y_{hij}$$

$$\hat{\beta}_i = \frac{p_1}{n} \sum_{h=1}^{p_1} \sum_{j=1}^{J} y_{hij} - \frac{1}{n} \sum_{h=1}^{p_1} \sum_{i=1}^{p_2} \sum_{j=1}^{J} y_{hij}$$

By defining

$$\overline{y} = \frac{1}{n} \sum_{h=1}^{p_1} \sum_{i=1}^{p_2} \sum_{j=1}^{J} y_{hij}$$

$$\overline{y}_{h..} = \frac{1}{p_2 J} \sum_{i=1}^{p_2} \sum_{j=1}^{J} y_{hij}$$

and

$$\overline{y}_{.i.} = \frac{1}{p_1 J} \sum_{h=1}^{p_1} \sum_{j=1}^{J} y_{hij}$$

the normal equations become

$$\hat{\mu} = \overline{y}$$

$$\hat{\alpha}_h = \overline{y}_{h..} - \overline{y}$$

$$\hat{\beta}_i = \overline{y}_{.i.} - \overline{y}$$

The sum of squares of deviations of all measurements with respect to the general average is given by

$$\sum_{h=1}^{p_1}\sum_{i=1}^{p_2}\sum_{j=1}^{J}(y_{hij}-\bar{y})^2=\sum_{h=1}^{p_1}\sum_{i=1}^{p_2}\sum_{j=1}^{J}y_{hij}^2-n\bar{y}^2$$

It can easily be shown that this sum, also called the total sum of squares (SS_t), can be broken down in the following way

$$\sum_{h=1}^{p_1}\sum_{i=1}^{p_2}\sum_{j=1}^{J}y_{hij}^2-n\bar{y}^2=\sum_{h=1}^{p_1}p_1J(\bar{y}_{h..}-\bar{y})^2+\sum_{i=1}^{p_2}p_2J(\bar{y}_{.i.}-\bar{y})^2$$
$$+\sum_{h=1}^{p_1}\sum_{i=1}^{p_2}\sum_{j=1}^{J}(y_{hij}-\bar{y}_{h..}-\bar{y}_{.i.}+\bar{y})^2$$

The following notations are used to describe this breakdown into sums of squares

$$SS_t=\sum_{h=1}^{p_1}\sum_{i=1}^{p_2}\sum_{j=1}^{J}(y_{hij}-\bar{y})^2$$

$$SS_A=\sum_{h=1}^{p_1}p_1J(\bar{y}_{h..}-\bar{y})^2$$

$$SS_B=\sum_{i=1}^{p_2}p_2J(\bar{y}_{.i.}-\bar{y})^2$$

$$SS_r=\sum_{h=1}^{p_1}\sum_{i=1}^{p_2}\sum_{j=1}^{J}(y_{hij}-\bar{y}_{h..}-\bar{y}_{.i.}+\bar{y})^2$$

This gives the following equality

$$SS_t=SS_A+SS_B+SS_r$$

Considering the total sum of squares, SS_t, it can be observed that the sum of its components is always zero

$$\sum_{h=1}^{p_1} \sum_{i=1}^{p_2} \sum_{j=1}^{J} (y_{hij} - \bar{y}) = 0$$

Therefore, and as all y_{hij} are independent variables, SS_t has n-1 degrees of freedom. As all y_{hij} have an $N(\bar{y},\sigma^2)$ distribution, the terms $(y_{hij} - \bar{y})$ have an $N(0,\sigma^2)$ distribution, which implies that $\frac{SS_t}{\sigma^2}$ has an χ^2_{n-1} distribution. In the same way, it can be shown that $\frac{SS_A}{\sigma^2}$ has a $\chi^2_{p_1J-1}$ distribution, $\frac{SS_B}{\sigma^2}$ has a $\chi^2_{p_2J-1}$ distribution and $\frac{SS_r}{\sigma^2}$ has a $\chi^2_{p_1p_2J-p_1-p_2+1}$ distribution. The two quotients

$$\frac{SS_A/\sigma^2(p_2J-1)}{SS_r/\sigma^2(p_1p_2J-p_1-p_2+1)} = \frac{SS_A(p_1p_2J-p_1-p_2+1)}{SS_r(p_2J-1)}$$

and

$$\frac{SS_B/\sigma^2(p_1J-1)}{SS_r/\sigma^2(p_1p_2J-p_1-p_2+1)} = \frac{SS_B(p_1p_2J-p_1-p_2+1)}{SS_r(p_1J-1)}$$

have F distributions with $(p_2J-1,\ p_1p_2J-p_1-p_2+1)$ and $(p_1J-1,\ p_1p_2J-p_1-p_2+1)$ degrees of freedom, respectively. This makes it possible to test the hypothesis of the effects of factor A or of factor B by using a one-sided F test.

REFERENCES

J.S.A. Amenta, Amer. J. Clin. Pathol., 49 (1968) 842.
G. Barrie Wetherill, Elementary Statistical Methods, Chapman and Hall, London, 2nd ed., 1972.
W.J. Dixon, Biometrics, 9 (1953) 74.
K. Doerffel, Z. anal. Chem., 185 (1962) 1.
J.A.H. Gooszen, Clin. Chim. Acta, 5 (1960) 431.
G. Gottschalk, Z. anal. Chem., 276 (1975) 81.
R.F. Hirsch, Anal. Chem., 49 (1977) 691A.
P. Jonckheere, A. Vanhoutte, Beginselen van Wiskundige Statistiek (Op Middelhoog Niveau), Standaard, Antwerpen, 1966.
M.G. Kendall and A. Stuart, The advanced theory of statistics, Vol I, 3rd ed., 1969, Vol II, 2nd ed., 1967, Vol III, 2nd ed., 1968, Charles Griffin, London.
H.R. Lindman, Analysis of Variance in Complex Experimental Designs, Freeman, San Francisco, 1974.

M.J. Maurice, Z. anal. Chem., 158 (1957) 271.
J.H. Riddick, R. Flora and Q.L. Van Meter, Clin. Chem., 18 (1972) 250.
H. Scheffé, The analysis of variance, Wiley, New York, 1959.
T.I. Walker, Int. J. Environ. Anal. Chem., 5 (1977) 25.
G. Wernimont, Anal. Chem., 23 (1951) 1572.
W.J. Youden and E.H. Steiner, Statistical Manual of the Association of Official Analytical Chemists, The Association of Official Analytical Chemists, Washington, D.C., 1975.

Chapter 5

RELIABILITY AND DRIFT *

Reed and Henry (1974) defined the reliability of a test as its ability to maintain accuracy and precision into the future. If a test has maintained a steady state of these characteristics over a long period of time, then the test is said to be reliable.

The reliability of a method can be studied in several ways. One technique is to follow the method over a long period of time and to reach a conclusion about the reliability *a posteriori*, i.e., from experience. This is the technique adopted in modern routine laboratories, where it is called quality control. The other approach is to try and predict the reliability, which can be done more or less, although not completely, by the determination of what Youden and Steiner (1975) called the "ruggedness" of a test. We shall discuss both aspects in the following sections. The notion of reliability is related to the notion of drift, which can be defined as a systematic trend in the results as a function of time. Drift has been found to occur in many instances, particularly in automatic apparatus in which many determinations per hour are carried out. An example was given by Bennet et al. (1970). It should not be concluded that automatic methods are more prone to show drift than manual methods, but rather that the larger series of determinations carried out with the former methods allows easier detection of drift.

5.1. THE A *POSTERIORI* APPROACH ; QUALITY CONTROL

5.1.1. Control chart methods for detection of drift

Many methods for quality control are currently in use, among which control

* This chapter has been written with the collaboration of Y. Michotte, Pharmaceutical Institute, Vrije Universiteit Brussel, Belgium.

chart methods, in clinical chemistry also called Levey-Jennings chart methods and in industrial analytical chemistry Shewhart charts, are the most common (Levey and Jennings, 1950 ; Shewhart, 1931 ; Koehler, 1960 ; Grannis and Caragher, 1977).

Reference samples are analysed every day or with each run and their values are plotted on a chart as depicted in Fig. 5.1. A solid line depicts the mean value and the broken lines limits of $\pm$ 2s and $\pm$ 3s. These limits have to be determined before starting the quality control scheme. The 2s limit is usually called the warning limit and the 3s limit the action limit. The laboratory under control follows rules such as : "If one point falls outside the action limit or two consecutive points outside the warning limit but within the action limits, the results are accepted but the procedure is nevertheless investigated", etc.

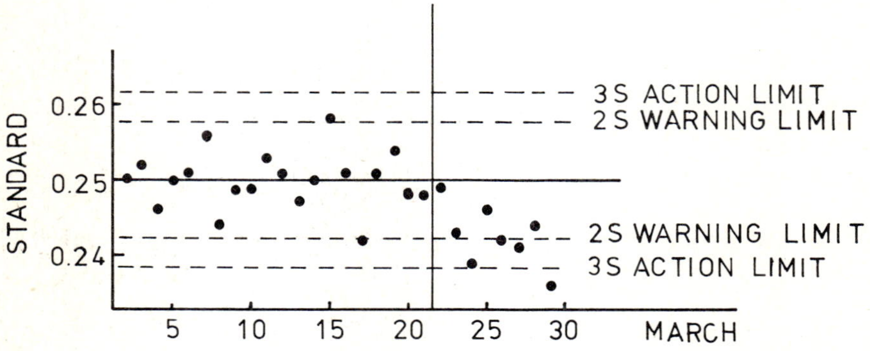

Fig. 5.1. Control chart for reference sample (from Reed and Henry, 1974).

Clearly, the emphasis is on the detection of time-dependent systematic errors, i.e., errors that influence the accuracy. This is also true for the whole of this chapter. It is also possible, however, to evaluate a trend in the precision, which necessitates the analysis of at least two controls per run.

There are several variants to these methods, and readers are referred to books on clinical chemistry such as that already cited for a full account of these procedures. Control chart methods are relevant to our purpose only in that they permit one to observe whether a method remains acceptable or not ; they do not allow a quantitative measure of the reliability.

5.1.2. Operational research methods

These methods are not as widely used as control chart methods but they are becoming increasingly popular. As one of the aims of this book is to stimulate the introduction of operational research and other modern mathematical techniques in analytical chemistry, we shall discuss these methods in some detail.

Two operational research techniques have been proposed, namely the Cusum technique and Trigg's monitoring method. The former has been used most in routine applications so far.

5.1.2.1. The Cusum technique (Lewis, 1971 ; Whitby et al., 1967)

For a series of control measurements $x_0, x_1, x_2, \ldots, x_t$, one determines the cumulative sum of differences between the observed value and the previously determined mean value, $\bar{x}$

$$C_1 = x_1 - \bar{x} \qquad (5.1)$$

$$C_2 = C_1 + (x_2 - \bar{x}) \qquad (5.2)$$

$$C_3 = C_2 + (x_3 - \bar{x}) \qquad (5.3)$$

These values are displayed on a chart such as that shown in Fig. 5.2. If the deviations from $\bar{x}$ are random, then the C values oscillate around the line at zero, at least if the mean $\bar{x}$ is an accurate estimate of the true mean value. If not, they will veer away from this line.

The interpretation of the results obtained is not as straigthforward as in the control chart methods. In particular, it is not evident from the Cusum results how one should decide when a trend is significant and when it is not. The most general means of coming to such a decision appears to be the use of a V-mask. This is illustrated in Fig. 5.2, adapted from Lewis'paper.

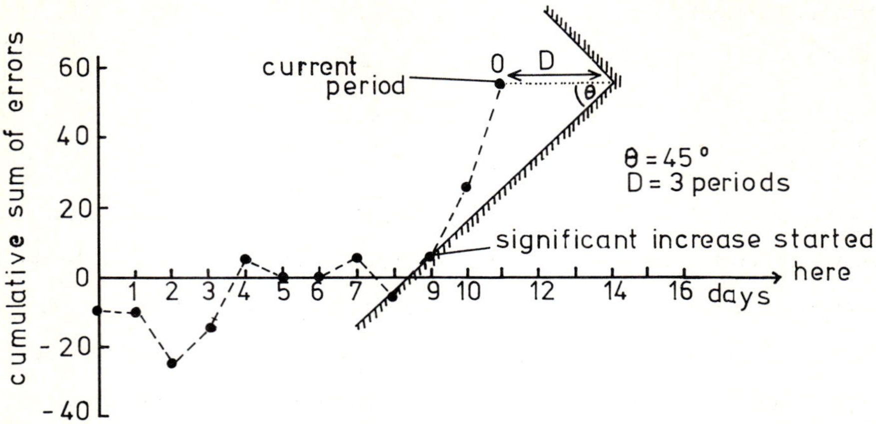

Fig. 5.2. Illustration of the Cusum method (adapted from Lewis, 1971).

When one wishes to evaluate a possible trend at time t using C_t, one places the mask so that point 0 coincides with C_t. If the Cusum line cuts one of the limits of the mask, then the trend is considered to be significant. The difficulty resides in the choice of the angle θ and the distance D, which is somewhat intuitive. Taylor (1968) gave some rules for choosing these parameters and Bissell (1969) surveyed the many related methods that have been proposed.

5.1.2.2. Trigg's monitoring technique (Trigg, 1964 ; and Batty, 1969)

To determine whether or not there is a drift, one makes observations at regular times. Such a series of observations made at specified times is called a time series, and the analysis of time series is a classical statistical problem (see Kendall, 1973) used for example in the evaluation of economic trends. In Chapters 10 and 26, the application of time series concepts in the characterization of continuous processes in analytical chemistry is discussed in more detail.

The main difficulty in the analysis of time series as applied here is to separate the long-term effects from irregular, random effects, and one of the methods used to do this is the application of moving averages. For a series of control measurements x_1, x_2, ..., x_t, one defines

$$\frac{x_1 + x_2 + \dots x_n}{n}, \quad \frac{x_2 + x_3 + \dots x_{n+1}}{n}, \quad \frac{x_3 + x_4 + \dots x_{n+2}}{n}$$

as the moving averages of order n. Moving averages have the effect of reducing the random variations, thereby smoothing the time series. To avoid too large effects from the extreme values, one often uses weighted moving averages, which are obtained by giving larger weights to the central values than to the extreme values. It appears that these simple methods have not been used (or at least have been used only unfrequently) for quality control in analytical chemistry. A more complex technique, called Trigg's monitoring technique, has been proposed by Cembrowski et al. (1975), who gave an interesting discussion and comparison of Levey-Jennings control charts, the Cusum technique and Trigg's method.

In Trigg's method, one calculates an exponentially weighted average, C_t, of the observations

$$C_t = \alpha\, x_t + (1 - \alpha)\, C_{t-1} \tag{5.4}$$

As in the same way

$$C_{t-1} = \alpha\, x_{t-1} + (1 - \alpha)\, C_{t-2} \tag{5.5}$$

C_t is given by

$$C_t = \alpha\, x_t + \alpha\,(1 - \alpha)\, x_{t-1} + (1 - \alpha)^2\, C_{t-2} \tag{5.6}$$

Eqn. 5.4 is equivalent to

$$C_t = \sum_{n=0}^{t-1} \alpha\,(1 - \alpha)^n\, x_{t-n} + (1 - \alpha)^t\, C_0 \tag{5.7}$$

In eqn. 5.7, α is a constant which is generally chosen to be 0.1 or 0.2. If $\alpha = 0.2$, the average at time t is calculated with a weight of 0.2 for the current observation, 0.2 x (1 - 0.2) = 0.16 for the last but one, and 0.128, 0.102, etc., for the preceding ones in order of decreasing time. The number of observations that should be included depends on α. For $\alpha = 0.2$, the number of observations to be taken into account may be limited to 9. In fact, C_t is a moving average, where the weights of the observations in the calculation decrease with time. The moving average C_t is considered as a predictor for the next observation, x_{t+1}, and the difference e_t between x_{t+1} and C_t is considered to be the error of the prediction

$$e_t = x_{t+1} - C_t \tag{5.8}$$

The smoothed error $\overline{e}_t$ is then calculated according to the same principle as in eqn. 5.4

$$\overline{e}_t = \alpha\, e_t + (1 - \alpha)\, \overline{e}_{t-1} \tag{5.9}$$

When $\overline{e}_t$ changes continuously in one direction as a function of time, a changing trend in the results is indicated.

Instead of observing directly the trend in $\overline{e}_t$, one compares $\overline{e}_t$ with the mean absolute deviation, MAD_t

$$MAD = \alpha.\text{latest absolute error} + (1 - \alpha)\text{ previous MAD}$$

or

$$MAD_t = \alpha\, |e_t| + (1 - \alpha)\, MAD_{t-1} \tag{5.10}$$

by means of Trigg's tracking signal, T_t

$$T_t = \frac{\overline{e}_t}{MAD_t} \tag{5.11}$$

The tracking signal oscillates between +1 and -1 and the more different it is from zero the more significant the trend is. For example, for $\alpha = 0.2$ a value of $T_t = 0.4$ indicates an 80% confidence level, i.e., that there is an 80% probability that a significant change has taken place. The tracking signal can therefore be used as a criterion to describe drift.

In Table 5.II a worked example is given. To be able to apply eqns. 5.4, 5.9 and 5.10, initial values of C_{t-1}, $\bar{e}_{t-1}$ and MAD_{t-1} must be determined. The smoothed forecast error is initially set to zero and the initial mean absolute deviation, MAD, is set to $MAD_{t-1} = (\frac{2}{\pi})^{1/2}$ s (Cembrowski et al., 1975). The standard deviation, s, was calculated in preceding experiments and was found to be 10.027. C_{t-1}, the immediate past exponentially weighted average, is initially set to the average value of previous determinations.

In the example given, the values of C_{t-1}, $\bar{e}_{t-1}$ and MAD_{t-1} at time zero are 50, 0 and 8, respectively. All calculations to be effected in order to obtain the tracking signal are described in Table 5.II. α is chosen to be 0.2.

As up to t = 8 the value of T_t does not exceed 0.50, which, according to Table 5.I, corresponds to a 90% confidence level (the lowest level one could use in practice), one could not be confident that a statistically significant change occurs up to that time. At t = 9, however, the tracking signal reaches 0.60, which corresponds to a 95% confidence level. At t = 10, $T_t = 0.68$, which means that at this time an increase has taken place with a 98% level

Table 5.I

Tracking signal values (Cembrowski et al., 1975)

Confidence level (%)	Tracking signal	
	$\alpha = 0.1$	$\alpha = 0.2$
70	0.24	0.33
80	0.29	0.40
85	0.32	0.45
90	0.35	0.50
95	0.42	0.58
96	0.43	0.60
97	0.45	0.62
98	0.48	0.66
99	0.53	0.71
100	1.00	1.00

Table 5.II

Example of calculation of Trigg's tracking signal

	$t = 0$	1	2	3	4	5	6	7	8	9	10
	$x_t = 40$	52	36	55	47	61	57	49	60	65	64
C_{t-1}	50	48	48.8	46.24	47.99	47.79	50.43	51.74	51.19	52.95	55.36
$e_t = x_t - C_{t-1}$	-10	4	-12.8	8.76	-0.99	13.21	6.57	-2.74	8.81	12.05	8.64
αx_t	8	10.4	7.2	11	9.4	12.2	11.4	9.8	12	13	12.8
$(1-\alpha)\ (C_{t-1})$	40	38.4	39.04	36.99	38.39	38.23	40.34	41.39	40.95	42.36	44.29
$C_t = \alpha x_t + (1-\alpha)\ C_{t-1}$	48	48.8	46.24	47.99	47.79	50.43	51.74	51.19	52.95	55.36	57.09
$\alpha\ e_t$	-2	0.8	-2.56	1.75	-0.2	2.64	1.31	-0.55	1.76	2.41	1.73
$(1-\alpha)\ \overline{e}_{t-1}$	0	-1.6	-0.64	-2.56	-0.65	-0.68	1.57	2.30	1.40	2.53	3.95
$\overline{e}_t = \alpha\ e_t + (1-\alpha)\ \overline{e}_{t-1}$	-2	-0.8	-3.2	-0.81	-0.85	1.96	2.88	1.75	3.16	4.94	5.68
$\alpha\ \lvert e_t \rvert$	2	0.8	2.56	1.75	0.2	2.64	1.31	0.55	1.76	2.41	1.73
$(1-\alpha)\ MAD_{t-1}$	6.4	6.72	6.02	6.86	6.89	5.67	6.65	6.37	5.54	5.84	6.60
$MAD_t = \alpha\ \lvert e_t \rvert + (1-\alpha)\ MAD_{t-1}$	8.4	7.52	8.58	8.61	7.09	8.31	7.96	6.92	7.30	8.25	8.33
$T_t = \dfrac{\overline{e}_t}{MAD_t}$	-0.24	-0.11	-0.37	-0.09	-0.12	0.24	0.36	0.25	0.43	0.60	0.68

of confidence. A level of 97% is in practice always high enough to infer that a significant change has occurred (Lewis, 1971).

5.1.3. Other statistical methods

Various other statistical techniques have been applied to test whether or not there is a drift in the results. For example, Gindler et al. (1971) used the chi-square test. This test, which is discussed in more detail in section 8.6, is used to discriminate between different distributions of data. According to Gindler et al., the chi-square test easily demonstrates changes in patient distribution data from day to day, even when the means are constant. Laboratory error caused, for example, by evaporation of samples is cited as only one possible source of such changes.

Other possible causes are not within the scope of analytical chemistry. They include changes of population, medical treatment, diet, etc. Gooszen (1960) described a so-called run test for application in the clinical chemical laboratory. One uses the results of, for example, 20 control determinations and determines the median, the observations with a lower result being denoted by a minus sign and those with a higher result by a plus sign. A sequence of similar signs is called a run. In the following example, adapted from Gooszen, the median is situated between 7 and 9 and there are 9 runs.

1	2	3	4	5	6	7	8	9	10	11	12	13	14	15	16	17	18	19	20	result-number
5	9	9	10	7	3	7	10	9	7	9	9	4	9	10	9	4	6	2	3	result
-	+	+	+	-	-	-	+	+	-	+	+	-	+	+	+	-	-	-	-	sign
—	———			———			——		—	——		—	———			————				run

A table given by Dixon and Massey (1957) indicates that in this instance the probability of finding 9 or fewer runs is 0.242 and that the critical value for rejecting the hypothesis that there is no drift at the 0.05 probability level is 6 runs. It is not our intention to give a complete review of all of the applications of classical statistics here. Articles by Glick (1972) and

Thiers et al. (1976) and the ANOVA method of Riddick et al. (1972), which was discussed in section 4.1.5 and also allows drift detection, can be cited. It should also be observed that these methods usually do not give a rapid warning of the existence of a systematic trend, as is the case with the operational research methods.

5.2. THE A *PRIORI* APPROACH ; RUGGEDNESS OF A METHOD

The reason for incomplete reliability of a method with time is that it is sensitive to minor changes in procedure, such as variations in concentrations of reagents or heating rates. One can, of course, try new methods and see how they behave over a long period of time in order to test their reliability (see section 5.1). However, it is preferable to have an idea of the reliability to be expected, which can be obtained by measuring the sensitivity of the method to small variations.

It is clear that insensitivity to changes in procedure is an important asset for an analytical method. Therefore, this property, which has been called "ruggedness" (Youden and Steiner, 1975), can be considered as an evaluation criterion. An insufficiently "rugged" method is also subject to large laboratory biases. As we have already stated, it is unfortunate but well known that methods proposed in the literature do not always yield the expected good results. Laboratory bias is estimated (see Chapters 2, 3 and 4) by collaborative research programmes using the two-sample method or analysis of variance. These collaborative programmes require important organizational efforts, so that it is out of the question to subject all promising methods in an early stage of development to such programmes. Here again, an *a priori* approach, permitting a prediction of the laboratory bias to be expected, would be useful. It can therefore be concluded that a measure of the ruggedness gives an idea of the day-to-day or between-laboratory variations to be expected. To study the effect of minor and inevitable variations, one could carry out a factorial experiment (see Chapter 12). In this instance, one would use a two-level experiment, one of these levels being that given in the proposed procedure and the other a level which deviates from the former level to an extent that can be reasonably conceived to occur in practice. Denoted by plus and minus signs, and following the practice introduced by Plackett and Burman (1940), these levels are called the nominal and extreme values, respectively. In this type of investigation, one must consider as many parameters as possible, and therefore introduce a large number of factors in the factorial experiment. If this number is n, then the number of experiments to be carried out in a complete design is 2^n. As the typical number of factors is between 5 and 10, it is clear that complete factorial experiments are often impractical for this purpose. Several designs have been proposed to obtain an estimate in a much smaller number of experiments. To understand this, let us first consider the design of Table 5.III proposed by Youden and Steiner for seven factors using eight experiments.

Table 5.III

Partial factorial experiment for seven factors

Experiment	Factors							Measurement obtained
	A	B	C	D	E	F	G	
1	+	+	+	+	+	+	+	y_1
2	+	+	-	+	-	-	-	y_2
3	+	-	+	-	+	-	-	y_3
4	+	-	-	-	-	+	+	y_4
5	-	+	+	-	-	+	-	y_5
6	-	+	-	-	+	-	+	y_6
7	-	-	+	+	-	-	+	y_7
8	-	-	-	+	+	+	-	y_8

This means that eight experiments are carried out, each yielding a result, y_1, ..., y_8. The third experiment, for example, is carried out in such a way that factors A, C and E take their nominal values while the others are at the extreme level. Note that Table 5.III is constructed in such a way that each factor occurs four times at the nominal and four times at the extreme level. To determine the effect of changing factor A from the nominal + level to the extreme - level, one compares the mean value of the results obtained at both levels. In this instance, this means carrying out the operation

$$D_A = (y_1 + y_2 + y_3 + y_4)/4 - (y_5 + y_6 + y_7 + y_8)/4$$

For factor C, the following calculation should be carried out

$$D_C = (y_1 + y_3 + y_5 + y_7)/4 - (y_2 + y_4 + y_6 + y_8)/4$$

One observes that in doing this, one divides the experiments into two groups for each factor. In one of these groups, one factor (that being investigated) is at the + level. All the other factors, however, are twice at the - level and twice at the + level, in each group. When carrying out the comparison of averages, the effects of all of the other factors cancel out. In fact, this is completely true only when there is no interaction (see Chapter 4), but as the variations introduced in the factor levels are small this will be of relatively little importance.

The obtaining of the differences D_A, ..., D_G is not sufficient in itself, and one must determine whether these differences are significantly greater than the experimental error determined by carrying out replicate measurements at the nominal level. These replicate measurements do not involve extra work as the author of the method probably carried them out already in order to measure the repeatability of the proposed procedure. If this is characterized by a standard deviation, s, then when there is no significant factor, the standard deviation on the mean of four measurements is $s/\sqrt{4}$ and the standard deviation, s_D, on the difference between two averages, $\sqrt{2s^2}/\sqrt{4} = s/\sqrt{2}$. The expected mean of the D-distribution being zero (again, when there is no significant factor), one can consider that a factor is significant when D is larger than $2s_D = \sqrt{2}.s$. When, in actual experimentation, a significant factor is noted, steps should be taken to eliminate it or, as this is often impossible, the procedure should state clearly the limits between which the parameter may be allowed to vary.

This design is very elegant but unfortunately it is impossible to propose an analogous device for, for example, six factors with seven experiments. The solution to this difficulty is to continue carrying out the above design where one of the variables is now a dummy one. As Youden and Steiner (1975) stated, one should "associate with factor G some meaningless operation such as solemmly picking up the beaker, looking at it intently and setting it down again". This means, in fact, that one uses the design of Table 5.IV.

Table 5.IV

Partial factorial experiment for six factors

Experiment	Factors					
	A	B	C	D	E	F
1	+	+	+	+	+	+
2	+	+	-	+	-	-
3	+	-	+	-	+	-
4	+	-	-	-	-	+
5	-	+	+	-	-	+
6	-	+	-	-	+	-
7	-	-	+	+	-	-
8	-	-	-	+	+	+

A design is also available for the three-factor situation, and is given by Table 5.V.

Table 5.V

Partial factorial design for three factors

Experiment	Factors			Measurement obtained
	A	B	C	
1	+	+	+	y_1
2	-	+	-	y_2
3	+	-	-	y_3
4	-	-	+	y_4

So-called cyclical designs with the same properties have been proposed for higher numbers of factors by Plackett and Burman (1946). Consider, for example, the 11-factor, 12-experiment design. The design is obtained from a first line given in their paper and which in this instance is

+ + - + + + - - - + -

and describes experiment 1. Experiments 2 - 11 are obtained by writing down all cyclical permutations of this line and the last, experiment 12, always contains only minus signs. The complete design is therefore given by Table 5.VI.

Table 5.VI

Partial factorial design for eleven factors

Experiment	Factors											Measurement obtained
	A	B	C	D	E	F	G	H	I	J	K	
1	+	+	-	+	+	+	-	-	-	+	-	y_1
2	-	+	+	-	+	+	+	-	-	-	+	y_2
3	+	-	+	+	-	+	+	+	-	-	-	y_3
4	-	+	-	+	+	-	+	+	+	-	-	y_4
5	-	-	+	-	+	+	-	+	+	+	-	y_5
6	-	-	-	+	-	+	+	-	+	+	+	y_6
7	+	-	-	-	+	-	+	+	-	+	+	y_7
8	+	+	-	-	-	+	-	+	+	-	+	y_8
9	+	+	+	-	-	-	+	-	+	+	-	y_9
10	-	+	+	+	-	-	-	+	-	+	+	y_{10}
11	+	-	+	+	+	-	-	-	+	-	+	y_{11}
12	-	-	-	-	-	-	-	-	-	-	-	y_{12}

Plackett and Burman gave designs for 8, 12, 16, 20, ..., 100 experiments. As a number of factors exceeding 15 seems improbable for analytical chemistry purposes, only the first line for the 15 factor experiment is given below

\+ + + + - + - + + - - + - - -

In the same way as for the 6-factor example using Youden and Steiner's design, arrangements for 8 - 10 factors can be derived from the 11-factor design and for 12 - 14 factors from the 15-factor design, using dummy factors. These methods have been called partial factorial experiments and can therefore be considered to be simple factorial experiments. As discussed later, they make it possible to investigate only main factors and are less useful for optimization purposes when interaction occurs.

REFERENCES

M. Batty, Oper. Res. Q., 20 (1969) 319.
A. Bennet, D. Gartelmann, J.I. Mason and J.A. Owen, Clin. Chim. Acta, 29 (1970) 161.
A.F. Bissell, Appl. Statistics, 18 (1969) 1.
G.S. Cembrowski, J.O. Westgard, A.A. Eggert and E. C. Toren, Jr., Clin. Chem., 21 (1975) 1396.
W.J. Dixon and F.J. Massey, Introduction to Statistical Analysis, McGraw-Hill, New York, 1957.
E.M. Gindler, D.T. Forman and K. Derstine, Clin. Chem., 17 (1971) 646.
J.H. Glick, Jr., Clin. Chem., 18 (1972) 1504.
J.A.H. Gooszen, Clin. Chim. Acta, 5 (1960) 341.
G.F. Grannis and T.E. Caragher, CRC Critical Reviews in Clinical Laboratory Science, 7 (1977) 327.
M.G. Kendall, Time Series, Charles Griffin, London, 1973.
T.L. Koehler, Chem. Eng., 25 (1960) 142.
S. Levey and E.R. Jennings, Am. J. Clin. Pathol., 20 (1950) 1059.
C.O. Lewis, Med. Biol. Eng., 9 (1971) 315.
R.L. Plackett and J.P. Burman, Biometrika, 23 (1946) 305.
A.H. Reed and R.J. Henry, in R.J. Henry, D.C. Cannon and J.W. Winkelman (Editors), Clinical Chemistry : Principles and Techniques, 2nd ed., Harper and Row, 1974.
J.H. Riddick, R. Flora and Q.L. Van Meter, Clin. Chem., 18 (1972) 250.
W.A. Shewhart, Economic Control of the Quality of Manufactured Product, Macmillan, New York, 1931.
H.M. Taylor, Technometrics, 10 (1968) 479.
R.E. Thiers, G.T. Wu, A.H. Reed and L.K. Oliver, Clin. Chem., 22 (1976) 176.
D.W. Trigg, Oper. Res. Q., 15 (1964) 271.
L.G. Whitby, F.L. Mitchell and D.W. Moss, Advan. Clin. Chem., 10 (1967) 102.
W.J. Youden and E.H. Steiner, Statistical Manual, Association of Official Analytical Chemists, Benjamin Franklin Station, Washington, D.C., 1975.

Chapter 6

SENSITIVITY AND LIMIT OF DETECTION

6.1. INTRODUCTION

There is no doubt that the detection limit is one of the most important performance characteristics of an analytical procedure. Progress in analytical chemistry might well be measured by the shift of the detection limit towards lower values. Of course, the picture emerging would reflect only part of the progress. However, it cannot be denied that many problems in analytical chemistry are problems of detecting and determining elements or compounds in small amounts of sample (micro-analysis), of determining very low concentrations or small amounts in larger samples (trace analysis) or even of determining low concentrations in small samples.

Comparing analytical procedures by their limits of detection is not easy. In many papers describing analytical procedures no detection limits are given, and to the analyst facing the problem of choosing a procedure from several alternatives, this omission is very disappointing. Even more disappointing is the lack of uniformity in describing performances with respect to the smallest amounts or concentrations that can be detected or determined.

Often a procedure is said to be very sensitive when the limit of detection is low, and the limit of detection and sensitivity in many instances are regarded as synonymous. However, in other branches of science sensitivity is defined as the slope of the curve that is obtained when the result of the measurement is plotted against the amount that is to be determined. In analytical chemistry, sensitivity defined in this way is equal to the slope of the analytical calibration curve (Kaiser, 1965) and throughout this book this definition of sensitivity will be used. The lower limit of detection literally is to be understood as the limit below which detection is impossible. Although this clarifies the meaning of the term, it certainly is not sufficient when the detection limit is to be used as a performance characteristic of an analytical procedure. It appears that quantification of this characteristic gives rise to considerable confusion, as has been clearly demonstrated by Currie (1968). Fig. 6.1, from the paper of Currie, gives several values of limits of detection for a specific radioactivity measurement process. These values were calculated by using different definitions. These differences can be partly ascribed to differences in formulating the problem (when is a component detected and with what certainty ?).

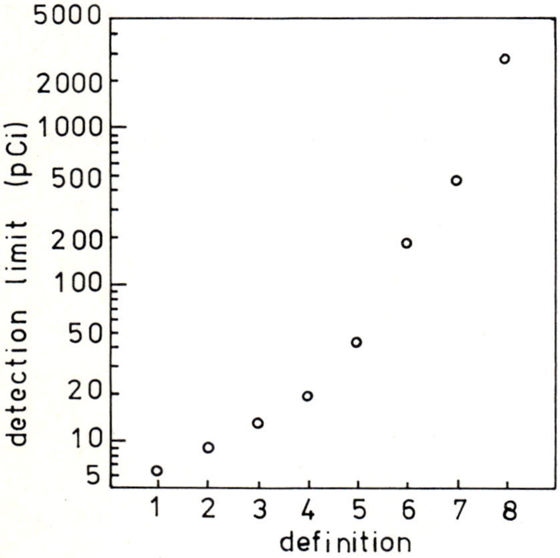

Definitions :

1-background standard deviation (σ_B)
2-10% of the background
3-$2\sigma_B$
4-$3\sigma_B$
5-$3\sigma_B+3\sigma_D$ (σ_D=sample standard deviation)
6-twice the background
7-1000 dpm
8- 100 dps

Fig. 6.1. "Ordered" detection limits ; literature definitions. The detection limit for a specific radioactivity measurement process is plotted in increasing order, according to commonly-used alternative definitions (from Currie, 1968). Reprinted with permission ; copyright American Chemical Society

Lower limits to the detection of elements and compounds are set because of the presence of errors (noise). Therefore, a definition and quantification of detection limits must be based upon statistics (Kaiser, 1947). In this chapter on the discussion of the detection limit and related quantities, we shall partly follow Currie (1968), who introduced a decision limit, a detection limit and a determination limit.

6.2. SENSITIVITY AND THE ANALYTICAL CALIBRATION FUNCTION

The sensitivity of a procedure designed for a quantitative analysis can be defined as the slope of the analytical calibration function $y = f(x)$. This calibration function relates the result (y) of the measuring process (output, analytical signal) to the concentration or amount (x) of the component to be determined. The output can be a meter reading, an electric current or voltage, a weight, etc. The sensitivity (S) can be written as the differential quotient

$$S = dy/dx \tag{6.1}$$

For linear relationships between x and y and in the absence of a blank, the sensitivity is simply the ratio between y and x. Fig. 6.2, from a paper of Specker (1968) clearly illustrates the concept of sensitivity.

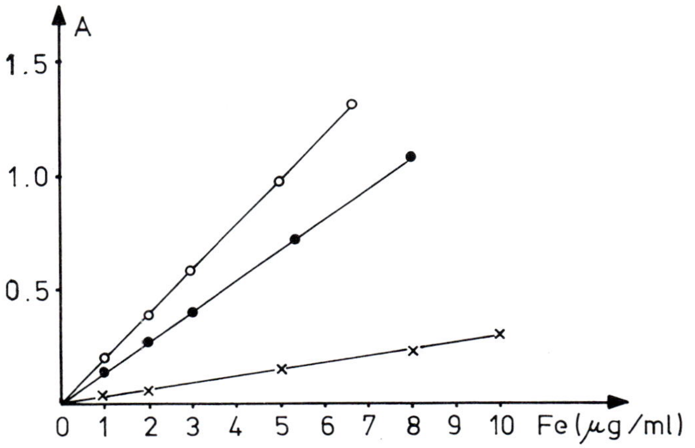

Fig. 6.2. Calibration lines for the photometric determination of iron (definition of sensitivity). Line 1, with 2-pyridinealdoxime, $\Delta A/\Delta c = 0.18$; line 2, with *o*-phenanthroline, $\Delta E/\Delta c = 0.14$; line 3, with 2,6-pyridinedicarboxylic acid, $\Delta A/\Delta c = 0.028$. Thickness of cell, 1 cm. Ordinate, absorbance ; abscissa, iron concentration. (From Specker, 1968).

These calibration graphs were obtained by plotting the absorbance against the iron concentration. It appears that the determination of iron with 2-pyridine aldoxime has a greater sensitivity than the procedures with *o*-phenanthroline and with 2,6-pyridinedicarboxylic acid.

For purposes of characterizing analytical procedures, the sensitivity is of limited importance. For instance, the sensitivity can easily be influenced without altering the procedure significantly. Connecting an amplifier to the output of an instrument can easily bring the output from the millivolts to the volts range and, according to the definition, the sensitivity is then increased by a factor of 1000. Similarly, the sensitivity of a photometric determination can be increased by increasing the optical path length.
However, errors (noise) are usually magnified to the same extent.

Sensitivities are seldom constant over large concentration ranges and sensitivities are therefore meaningful only when concentrations or concentration ranges are specified. Here again, sensitivities can be easily manipulated ; a wide variety of linearizing devices are available. This does not mean, of course, that any calibration graph is acceptable to the analytical chemist, and one must at least be cautious about non-linear calibration graphs. In some instances, theoretical considerations can lead to linearization of calibration graphs that are fully justified, and the use of logarithms in spectrophotometry (Beer-Lambert laws) and potentiometry (Nernst) is well known in this respect. In other instances, non-linear graphs are due to saturation effects, sometimes even resulting in a change of slope from positive to negative.

The range over which the sensitivity can be considered to be constant has, of course, a lower and an upper limit. By definition, the lower limit will be the detection limit (as defined in section 6.3) and the concentration where the sensitivity begins to change (going from lower to higher concentrations) can be regarded as the upper limit. In general, such a change will be gradual and the upper limit cannot be specified unless a specification is given of what is to be considered to be a straight calibration graph. A definition of the upper limit might be the concentration where the response differs by a certain percentage (for instance, 3%) from the response that might be expected from the sensitivity near the detection limit. As far as we know, no generally accepted definition of the upper limit and thus of the linear (dynamic) range for characterizing a procedure has been proposed. The concept of the linear range is illustrated in Fig. 6.3. The linear range is usually expressed as the number of decades between the lower and upper limit.

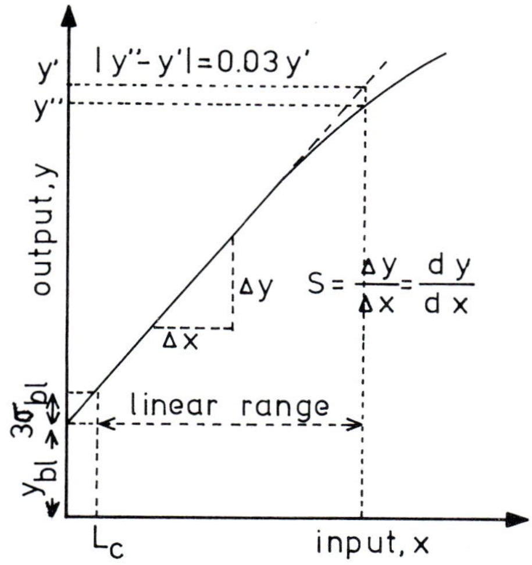

Fig. 6.3.

As has been stated, the sensitivity as a means of characterizing procedures is of limited value, and this is particularly true for calibration graphs near the (lower) limit of detection. However, for a good understanding of the general nature of analytical procedures, the sensitivity is a useful parameter (see Part V), and it is also a useful parameter when discussing the selectivity and specificity of analytical procedures (Chapter 7).

6.3. DECISION LIMIT

When the analytical chemist accepts that random errors are unavoidable, he also has to accept that there are limits to the detection (and thus to the determination) of elements and compounds. He intuitively may feel that it makes no sense to detect or determine amounts that are smaller than the random errors inherent to the procedure used. In fact, a rough estimate of the detection limit could be made by taking the value of the standard deviation (in units of concentration or amount). However, this rough picture needs some refinement.

The concentration or amount of the component to be determined (x) can be calculated from the measurement (y) by making use of the calibration function ($y = f(x)$). The discussion will be given in terms of signals. Usually y is regarded as the difference between two measurements, i.e., a measurement of the unknown sample (y_u) and a measurement of the blank (y_{bl}). Then the problem can be formulated in two ways which are essentially the same : it can be questioned whether $y_u - y_{bl}$ differs significantly from zero or whether y_u differs significantly from y_{bl}. Of course, this problem can be attacked only by means of statistics. However, to make any statements at all some assumptions have to be made about the distribution of errors. The case of a normal distribution of the reading of the blank is represented in Fig. 6.4. The standard deviation is denoted by σ_{bl} and the true value or limiting mean of the blank by $\bar{y}_{bl}$.

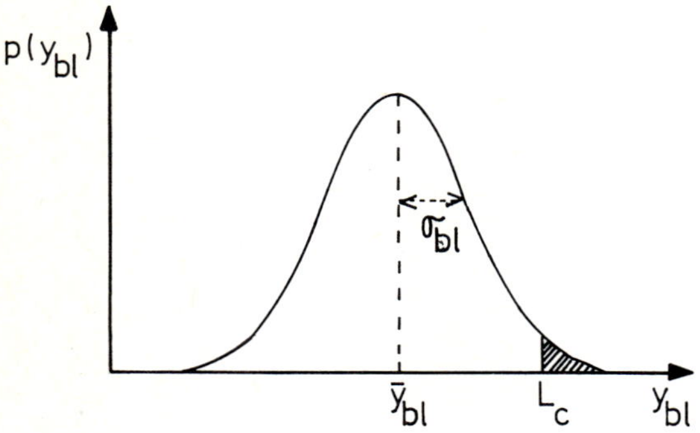

Fig. 6.4. Normal distribution of y_{bl}.

It is clear that the probability of measuring signals $y_{bl} > L_c$ will be

$$\alpha = \int_{L_c}^{\infty} p(y_{bl})\, d_{y_{bl}} \tag{6.2}$$

where $p(y_{bl})$ represents the distribution function of y_{bl}. If signals larger than the decision limit, L_c, are interpreted as "component present", a fraction α of the measurements of the blank is misinterpreted.

The decision limit can be expressed in terms of signals by

$$L_c = \bar{y}_{bl} + k_c\sigma_{bl} \tag{6.3}$$

Conversion into concentrations is easily possible by multiplication with the calibration constant.

Introducing a value of $k_c = 3$ leads to $1-\alpha = 99.86\%$. Then L_c is equal to Kaiser's detection limit (Kaiser, 1947). The choice of k_c, of course, is arbitrary and depends on the confidence that is required for the answer to the question of whether the component is detected or not. The decision limit, L_c, cannot, in principle, be used as a quality criterion for the analytical procedure (Currie, 1968 ; Svoboda and Gerbatsch, 1968 ; Wilson, 1970). This is illustrated by Fig. 6.5, where the two probability distribution functions of y_{bl} and y_u overlap. The distribution of y_u is chosen to have a maximum at $\bar{y}_u = L_c$. Thus, Fig. 6.5 represents a situation of a large number of repeated measurements on a sample with a concentration corresponding (via the calibration constant) to the decision limit, L_c.

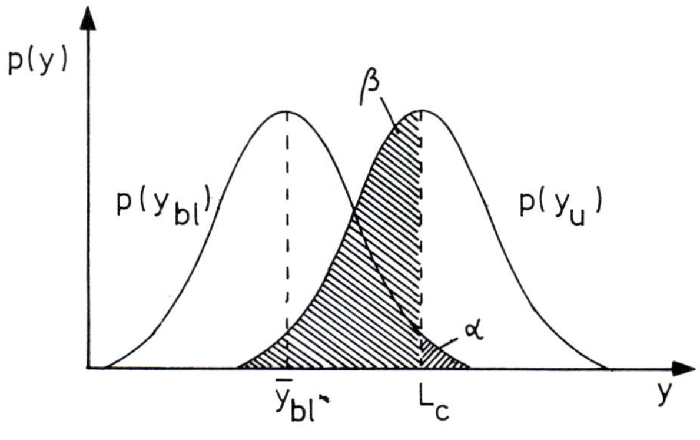

Fig. 6.5. Illustration of decision limit.

The standard deviations σ_{bl} and σ_u are considered to be equal (which is usually the case for small concentrations). Signals larger than L_c can be interpreted by "component present". However, a fraction β of the measurements on a sample with a content L_c of the component to be detected will yield signals smaller than L_c. β is given by

$$\beta = \int_{-\infty}^{L_c} p(y_u)\ dy_u \qquad (6.4)$$

From Fig. 6.5, it appears that $\beta = 0.5$ and the statement about the absence of the component is very unreliable. To express this differently : the error of the first type (deciding that the component is present when it is not) is small (α), whereas the error of the second type (deciding that the component is absent when it is present) is large (β) (see also 3.2.1 and Chapter 2). Signals larger than L_c can be interpreted as the detection of the component with quasi certainty, whereas signals smaller than L_c allow no decision to be made about the absence of the component.

6.4. DETECTION LIMIT

The *a posteriori* decision about the presence of a component from a measured signal has resulted in a definition of the decision limit as given above. In order to characterize an analytical procedure, it is necessary to define a level L_D specifying the detection capabilities of the analytical procedure. This level, the detection limit, should correspond to a concentration that, with great probability, will yield signals that can be distinguished from the signals obtained from the blank. This, of course, corresponds to reducing the error of the second type, and thus of reducing β. In Fig. 6.6 a situation is represented where $\alpha = \beta$. Here the limiting mean of y_u can be used for defining a detection limit, L_D

$$L_D = \overline{y}_{bl} + k_d\ \sigma_{bl} = L_c + k'_d\ \sigma_{bl} \qquad (6.5)$$

Here again the standard deviations of the distributions $p(y_{bl})$ and $p(y_u)$ have been assumed to be identical. The detection limit as defined by eqn. 6.5 is equal to the limit of guarantee of purity as defined by Kaiser (1965) when

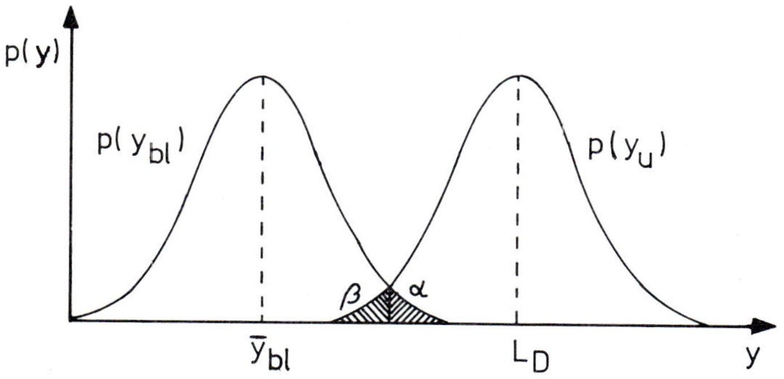

Fig. 6.6. Illustration of detection limit.

$k_d = 6$ and $k'_d = 3$. If a concentration is equal to the detection limit, it can be detected with 99.86% certainty. Smaller concentrations cannot be detected unless a smaller confidence is accepted.

6.5. DETERMINATION LIMIT

A determination limit can be defined as the limit at which a given procedure will be sufficiently precise to yield a satisfactory quantitative estimate of the unknown concentration. Such a limit, L_q, can be defined in terms of $\bar{y}_{bl}$ and σ_{bl}, again assuming that the standard deviations for blank and unknown are identical. One can write that the corresponding signal is

$$L_q = \bar{y}_{bl} + k_q \sigma_{bl}$$

and it can easily be shown that the relative standard deviation obtained from measurements at this level is $1/k_q$.
The relative standard deviation of the "quantitative" measurement at the decision level L_c will be $33\frac{1}{3}$%, and at L_D $16\frac{2}{3}$%.

6.6. DISCUSSION

No attempt will be made to discuss in detail the several aspects of the decision limit, the detection limit and the determination limit, and the reader is referred to the literature already cited and to the contributions of Kaiser (1966), Ehrlich (1969), Liteanu and Rica (1973, 1975) and Ingle (1974).

However, it is necessary to make some remarks about the use of the concepts introduced in this chapter. Definitions have been formulated in terms of the limiting mean of the blank ($\bar{y}_{bl}$) and the standard deviation. In practice, only a limited number of experiments will be available for the estimation of these quantities. If the estimates are used for calculation of the limits of decision and detection, an uncertainty is introduced, and deciding whether a measurement of the unknown differs significantly from the blank should therefore be carried out with the t-test (Currie, 1968 ; Gabriels, 1970 ; Plesch, 1975). This means that the constants k_c and k_d in eqns. 6.3 and 6.5 should be replaced by t-factors derived from Student's t-distribution. Surely the detection limit does not change by orders of magnitude if a reasonable number of measurements of the blank have been made. As long as there is no consensus about definitions, there should never be doubt about the way in which limits have been calculated. In a way we agree with Wilson (1973), who doubts the usefulness of the detection limit as a performance characteristic. Wilson proposes to supply information on the standard deviation of the blank. The detection limit is easily calculated from the standard deviation and its reliability can be taken into account when the number of degrees of freedom is known. It should be noted that in general it is not permissible to calculate detection limits from standard deviations obtained from measurements at concentration levels much higher than the detection limit, or at least the analyst has to be aware of the pitfalls in doing so. Standard deviations are usually a function of concentration. Complications can also arise from non-linearity of the calibration function. For details, the reader is referred to the papers by Ingle and Wilson (1976) and Liteanu et al. (1976).

Detection limits should be regarded as characteristics of well described analytical procedures. It makes no sense, for instance, to specify the detection limit for titrations in general. A change in conditions will lead to a change in the procedure and possibly to a change in the limits.

The nature of the procedure (usually the measurement) will either lead to a formulation of the detection limit in terms of amounts of the component to be detected or in terms of concentrations. With specified amounts of sample, concentrations can easily be converted into amounts, and *vice versa*. Of course, similar reasoning applies to the sensitivity and the linear range. However, it is essential to specify the units when quoting values for the performance characteristics without converting, for instance, concentrations into amounts. Such a conversion can easily obscure the merits of the procedure.

6.7. GAS CHROMATOGRAPHIC DETECTORS

The description of a number of gas chromatographic detectors by Hartmann (1971) will serve as an illustration of the (use of the) concepts introduced in this chapter. The set of characteristics as given in Table 6.I (taken from the paper by Hartmann) consists of the sensitivity, the noise and the linear range. In addition, some of the operating conditions have been specified. The reader should be aware of the fact that for purposes of selection of the "best" detector, the information gathered in this table is incomplete. The figures apply to a set of specific operating conditions, although it can be assumed that these operating conditions have been optimized. It should also be noted that the figures will be different for different compounds.

The sensitivity used by Hartmann is essentially the same characteristic as defined in section 6.2. However, a direct measurement of the sensitivity of a gas chromatographic detector would involve a feed of known concentration and measurement of the output. The values in the table apparently are derived from peak areas, the flow velocity of the carrier gas and the weight of the sample injected. The figures derived in this way are average sensitivities, which, of course, are identical with the sensitivities provided that they can be considered to be constant in the concentration range covered by the

Table 6.I.

Characteristics of gaschromatographic detectors

	Thermal conductivity detector 20 ml/min He	Flame ionization detector 20 ml/min N_2	Helium ionization detector 60 ml/min He	Alkali flame ionization detector 20 ml/min N_2	Electron capture detector 35 ml/min N_2 ^{3}H	Electron capture detector 35 ml/min N_2 ^{63}Ni	Flame photometric detector 80 ml/min N_2
Sensitivity = S	10.000mV.ml/mg	0.01 C/g	100 C/g	20 C/g	800A ml/g	40A ml/g	$4x10^{-10}$A
Noise = N	0.01 mV	10^{-14} A	$2x10^{-12}$A	$3x10^{-14}$A	$8x10^{-12}$A	$2x10^{-12}$A	$2x10^{-12}$g/sec
Detectability=2N/S	$2x10^{-9}$ g/ml	$2x10^{-12}$g/sec	$4x10^{-14}$g/sec	$3x10^{-15}$g/sec	$2x10^{-14}$g/ml	10^{-13}g/ml	$1.7x10^{-12}$g/sec
Linear range	10^{5}	10^{7}	$5x10^{3}$	10^{3}	500	50	None
Limit of detection	$2x10^{-5}$mg	$2x10^{-8}$mg	$4x10^{-10}$mg	$3x10^{-11}$mg	$2x10^{-10}$mg	10^{-9}mg	$1.7x1P^{-8}$mg

chromatographic peak. It is important to note that the units used in the expression for the sensitivities are dependent on the nature of the detector. For instance, the thermal conductivity detector responds to changes in concentration whereas the flame-ionization detector responds to the mass of the compound entering the detector per unit time. For this reason, amongst others (see section 6.2), it is difficult to compare detectors by their sensitivities.

The noise quoted by Hartmann (1971) is defined as the average peak-to-peak amplitude measured at the output. Depending on the nature of the noise, the average amplitude, N, can be put equal to the decision limit, L_C, as defined in this chapter (N roughly equals 4σ) and expressed in units of the output (y). The detectability, 2N/S, then roughly approximates the detection limit, L_D, expressed in units of the input (x) of the detector. Again, these units are different for the several detectors.

In order to compare these detectors in combination with a chromatographic column, the detection limit of the detector has to be converted into the detection limit of a gas chromatographic procedure. This can easily be done when the carrier gas velocity, the peak width and the amount of sample are known. Assuming that the characteristics would apply when the carrier gas velocity is 50 ml/min. in all instances and the peak width at half height is 12 sec. (or 10 ml), one can arrive at the detection limits of the procedure quoted in Table 6.I by simply multiplying the detection limit of the detector by the peak width. These detection limits necessarily are not exact and can serve only as rough figures for comparing detectors. A better comparison would be possible only when more details of the behaviour of the whole system, i.e., detector + column + injector + compound to be determined, are known.

REFERENCES

L.A. Currie, Anal. Chem., 40 (1968) 586.
G. Ehrlich, Wissenschaftl. Zeitschr., 11 (1969) 22.
R. Gabriels, Anal. Chem., 42 (1970) 1439.
C.H. Hartmann, Anal. Chem., 43(2) (1971) 113A.
J.D. Ingle, J. Chem. Educ., 51 (1974) 100.
J.D. Ingle and R.L. Wilson, Anal. Chem., 48 (1976) 1641.
H. Kaiser, Spectrochim. Acta, 3 (1947) 40.
H. Kaiser, Z. Anal. Chem., 209 (1965) 1.
H. Kaiser, Z. Anal. Chem., 216 (1966) 80.
C. Liteanu and I. Rica, Mikrochim. Acta, (1973) 745.
C. Liteanu and I. Rica, Mikrochim. Acta, (1975) 311.
C. Liteanu, E. Hopirtean and C. Popescu, Anal. Chem., 48 (1976) 2013.
R. Plesch, GIT (Glas-Instrum.-Tech.) Fachz. Lab., 19 (1975) 676.
H. Specker, Angew. Chem., Int. Ed. Engl., 7 (1968) 252.
V. Svoboda and R. Gerbatsch, Z. Anal. Chem., 242 (1968) 1.
A.L. Wilson, Talanta, 20 (1973) 725.

Chapter 7

SELECTIVITY AND SPECIFICITY

7.1. INTRODUCTION

A quantitative analysis of an element or a compound can be devised when a measurable property (y) that is dependent on the concentration or amount to be determined (x) can be found. Usually the quantity y also depends on several other parameters, such as temperature and amount of sample and reagents. In a well formulated procedure, these parameters are specified and have to be kept constant, although it must be accepted that they are subject to fluctuations that cannot be controlled. Apart from inherent (random) fluctuations, the relationship between x and y is deterministic. Thus, the analytical calibration function $y = f(x)$, which is preferably but not necessarily linear, should be regarded as characteristic of the analytical procedure.

However, analytical calibration functions are usually influenced by the presence of other components than that which is to be determined, and then the relationship $y = f(x)$ applies to only one kind of matrix. For instance, the relationship found for the determination of calcium in a "synthetic" solution will not necessarily hold for the determination in a real sample such as sea water. Parameters that describe the sample matrix must be specified when describing a procedure. This, of course, is a severe complication and much effort has to be devoted to circumventing these difficulties. This can be done either by divising suitable calibration methods or by developing selective and specific analytical procedures.

In both selective and specific analytical procedures, the measurement used for the determination is not influenced by the presence of other components. The difference between the two terms is to some extent artificial. A simple example concerning qualitative analysis can clarify the meaning of and the difference between the terms.

If a reagent gives a colour with only one ion, the reagent is said to be specific for that particular ion. If the reagent yields colours with many ions, but with a distinct colour for each ion, the procedure of the colour reaction might be called selective. In both instances the outcome of the detection of the ions would not be influenced by the presence of other ions. In other words, there are no interferences (matrix effects).

In the same sense, multi-component analysis by means of gas chromatography yielding well resolved peaks for all of the components of the sample can be regarded as a selective procedure. In contrast, X-ray fluorescence analysis might yield well resolved peaks for a set of elements, but the size of each peak usually depends on the content of the corresponding elements and on the entire matrix. This procedure clearly is not selective.

When considering the problem of selectivity (and of specificity) in more detail, the analyst will discover that a distinction between non-selective and selective is artificial. X-ray fluorescence can be made more selective when the sample is diluted with borax, for instance. Selectivity can thus be varied and hence there must be a basis for expressing the degree of selectivity (and/or of specificity) if selectivity is to be used as a characteristic of an analytical procedure. Moreover, it has to be a uniform basis if different procedures are to be compared.

At present there seems to be no uniformity in the analytical literature when describing selectivity, specificity, interferences and matrix effects. Well described analytical procedures usually apply to well defined samples (blood, steel, sea water, etc.). Papers that describe procedures in a more general way usually give some indication of the interferences that can be expected. In some instances maximum allowable concentrations of potentially interfering components are given (spectrometric techniques), while in other instances selectivity coefficients have been introduced (ion-selective or specific electrodes). In a way, the resolution as used in chromatographic procedures, falls in this category.

7.2. QUANTIFICATION OF SELECTIVITY AND SPECIFICITY

As has been stressed by Belcher (1965, 1966, 1976) and by Betteridge (1965), it is necessary to clarify the term selectivity and to avoid the use of terms such as highly selective and non-selective and a selectivity index was proposed for this purpose. This index gives some information about (the number of) possible interfering substances but permits no real quantification of the interferences. It is questionable whether the index proposed is more than a shorthand notation of information that usually is (or rather should be made) available when proposing a procedure. Wilson (1965) considered that the compression of the necessary information into one index would be confusing and might lead to ambiguity.

Another, more quantitative, approach was followed by Kaiser (1972). The concepts of selectivity and specificity as proposed by Kaiser are closely related to a more general form of the calibration function $y = f(x)$. When matrix effects or interferences are present, the analytical calibration function $y = f(x)$, all other parameters being kept constant, has to be extended to

$$y_i = f_i\ (x_1, \ldots\ldots\ldots, x_{i-1}, x_i, x_{i+1}, \ldots\ldots\ldots, x_n) \tag{7.1}$$

The concentration (or amount) of component i (x_i) can be derived from the measurement y_i, provided that concentrations of all of the other components present ($x_1, \ldots, x_{i-1}, x_{i+1}, \ldots, x_n$) are known. If these concentrations are known, they have to be determined even if one is not interested in the entire composition of the sample. (It is, of course, possible to choose a suitable calibration procedure in order to reduce eqn. 7.1 to the simpler calibration function $y = f(x)$. This is usually possible by using standards of almost the same composition as the unknown sample.)

If the entire composition is to be determined, whether one is interested in it or not, a set of measurements $y_1, y_2, \ldots, y_m$ is necessary. It is clear that m has to be equal to or greater than n for the problem to be solved. Thus the following set of equations is necessary

$$
\begin{aligned}
y_1 &= f_1(x_1, x_2, \ldots\ldots\ldots, x_n) \\
y_2 &= f_2(x_1, x_2, \ldots\ldots\ldots, x_n) \\
&\vdots \\
y_m &= f_m(x_1, x_2, \ldots\ldots\ldots, x_n)
\end{aligned}
\qquad (7.2)
$$

In practice, the set of equations for a number of components exceeding two or three can be handled only when the functions are linear. Either the functions can be linearized or a limited range of compositions with a linear dependence can be considered. Then eqn. 7.2 reduces to

$$
\begin{aligned}
y_1 &= S_{11}x_1 + S_{12}x_2, \ldots\ldots\ldots, S_{1n}x_n \\
y_2 &= S_{21}x_1 + S_{22}x_2, \ldots\ldots\ldots, S_{2n}x_n \\
y_m &= S_{m1}x_1 + S_{m2}x_2, \ldots\ldots\ldots, S_{mn}x_n
\end{aligned}
\qquad (7.3)
$$

For a full description of the system (set of equations), m needs not exceed n. Hence a set of $m.n$ (minimal n^2) constants is required. These can be obtained from a calibration with n samples of different composition, each yielding m measurements. For instance, a calibration for an n-component spectrophotometric analysis requires n samples to be measured at at least n wavelengths.

It can be observed that the constants S_{ji} in eqn. 7.3 can be regarded as partial sensitivities, i.e.

$$S_{ji} = (\partial y_j/\partial x_i)_{x_1, \ldots\ldots\ldots, x_{i-1}, x_{i+1}, \ldots\ldots\ldots, x_n} \qquad (7.4)$$

It is clear that the mathematical model of a multi-component analysis as given by eqn. 7.3 is an idealized model (linear dependence, no cross terms such as x_1x_2). However, it will show the possibilities and limitations of presenting selectivity and specificity in an efficient way.

The idealized model of an analysis of n components using n independent measurements as expressed by eqn. 7.3 (with m = n) represents a selective method if all but the n coefficients S_{ii}(i = 1, ..., n) are zero ; then each measurement depends on only one component in the sample. This model, for instance, represents a gas chromatographic determination of n components where the concentrations are derived from the areas of a set of n well resolved peaks.

Specificity is a special case of selectivity. Of all n^2 coefficients, only one (partial) sensitivity retains a value. Taking gas chromatography as an example again : the detector senses only one component if the procedure is specific.

Full selectivities (and specificities) are rare for analytical procedures. Therefore, Kaiser introduced a parameter to express the degree of selectivity (or specificity). Expressed in the same symbols as those used in the set of eqn. 7.3, the selectivity parameter Ξ is defined as

$$\Xi = \underset{j = 1, \ldots, n}{\mathrm{Min}} \frac{|S_{jj}|}{\sum_{i=1}^{n} |S_{ji}| - |S_{jj}|} - 1 \qquad (7.5)$$

For each equation of the set of n eqns. 7.3, the sum of the partial sensitivities S_{ji} with i ≠ j is determined $\left(\sum_{i=1}^{n} |S_{ji}| - |S_{jj}| \right)$. If this sum is small compared with S_{jj}, the expression in eqn. 7.5 is large, i.e., for the element i = j with measurement j the procedure is selective. Full selectivity corresponds to a value of infinity. The equation of the set yielding the smallest value of the expression in eqn. 7.5 is the weakest part of the procedure and, according to Kaiser, this minimum value determines the selectivity of the entire procedure. It is clear that, when reducing the set of n^2 partial sensitivities to one selectivity parameter Ξ, much information concerning the procedure is lost. Wilson (1974), in discussing the performance characteristics, considers this to be a serious drawback. Indeed, a procedure with a low value of Ξ is to be considered as a poor procedure. However, in the absence of the component responsible for the low value of Ξ, a poor procedure can be acceptable. It therefore appears to be necessary to quote partial sensitivities for all possibly interfering elements or compounds rather than compressing the required information into one parameter.

It is possible to define a parameter for the specificity in an analogous way. However, specificity is met even more infrequently than selectivity. The parameter expressing the degree of specificity has the same disadvantages as the selectivity parameter and therefore will not be discussed in this chapter. For a further discussion, we also refer to Pszonicki (1977) and Pszonicki and Lukszo-Biénkowska (1977) who used a somewhat more complex model to define (non)specificity. The concepts introduced by Kaiser (1972) are useful in an entirely different context, and we shall return to these aspects in Chapter 17.

7.3. SOME EXAMPLES

Ion-specific (ion-selective) electrodes are not as specific or selective as the term suggests. The electrode potential, normally represented by the Nernst equation, can be replaced with an equation of the type

$$E_j = E_{jo} + \frac{RF}{n_j F} \ln(a_j + \sum_{i \neq j} k_{ji} a_i^{n_j/n_i}) \qquad (7.6)$$

where E_j is the electrode potential, E_{jo} the (standard) potential (for activities $a_j = 1$ and all $a_i = 0$) and n_i the valency of the ion i. The constants k_{ji} are termed selectivity constants and are usually quantified in publications on ion-selective electrodes. The selectivity parameter defined by Kaiser can easily be calculated if eqn. 7.6 is transformed into a linear equation

$$y_i = \exp\{(E_j - E_{jo}) \frac{n_j F}{RT}\} = a_j + \sum_{i \neq j} k_{ji} a_i^{n_j/n_i} = \sum_{i=1}^{n} S_{ji} x_i \qquad (7.7)$$

Eqn. 7.7 is reduced to an equation similar to one of the set of eqns. 7.3. The selectivity coefficients can be considered as partial sensitivities if the potential measurements are transformed logarithmically and the activities are assumed to be equal or proportional to the concentrations.

This, of course, applies only under certain conditions. Also, the system has to show a linearity, which will seldom be the case. Again, we conclude that the concept of Kaiser is of limited value.

Another example illustrating a much better use of the selectivity parameter is its application in spectrophotometric determinations in general and the determination of chlorine and bromine in particular. In Table 7.I the absorption coefficients of Cl_2 and Br_2 in chloroform at six wavenumbers are given

Table 7.I

Absorptivities of Cl_2 and Br_2 in chloroform (Landolt-Börnstein, 1951)

wavenumber σ $(cm^{-1}) \times 10^{-3}$	absorptivities	
	a_{Cl_2}	a_{Br_2}
22	4.5	168
24	8.4	211
26	20	158
28	56	30
30	100	4.7
32	71	5.3

Obviously for the determination of chlorine and bromine, only two measurements are required. A combination of two possible wavelengths leads to a procedure with a certain selectivity. Each of the possible 15 combinations has a certain selectivity and the combination with the highest selectivity is most attractive for analytical purposes. The reader can easily verify that the combination of $\sigma = 24 \cdot 10^3\ cm^{-1}$ and $\sigma = 30 \cdot 10^3\ cm^{-1}$ leads to the best selectivity. The analytical calibration functions are

$$A_1 = 8.4\ C_{Cl_2} + 211\ C_{Br_2}$$

$$A_2 = 100\ C_{Cl_2} + 4.7\ C_{Br_2}$$

where A_1 and A_2 are the absorbancies at the two wavelengths. The selectivity parameter $\Xi = 16.6$. The example given here is rather simple. Inspection of the spectra might easily lead to the same conclusion as can be shown by Fig. 7.1. However, the concepts illustrated here can be used for situations where judgement by eye is not easy. The use of these principles in some optimization problems will be discussed in Chapter 17.

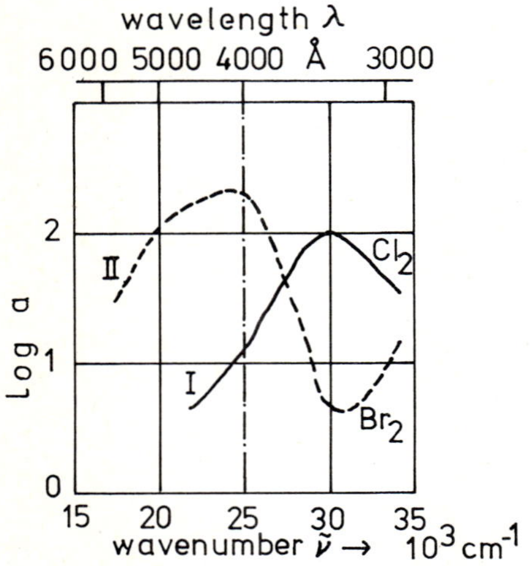

Fig. 7.1. Spectra of Cl_2 and Br_2 in Chloroform (Landolt-Börnstein, 1951).

REFERENCES

R. Belcher, Talanta, 12 (1965) 129.
R. Belcher, D. Betteridge, Talanta, 13 (1966) 535.
R. Belcher, Talanta, 23 (1976) 883.
D. Betteridge, Talanta, 12 (1965) 129.
H. Kaiser, Z. Anal. Chem., 260 (1972) 252.
Landolt-Börnstein, Zahlenwerte und Funktionen, 3. Teil, Atom- und Molekularphysik, Springer Verlag, Berlin, 1951, p. 232.
L. Pszonicki, Talanta, 24 (1977) 613.
L. Pszonicki and A. Lukszo-Biĕnkowska, Talanta, 24 (1977) 617.
A.L. Wilson, Talanta, 12 (1965) 701.
A.L. Wilson, Talanta, 21 (1974) 1109.

Chapter 8

INFORMATION

8.1. INTRODUCTION

In the analytical chemical literature, qualitative analytical methods are often referred to as "good", "valuable", "excellent", "specific", etc., with no further explanation of these terms. An objective interpretation of such terms is not easy and therefore the resulting choice of methods often does not have a completely rational basis. Whereas quantitative analytical methods can be evaluated by using criteria such as precision, accuracy and reliability, and other criteria discussed in the preceding chapters, no comparable and generally accepted criteria exist for qualitative analysis.

Information theory, introduced in analytical chemistry some years ago (see for instance Kaiser, 1970), permits a mathematical evaluation of qualitative methods by calculation of the expected or average amount of information obtained from the analysis. Quantitative methods can also be evaluated on the basis of principles of information theory (see for instance Doerffel and Hildebrandt, 1970 ; Eckschlager, 1971, 1972 a, b, 1973 a, b, 1975 ; Griepink and Dijkstra, 1971). However, the application of information theory is clearly more important for qualitative analysis, where it fulfils a need for criteria. In explaining the use of information theory in analytical chemistry we shall therefore confine the discussion to qualitative analysis.

The aim of an analysis is to reduce the uncertainty with respect to the sample to be analysed. It will be appreciated that the reduction of uncertainty is considered to be equivalent to obtaining information. This corresponds with the common use of the terms uncertainty and information. A newspaper offers only news (information) if the reader has not yet been informed about the events through other communication channels. If he *has* been informed, he is (almost) certain about the contents of the pages of the newspaper. The same is true of qualitative analysis : the analysis is carried out because there is an uncertainty about the identity of the components in the sample. After the analysis, the state of uncertainty is (hopefully) turned into a state of certainty (or, at least, of less uncertainty) ; in other words, the analysis has yielded a certain amount of information.

8.2. INFORMATION CONTENT

In order to use information as an evaluation criterion, the uncertainty before and after analysis, and thus the information, has to be quantified. For this quantification we associate large and small uncertainties with large and small numbers of possible identities of the components in the sample. In the case of certainty there is only one possible identity.

Information theory is related to classical probability theory. For a large number of possible identities the probability of each will, in general, be small. Similarly, if there are a small number of possible identities, the probability of each will be large. Following this reasoning, we can arrive at an expression for the information obtained from an analysis. Before the experiment the uncertainty can either be expressed in terms of the number of possible identities, n_o, each having a probability $p_o = 1/n_o$. After the ith experiment the number of possible identities is reduced to n_i with probabilities $p_i = 1/n_i$. The information I_i obtained from the ith experiment can be defined by (Brillouin, 1960).

$$I_i = k \log_z (n_o/n_i) \qquad (8.1)$$

This expression can be replaced by

$$I_i = k \log_z (p_i/p_o) \qquad (8.2)$$

where $\log_z$ is the logarithm to the base z and k is a constant ; both z and k depend on the units used for expressing the information. Usually z is put equal to 2 and k equal to 1. Then I_i is expressed in bits (binary digits).

Strictly, the information expressed by eqns. 8.1 and 8.2 is the information obtained from one particular outcome of an experiment. This amount of information is also called specific information (Arbeitskreis Automation in der Analyse, 1974). However, if all possible outcomes of the experiment yield the same specific information (for instance, all melting points lead to the same uncertainty after analysis), the average information is equal to the specific information. Then eqns. 8.1 and 8.2 are also expressions for the information content of the procedure (for instance, identification by means of melting points).

The application of eqns. 8.1 and 8.2 assumes a simple model of the analytical problem, in which each of the possible identities has the same probability before analysis. However, usually some identities (substances) are more likely to be found than others. Further, it should be noted that the model applies only to the identification of pure substances.

A numerical example will illustrate the concepts introduced so far. Let us assume that in a qualitative analysis it is known that the sample to be analysed is one of 100 possible substances and that the measurement yields a signal corresponding to 10 possible identities. Then, the application of eq. 8.1 or 8.2 leads to the specific information $I = \log_2(100/10)$ or $I = \log_2(0.1/0.01) = 3.32$ bits. If all possible results of the experiment lead to the same reduction of possible identities, the information content of the procedure also is 3.32 bits. Such a situation can be met in thin-layer chromatography when only 10 groups of 10 substances each can be distinguished by their R_f values.

With different reduction factors, the information content clearly is not equal to the specific information for the several (results of the) experiments. Suppose that in the thin-layer chromatographic experiment 10 substances have identical R_F values (or R_F values that cannot be distinguished) and that all other substances have, for instance, R_F values of zero. Then in 10% of the experiments the information obtained will be 3.32 bits, whereas in 90% an information of only $I = \log_2(100/90) = 0.15$ bit is obtained. The weighted average for the specific information or the information content of such a thin -layer chromatographic procedure will be $I = 0.1 \times 3.32 + 0.9 \times 0.15 = 0.47$ bits (assuming that all 100 substances are to be found with the same probability).

In symbol form, the equation used can be written as

$$I = \sum_i - \frac{n_i}{n_o} \log_2 \left(\frac{n_i}{n_o}\right) = \sum_i p_i I_i \tag{8.3}$$

where I is the information content of the procedure, n_o the number of possible identities before the experiment (with equal probabilities) and n_i the number of possible identities after interpretation of the experiment with result y_i (signal y_i). I_i is the information obtained from the experiment with result y_i and p_i is the probability of measuring a signal y_i.

For general use, a more generally applicable model has to be introduced. This model will represent a set of possible identities before the experiment ($x_1, x_2, \ldots, x_j, \ldots, x_n$), each having a probability p_j. The uncertainty before the experiment, H, can be expressed by means of the equation of Shannon (Shannon and Weaver, 1949)

$$H = \sum_{j=1}^{n} - p_j \log_2 p_j \tag{8.4}$$

The uncertainty H is also called entropy (Shannon and Weaver, 1949 ; Eckschlager, 1975 ; Belyaev and Koveshnikova, 1972), because of its analogy with the entropy expression as used in thermodynamics. Similarly, after the experiment with result y_i

$$H_i = \sum_{j=1}^{n} - p_{j/i} \log_2 p_{j/i} \tag{8.5}$$

where $p_{j/i}$ is the (conditional) probability (also called Bayes' probability : see for instance Raeside, 1976 ; Chapter 25) of identity x_j, provided that the experiment has yielded a signal $y_i (i = 1, \ldots, m)$. The uncertainty or entropy

H_i depends, of course, on the signal measured. The difference $H-H_i$ is equal to the specific information. Clearly, in order to arrive at an equation for the information content, we have to take the weighted average of $H-H_i$, which leads to the expression

$$I = H - \sum_{i=1}^{m} p_i H_i \tag{8.6}$$

where p_i is the probability of measuring a signal y_i. By making use of Shannon's uncertainty equation, eqn. 8.6 can be written as

$$I = \sum_{j=1}^{n} - p_j \log_2 p_j - \sum_{i=1}^{m} p_i \sum_{j=1}^{n} - p_{j/i} \log_2 p_{j/i} \tag{8.7}$$

Calculation of the information content in general requires a knowledge of the following probabilities :

(a) The probabilities of the identities of the unknown substance before analysis (p_j). The first term on the right-hand side of eqn. 8.7 represents what is known about the analytical problem in a formal way, or the "pre-information". The analytical problem in terms of the probabilities p_j is essential for calculating the information content. An infinite number of possible identities each having a very small probability (approaching zero) represents a situation without pre-information. The uncertainty is infinitely large and solving the analytical problem requires an infinite amount of information.

(b) The probabilities of the several possible signals (p_i). These probabilities depend on the relationship between the identities and the signals (tables of melting points, R_F values, spectra, etc.), and also on the substances expected to be identified (p_j). If an identity is not likely to be found, the corresponding signal is not likely to be measured. It should be noted that one identity can lead to several signals because of the presence of experimental errors.

(c) The probabilities of the identities when the signal is known ($p_{j/i}$). In fact, these probabilities are the result of the interpretation of the measured signals in terms of possible identities. To this end the following "interpretation" relationship can be used

$$p_{j/i} = \frac{p_j \cdot p_{i/j}}{\sum_j p_j \cdot p_{i/j}} \qquad (8.8)$$

This relationship shows that the probabilities for the identities after analysis can be calculated from the pre-information (p_j) and the relationships between the identities and the signals ($p_{i/j}$). Equation 8.8 is found in the literature as Bayes'theorem (see for instance Raeside, 1976 ; Chapter 25). It should be observed that one particular signal can correspond with more than one identity.

A few final remarks will conclude this section. As has been shown, uncertainties and information can be related to probabilities. Shannon's equation is one of several possible equations that can be used to define uncertainty or entropy and information (Aczél and Daróczy, 1975 ; Eckschlager and Vadja, 1974). It must be stressed that the information content is a characteristic of an analytical procedure in relation to the analytical problem. The same procedure applied to different problems can have different information contents.

The following sections will serve as illustrations of the principles introduced so far. For more extended treatments the reader is referred to the literature already cited.

8.3. AN APPLICATION TO THIN-LAYER CHROMATOGRAPHY

In thin-layer chromatography (TLC), the signal that permits the identification of an unknown substance is an R_F value. If we assume that substances whose R_F values differ by 0.05 can be distinguished, the complete range of R_F values can be divided into 20 groups (0-0.05, 0.06-0.10, ...). Such a simplified model leads to a situation where substances with R_F values of, for instance, 0.05 and 0.06 are considered to be separated, which clearly is not real. However, the model allows an easy calculation of approximate values of the information content. Further, at least in this application, it is not important to distinguish exactly which substances are separated and which are not, as the purpose is rather to see how well the substances are spread out over the plate.

Each of the 20 groups of R_F values can then be considered as a possible signal (y_1, y_2, ..., y_{20}) and there is a distinct probability (p_1, p_2, ..., p_{20}) that an unknown substance will have an R_F value within the limits of one of the groups. Let us consider a TLC procedure that is used to identify a substance belonging to a set of n_o substances and that n_1 substances fall into group 1, n_2 into group 2, etc. If all substances have the same *a priori* probability to be the unknown compound, eqn. 8.3 can be used for calculation of the information content. To understand further the meaning of the information content, let us investigate some extreme conditions.

(a) All substances fall into the same group n_i.

In this instance $n_i/n_o = 1$ and thus $I_i = 0$. As all of the substances yield the same R_F value, the experiment does not indicate anything to the observer. No information is obtained because, in Brillouin's terminology, there is no uncertainty as to which event (signal, R_F value) will occur : whatever the unknown substance, the result will always be the same.

(b) All substances fall into different groups.

The information content is maximal as each substance will yield a different R_F value. The information content, from eqn. 8.3 with all $n_i = 1$, will now be equal to

$$I = n_o \cdot \frac{1}{n_o} \log_2 \frac{1}{n_o} = -\log_2 n_o \qquad (8.9)$$

It can be shown that this is indeed the maximum value which can be obtained. It is equal to the information necessary to obtain an unambiguous, complete identification of each substance. This information content is equal to the entropy before analysis, H.

From these extreme conditions, it follows that in order to obtain a maximum information content, the TLC system should cause an equal spread of the R_F values over the entire range. Of course, if there are more groups of R_F values than

Table 8.I

hR_F values of DDT and related compounds and information content of the proposed separations

R_F values were taken from Bishara et al. (1972)

Solvent system	p,p'-DDT	o,p'-DDT	p,p'-DDE	o,p'-DDE	p,p'-DDD	DDA	DDMU	DBP	Kel-thane	DPE	DBH	BPE	DDM	I
I	25	35	41	36	10	0	32	2	0	27	0	0	35	2.28
II	48	55	61	56	28	0	55	8	2	44	0	0	54	2.47
III	69	72	75	72	52	0	67	24	3	63	0	0	67	2.35
IV	76	76	77	75	64	0	74	45	8	72	2	4	75	2.50
V	67	69	72	69	52	0	70	69	7	66	1	2	71	1.98
VI	67	70	75	68	51	0	72	45	10	66	2	4	70	2.87
VII	70	70	75	70	63	0	74	61	23	66	6	13	69	2.78
VIII	78	79	83	79	77	6	83	75	53	77	27	39	80	2.56
IX	69	71	76	69	60	0	73	56	19	67	4	9	71	3.03
X	35	44	49	45	16	0	48	4	0	35	0	0	42	2.62
XI	58	63	65	62	42	0	62	18	3	55	0	0	61	2.41
XII	60	64	71	64	43	0	69	50	8	60	2	5	67	2.71
XIII	63	66	71	64	48	0	68	55	16	62	4	9	67	2.97
XIV	73	74	77	72	65	0	75	68	48	77	24	36	76	2.57
XV	83	84	85	83	78	5	83	81	58	82	46	55	83	1.47
XVI	84	84	84	83	80	0	84	82	75	84	71	74	83	1.70
XVII	87	88	89	87	81	20	87	83	64	86	59	63	86	1.47
XVIII	59	68	73	69	39	8	67	34	34	60	33	34	67	2.31
XIX	92	94	96	93	83	32	92	82	72	91	61	70	93	2.35
XX	78	80	80	77	69	17	78	67	43	74	24	35	78	2.50
XXI	82	82	82	80	81	0	81	81	81	81	80	81	82	0.39
XXII	80	80	80	77	79	5	78	78	79	79	79	79	80	0.39
XXIII	35	46	50	45	13	0	40	2	0	36	0	0	39	2.35
XXIV	40	51	52	48	16	0	44	3	0	42	0	0	42	2.28
XXV	71	74	77	72	59	0	74	64	21	70	5	13	72	2.72
XXVI	77	77	78	77	72	30	78	72	56	78	34	47	78	2.03
XXVII	27	40	43	42	10	0	35	2	0	32	0	0	35	2.08
XXVIII	70	75A	77	75	51	2	74	20	4	69	4	4	75	2.04
XXIX	65	69	76	68	48	0	72	57	7	64	2	5	70	2.93
XXX	85	85	85	84	84	4	80	83	84	84	83	84	84	0.77
XXXI	54	67	74	69	22	0	61	6	6	54	6	6	62	2.62
XXXII	100	100	100	100	94	35	100	94	92	100	83	92	100	1.57
XXXIII	93	94	96	94	90	7	96	93	75	94	67	70	92	1.88

there are substances, the maximum information content is related to the number of substances, while if there are more substances than groups, the maximum obtainable information content is limited by the number of groups, in the model introduced $I = \log_2 20 = 4.32$ bits. A model very similar to that described was used by Souto and Gonzales de Valesi (1970) and by Massart (1973) for comparison of TLC systems. The results of the application by Massart of information theory to the R_F values of DDT and related compounds determined by Bishara et al. (1972) are summarized in Table 8.I. It was concluded that systems V, XV, XVI, XVII, XXI, XXII, XXX, XXXII and XXXIII were of no interest. The best separations were obtained with solvents IX, XIII and XXIX and further investigations should be aimed at optimizing small changes in the best three solvents (for instance, by applying one of the techniques described in Part II).

The object of a qualitative analysis is to obtain an amount of information that is equal to the uncertainty before analysis. This, in practice, is often not possible with a single test and more experiments therefore have to be combined in order to achieve this aim. For example, in the toxicological analysis of basic drugs (Moffat, 1974), one will combine techniques such as UV and IR spectrometry, TLC and GLC or one will use two (or more) TLC procedures, etc., in order to obtain the necessary amount of information. Hence the next question which has to be answered is how to calculate the information content of two or more methods.

When two TLC systems are combined, one can consider the combination of the two R_F values both of which fall in the range 0.00-0.05 as one event (signal y_{11}), an R_F value of 0.00-0.05 for system 1 and of 0.05-0.10 for system 2 as a signal y_{12}, etc. As before one can define a probability p_{ij} for signal y_{ij}, so that $n_{ij}/n_o = p_{ij}$. For the general case of system 1 containing m_1 classes and system 2 containing m_2 classes, eqn. 8.3 can be converted into

$$I = \sum_{i=1}^{m_1} \sum_{j=1}^{m_2} - \frac{n_{ij}}{n_o} \log_2 \frac{n_{ij}}{n_o} \qquad (8.10)$$

At first sight one might assume that I is the sum of the information content of the systems 1 and 2

$$I = I(1) + I(2) \tag{8.11}$$

This is true only if the different information yielded by systems 1 and 2 is not correlated, i.e., if no part of the information is redundant. This can be understood more easily by considering a simple example represented by the R_F values for eight substances in three different solvents (Table 8.II).

Table 8.II

R_F values of eight substances in three different solvents

Substance	Solvent I	Solvent II	Solvent III
A	0.20	0.20	0.20
B	0.20	0.40	0.20
C	0.40	0.20	0.20
D	0.40	0.40	0.20
E	0.60	0.20	0.40
F	0.60	0.40	0.40
G	0.80	0.20	0.40
H	0.80	0.40	0.40
Information content	2	1	1

With solvent I one obtains 2 bits of information, while 3 bits are necessary for the complete identification of each possible substance. Solvents II and III each allow the acquisition of 1 bit. Running a plate first with solvent I and then with solvent II does indeed permit complete identification : 3 bits are obtained with this combination. Although solvents I and III have clearly different R_F values, the latter does not yield any more information than that obtained with solvent I. The information content of a procedure in which both solvents are used is still 2 bits. Both of these cases are, of course, extreme and the combination of two TLC procedures, and in general of any two procedures, will lead to an amount of information that is less than that which would be obtained by adding the information content of both procedures but equal to or higher than the information content of a single procedure. In practice, it is improbable that two chromatographic systems would yield completely "uncorrelated information" and even when combining methods such as chromatography and spectrophotometry some correlation must be expected.

The information content of combined procedures and the effects of correlation upon the information content are treated more extensively in Chapter 17.

The most important conclusions from this section are that the highest information content for individual systems is obtained when the substances are distributed evenly over the classes which can be distinguished, and that for combinations of methods, the "correlated information" should be kept as low as possible. This can be achieved by choosing unsimilar systems. It should be noted here that the amount of correlated information (also called mutual information) can be used as a similarity coefficient between systems (see Chapter 18).

Neither conclusion is surprising. Analytical chemists know that a TLC separation is better when the substances are divided over the complete R_F range and they also understand that two TLC systems in combination should not be too similar. Information theory allows one to formalize this intuitive knowledge and to quantify it, so that an optimal method can be devised.

The determination of the information content of spectral peaks (for instance, in mass spectrometry), is very similar to the application of information theory to TLC as described above. For binary coded peaks (peaks either absent or present, thus two intensity levels) the information content per peak position can be expressed by the simple equation

$$I = - p \log_2 p - (1-p) \log_2 (1-p)$$

where p is the probability of a peak being present at the peak location considered (Grotch, 1970 ; Erni, 1972 ; van Marlen and Dijkstra, 1976). The information content for combinations of peak locations is considered in Chapter 17.

8.4. AN APPLICATION TO GAS CHROMATOGRAPHY

Although the identification of substances by means of gas chromatographic retention indices is essentially the same as their identification by means of TLC R_F values, it is also possible to consider it in another way. This is necessary especially when the number of possible identities is large and when an assessment is to be made of the information content of a gas chromatographic identification by using more than one retention index.

As was remarked in section 8.2, the information content is calculated from the probabilities of the several possible identities before and after analysis. Actually, the identification is realized by first measuring the signal (retention index) and subsequently interpreting this signal in terms of possible identities. Fortunately, it is possible to convert eqn. 8.7 into an equation for the information content in terms of possible signals rather than possible identities (this is possible only when uncertainties have been expressed by the Shannon equation). It can be shown that the information content is

$$I = \sum_{i=1}^{m} - p_i \log_2 p_i - \sum_{j=1}^{n} p_j \sum_{i=1}^{m} - p_{i/j} \log_2 p_{i/j} \qquad (8.12)$$

where p_i is the probability of measuring a signal y_i and $p_{i/j}$ the (conditional) probability of measuring a signal y_i provided that the substance has the identity x_j. The values of p_i can be derived from the probabilities of the several identities and the relationship between the identities and signals (table of retention indices, taking into account that the signal is measured with a certain error). The values of $p_{i/j}$ represent the errors of the measurements because the $p_{i/j}$s represent the probabilities of the different signals y_i found when the identity of the component is known to be x_j. For different values y_i these errors can be, but need not to be, different.

Eqn. 8.12 for discrete signals (with discrete probabilities) can be converted into an equation for continuously variable signals (represented by probability distribution functions of the signals), as follows

$$I = \int - p(y) \log_2 p(y)dy - \sum_{j=1}^{n} p_j \int - p_{ej}(y) \log_2 p_{ej}(y)dy \qquad (8.13)$$

where p(y) is the distribution function of the expected signals (before analysis) and p_{ej} represents the error distribution function for substance x_j. If the errors are the same for all identities, eqn. 8.13 can be transformed into

$$I = \int - p(y) \log_2 p(y)dy - \int - p_e(y) \log_2 p(y)dy \qquad (8.14)$$

In work on the comparison of gas chromatographic columns (Dupuis and Dijkstra, 1975 ; Eskes et al., 1975) both p(y) and $p_e(y)$ were assumed to be Gaussian. For this situation, eqn. 8.14 reduces to the simple expression

$$I = \frac{1}{2} \log_2 \left(\frac{s_m^2}{s_e^2}\right) = \frac{1}{2} \log_2 \left(\frac{s_t^2 + s_e^2}{s_e^2}\right) \qquad (8.15)$$

where s_m^2 is the estimate of the variance of the measured signals (retention index + error), s_t^2 is the estimated variance of the "true" signals (retention index without errors) and s_e^2 is the estimate of the variance of the errors. As expected, the information content will be large when the retention indices differ widely and the experimental error of measuring the retention index is small. The information contents calculated by Eskes et al. (1975) are about 6.3 bits, depending slightly on the stationary phase and the nature of the substances that were studied (alcohols, ethers and carboxyl compounds). To stress once more : the information content is a criterion of a procedure in relation to the analytical problem (group of substances likely to be identified). The information calculated with eqn. 8.15 using a limited number of retention indices applies to a large number of substances provided that the value s_m^2 is a reliable estimate of the population.

The conclusions to be drawn in this section are the same as those drawn in section 8.3, and the remarks made in section 8.3 about combining procedures also apply to chromatography. The combination of stationary phases is discussed in Chapter 17.

8.5. DISCUSSION

The information content of an analytical procedure is one of several possible information parameters. Instead of Shannon's uncertainty function, other uncertainty functions which lead to other information contents can be used. Apart from the choice of the uncertainty function, one other fundamentally different information parameter has been used in analytical chemistry. This parameter has been called the "informing power" and was used, for instance, by Kaiser (1970), Huber and Smit (1969), Palm (1971), Massart and Smits (1974) and Eckschlager (1976 a, b) for characterizing spectrometric and chromatographic methods. The informing power is closely related to the structural information and the metric information defined by MacKay (1950), whereas the information content might be compared with MacKay's selective information. For the calculation of the informing power use can be made of the sampling theorem of Shannon and Weaver (1949) and the signal-to-noise ratio. The value of this characteristic increases when the number of independent measurements (required for reconstruction of the entire spectrum) is increased and when the signal-to-noise ratio is increased. Kaiser (1970) has shown that the number of independent measurements is proportional to the resolution. The higher the resolution, the greater is the number of peaks that can be resolved or distinguished. An increase in the signal-to-noise ratio permits a better discrimination between peaks of different magnitude. Hence the combination of resolution and signal-to-noise ratio leads to a measure of the number of different spectra or chromatograms that might be envisaged. Therefore, the informing power usually will be larger than the information content, because the number of spectra that might be discriminated by employing fully the power of the spectrometer is usually larger than the number of different spectra that occur in practice. Nature has set limits to the differences in spectra. The width of the peaks measured is often determined by the natural width of the peaks rather than by the resolution, and also the number of different spectra is limited owing to correlations (simultaneously occurring peaks) or to empty spectral regions. Hence the informing power is of limited use in analytical chemistry as the potential of the procedure is equally well described by the resolution and signal-to-noise ratio. However, the informing power as a composite characteristic might serve as a criterion when balancing signal-to-noise ratio and resolution. It also is a useful tool for problems related to data handling.

Some remarks should be made about the models used for the calculation of information contents. Although these models were used for qualitative analysis they are equally applicable to quantitative analysis. All models should take into account the occurrence of errors : in the TLC model considered in section 8.3, the error is taken into account to a certain extent by dividing the entire R_F range into a limited number of R_F sub-ranges (of, for instance, 0.05 unit). Nevertheless, in an actual experiment it may happen that a certain substance will fall into the wrong sub-range, the probability of such a wrong result has to be taken into account and in order to obtain an exact value of the information content a correction should be introduced. In this particular application the corrections are small and approximately identical for all systems. Therefore, a comparison of systems using information contents that have not been corrected for the error is permissible.

Another aspect of the modelling is the assumption of a probability distribution function for a large number of measurements from a limited number of such measurements. In the application to gas chromatography in section 8.4, a Gaussian (normal) distribution was assumed. The validity of such an assumption can be tested with the χ^2 test for goodness of fit, which is defined as

$$\chi^2 = \sum_{i=1}^{n} \frac{(O_i - A_i)^2}{A_i} \qquad (8.16)$$

where O_i is the value of the ith measurement and A_i is the calculated value on the basis of the assumption of the distribution. Obviously, smaller values of χ^2 correspond to a better agreement between the observed and the calculated values and thus correspond to a greater probability of the validity of the assumption.

The χ^2 test was used by Dupuis and Dijkstra (1975) and by Eskes et al. (1975) for verifying the assumption of a Gaussian distribution for the retention indices. It can be used for testing the validity of any distribution, and therefore also for rectangular distributions. In section 8.3 it was concluded that the highest information content is obtained when the substances are distributed regularly over the (R_F) range of measurements. It is obvious that evaluating (TLC) systems by means of their χ^2 values (for a rectangular distribution) runs parallel with the evaluation by the information contents. A small χ^2 indicates a nearly rectangular distribution and therefore a rather good separation. De Clercq and Massart (1975) compared the χ^2 criterion and the information content for the classification of TLC systems used for the qualitative analysis of 100 basic drugs. In this study a comparison of the discriminating power introduced by Moffat was also made. Moffat et al. (1974) regard two compounds as being discriminated in a particular system (i) if the difference between their characteristic values exceeds a certain critical value, which is termed the error factor (E_i). For example, when the qualitative technique used is UV spectrophotometry, two substances are considered to be separated if the wavelengths of two substances differ by 2 nm (E_i = 2 nm). For thin-layer chromatography, E_i is equal to a certain number of R_F units. For $E_i = 0.05$, two substances with R_F values of 0.08 and 0.10 are considered to be undiscriminated while those with R_F values of 0.08 and 0.16 are discriminated. The discriminating power (DP) is defined as the probability that two compounds selected at random from a large population would be discriminated. To calculate the DP of a system in which N compounds are investigated, the total number, M, of undiscriminated pairs of compounds (within the limits of E_i) is counted. The value of DP is then given by

$$DP = 1 - \frac{2M}{N(N-1)} \qquad (8.17)$$

This criterion was applied with success to the selection of optimal thin-layer and paper chromatographic systems for the qualitative analysis of 100 basic drugs. It was also applied to a comparison of the discriminatory effectiveness of different techniques, such as thin-layer chromatography, gas-liquid chromatography, UV and IR spectrophotometry and mass spectrometry. Finally, its application can be extended to the evaluation of combinations of methods. The DP and χ^2 criteria have the merit of simplicity and in many instances are equally useful as the information content. However, they are of a less fundamental nature and therefore less amenable to theoretical considerations.

8.6. TESTS OF FIT

In this section, tests to determine the underlying distribution of a set of observations will be examined.

Let $x_1, x_2, \ldots, x_n$ be independent random observations of a variable with an unknown probability distribution function f(x). The problem of testing whether f(x) is equal to some particular distribution function $f_0(x)$ is called a goodness-of-fit problem (see Kendall and Stuart, 1973). The test can be written as

$$H_0 : f(x) = f_0(x)$$
$$H_1 : f(x) \neq f_0(x)$$

When $f_0(x)$ is completely specified, i.e., all of its parameters are known, H_0 will be called a simple hypothesis, whereas if some parameters are unknown it will be called composite. Let us consider the simple hypothesis problem and suppose that the set of values which can be taken by the random variable x is divided into k classes. Using the completely known distribution function $f_0(x)$, it is possible to calculate the probability for an observation to be within each of the classes. These probabllities will be called $p_1, p_2, \ldots, p_k$ and their sum must be unity. The n observations can also be divided into the k classes. The observed frequencies of the classes will be called $f_1, f_2, \ldots, f_k$ $(\sum_{i=1}^{k} f_i = n)$.

These probabilities are estimators of the probabilities of the "true" distribution. The test proposed by Pearson (1900) for hypothesis H_0 is based on the statistic χ^2

$$\chi^2 = \sum_{i=1}^{k} \frac{(f_i - np_i)^2}{np_i} \tag{8.18}$$

which is analogous to 8.16. He showed that χ^2 has an approximately χ^2_{k-1} distribution. This approximation is very accurate when the np_i are almost equal. Values for the χ^2 distribution are given in Table II of the Appendix. Using this result, a value χ^2_α can be found in statistical tables with the property that

$$p(\chi^2 < \chi^2_\alpha) = 1 - \alpha$$

If

$$\chi^2 < \chi^2_\alpha$$

the hypothesis H_0 will be accepted.

The generality of this test is due to the weakness of the underlying assumptions. It can be used for any type of statistical scales provided that the observations can be divided into classes to which theoretical hypothetical probabilities can be associated.

Example

Thin layer chromatography of 200 compounds was performed. The results are grouped in 10 classes and the class width is 0.1 R_f units. The observed frequencies are

$$f_1 = 17 \; ; \; f_2 = 22 \; ; \; f_3 = 25 \; ; \; f_4 = 16 \; ; \; f_5 = 15 \; ;$$
$$f_6 = 21 \; ; \; f_7 = 12 \; ; \; f_8 = 23 \; ; \; f_9 = 29 \; ; \; f_{10} = 20$$

The question is asked whether or not the observed frequencies correspond to a rectangular distribution.

The null hypothesis H_0 = the distribution is restangular or $f_i = 20$. The alternative hypothesis $H_1 = f_i \neq 20$.

Using equation 8.16 χ^2 is calculated as follows

$$\chi^2 = \frac{(17-20)^2}{20} + \frac{(22-20)^2}{20} + \frac{(25-20)^2}{20} + \frac{(16-20)^2}{20} + \frac{(15-20)^2}{20}$$

$$+ \frac{(21-20)^2}{20} + \frac{(12-20)^2}{20} + \frac{(23-20)^2}{20} + \frac{(29-20)^2}{20} + \frac{(20-20)^2}{20}$$

$$\chi^2 = \frac{1}{20} \quad (9 + 4 + 25 + 16 + 25 + 1 + 64 + 9 + 81) = 11.70$$

A 5% significance level is chosen. The value of χ^2 at 9 degrees of freedom and $\alpha = 0.05$ equals 16.919. The null hypothesis is accepted which means that the distribution is not significantly different from a rectangular distribution.

REFERENCES

J. Aczél and Z. Daróczy, On measures of Information and their Characterisations, Academic Press, New York, 1975.

Arbeitskreis Automation in der Analyse, Z. anal. Chem., 272 (1974) 1 ; translated in I.L. Marr, Talanta, 22 (1975) 597.

Y.I. Belyaev and T.A. Koveshnikova, Zh. Anal. Khim., 27 (1972) 375.

R.H. Bishara, G.B. Born and J.E. Christian, J. Chromatogr., 64 (1972) 135.

L. Brillouin, Science and Information Theory, Academic Press, New York, 2nd ed., 1960.

H. De Clercq and D.L. Massart, J. Chromatogr., 115 (1975) 1.

K. Doerffel and W. Hildebrandt, Wissenschaffl. Zeitschr. T.H. "Carl Schoorlemmer" Leuna, 11 (1969) 30.

P.F. Dupuis and A. Dijkstra, Anal. Chem., 47 (1975) 379.

K. Eckschlager, Collect. Czech. Chem. Commun., 36 (1971) 3016.

K. Eckschlager, Collect. Czech. Chem. Commun., 37 (1972 a) 137.

K. Eckschlager, Collect. Czech. Chem. Commun., 37 (1972 b) 1486.

K. Eckschlager, Collect. Czech. Chem. Commun., 38 (1973 a) 1330.

K. Eckschlager, Z. anal. Chem., 277 (1975) 1.

K. Eckschlager, Z. Chem., 16 (1976 a) 111.

K. Eckschlager, Collect. Czech. Chem. Commun., 41 (1976 b) 1875.

K. Eckschlager and I. Vajda, Collect. Czech. Chem. Commun., 39 (1974) 3076.

F. Erni, Beitrag zur Computerunterstützten Strukturaufklärung, Thesis nr 4296, Eidgenössischen Technischen Hochschule Zürich, 1972.

A. Eskes, P.F. Dupuis, A. Dijkstra, H. De Clercq and D.L. Massart, Anal. Chem., 47 (1975) 2168.

B. Griepink and G. Dijkstra, Z. anal. Chem., 257 (1971) 269.

S.L. Grotch, Anal. Chem., 42 (1970) 1214.

J.F.K. Huber and H.C. Smit, Z. anal. Chem., 245 (1969) 84.
H. Kaiser, Anal. Chem., 42 (1970) 24A.
M.G. Kendall and A. Stuart , The advanced theory of statistics, Vol II, Charles Griffin, London, 1973.
D.M. MacKay, Philosophical Mag., 41 (1950) 289.
G. van Marlen and A. Dijkstra, Anal. Chem., 48 (1976) 595.
D.L. Massart, J. Chromatogr., 79 (1973) 157.
D.L. Massart and R. Smits, Anal. Chem., 46 (1974) 283.
A.C. Moffat, The Poisoned Patient : The Role of the Laboratory, CIBA Foundation Symposium 26 (new series), ASP (Elsevier-Excerpta Medica - North Holland), Amsterdam, 1974, p. 83.
A.C. Moffat, K.W. Smalldon and C. Brown, J. Chromatogr., 90 (1974) 1.
E. Palm, Z. anal. Chem., 256 (1971) 25.
K. Pearson, Phil. Mag., 5 (50) (1900) 157.
C.E. Raeside, Med. Phys., 3 (1976) 1.
E. Shannon and W. Weaver, The Mathematical Theory of Information, Univ. Illinois Press, Urbana, Ill., 1949.
J. Souto and A. Gonzalez de Valesi, J. Chromatogr., 46 (1970) 274.

Chapter 9

PRACTICABILITY

9.1. INTRODUCTION

In the preceding chapters a series of characteristics describing the quality of the analytical procedure or the quality of the analytical results have been discussed. The Committee on Standards of the International Federation of Clinical Chemistry (Büttner, 1976) used these characteristics to describe the reliability of an analytical method for routine use. In addition to this reliability as determined by the specificity, accuracy, precision and detection limit, there is a set of parameters that determine the practicability, which comprises speed, cost, technical skill requirements, dependability and laboratory safety. The division into two classes of parameters is artificial to some extent. However, if one considers the reliability as a measure of the quality of (the results obtained by) the procedure, practicability in a sense is the price that has to be paid for this quality. With this statement we also touch upon the fact that in a way all of the characteristics are interrelated ; we return to this point later in this chapter.

In discussing the practicability of analytical procedures we meet some difficulties, the most important of which is probably the impossibility or difficulty of quantifying all of the characteristics that determine the practicability. Therefore, it is difficult to use these characteristics as criteria in formal optimization procedures. Apart from the impossibility of a quantification, it is sometimes impossible even to define these parameters in a satisfactory way.

Although the analytical literature has little data on the cost of analyses, it is probably the most important characteristic that governs the practicability, as the application of an analysis is usually governed by economic factors.

Although a cost-benefit analysis is seldom explicitly made, it is usually found to be present in an implicit way. Of course, it would be much better to be clear on this point and, whenever possible, to make a cost-benefit analysis. Several aspects of such a cost-benefit analysis were clearly illustrated in a paper on the economic benefits to Australia of atomic-absorption spectroscopy (Brown, 1969), but it is beyond the scope of this chapter to discuss this paper in detail (see for some aspects also Part IV).

There are some properties or characteristics of analytical procedures (or of instruments) which will influence the choice of the analytical procedure but which will not be discussed, and aspects that are beyond the scope of this book are those related to the safety of (re)agents and apparatus, transportability, ease of handling procedures and instruments. With the last aspect we enter the area of a more subjective choice mechanism although we do not deny the importance of ergonomic studies related to this aspect. Even more subjective is the judgement of whether a procedure or apparatus is nice to look at or not ! Again, it cannot be denied that even these aspects will influence a non-rational choice.

9.2 COST

The cost of analysis depends on a large number of factors. To a large extent it depends on the nature of the analytical procedure, but possibly it also depends to a similar extent on the organization of the analytical laboratory. Analysis involves the use of labour, instruments, chemicals and energy, and all of these factors can be expressed in terms of money. Apart from these cost-determining factors, there are other aspects that have to be taken into account when the total amount that has to be paid is calculated. Analytical procedures have to be developed before it is possible to introduce them in the (routine) laboratory ; maintenance of laboratories, machines, etc., has to be provided ; efficient organization of the tasks to be done and appropriate communication channels are required in order to produce analytical results in a satisfactory way.

Rather than giving a detailed discussion of all of these factors, and rather than supplying accurate absolute figures on the cost of analysis, we prefer to present some general aspects. For this purpose we can consider a graph published some years ago that represents the relationship between the cost of analysis and the number of analyses to be carried out per day (Leemans, 1971). Although the absolute figures no longer apply, the trends emerging are still valid. In the graph (Fig. 9.1), the cost of seven procedures for determining the total nitrogen content of fertilizers is plotted.

Table 9.I.

Some characteristics of procedures for the determination of nitrogen (Leemans, 1971)

Analytical procedure	Dead time of analysis (min)	Standard deviation of analysis (% N)
Total N, classical distillation	75	0.17
Total N, DSM automated analyzer	12	0.25
NO_3-N, Technicon AutoAnalyzer	15	0.51
NO_3-N, ion-specific electrode	10	0.76
NH_4NO_3 : $CaCO_3$ ratio, X-ray diffraction	8	0.8
Total N, fast neutron -activation analysis	5	0.17
Specific gravity γ-ray absorption	1	0.64

One should bear in mind that only prices are compared in Fig. 9.I and other characteristics (some of which are given in Table 9.I) can be different. From the graph, the following trends can be discerned.

(1) The cost per analysis is almost independent of the number of analyses to be carried out when the procedure requires much labour and cheap instruments (or no instruments). This situation usually applies to "classical" analysis (manual titration, gravimetric analysis, etc.), such as, in the nitrogen determination, to the classical distillation procedure. The cost of chemicals is usually negligible when applying such procedures.

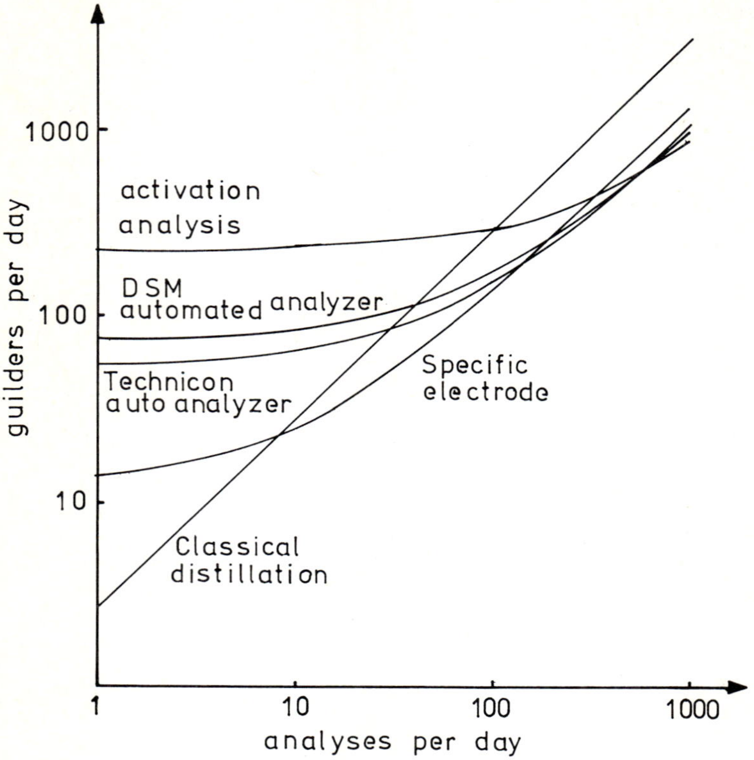

Fig. 9.1. Costs of some off-line analytical techniques for the analysis of inorganic nitrogen (1968) (Leemans, 1971).

(2) Fully automated (either laboratory or on-line) equipment delivers results at an almost constant price per unit time. Up to the (maximum) capacity of the equipment, the cost per analysis decreases as the inverse of the number of analyses per unit time.

(3) The actual situation is usually intermediate between (1) and (2). Expressed as an equation, the total cost K_t per analysis for a series of n analyses per unit time is

$$K_t = \frac{a}{n} + b \qquad (9.1)$$

where a and b are constants. Essentially the same equation was given by Bechtler (1970). The constant a consists of the cost of apparatus, investment, maintenance, amortization and the research and development required prior to introducing the procedure in the analytical laboratory. Labour, chemicals and energy are included in the constant b. Labour, of course, includes the efforts required for additional tests such as calibrations, in addition to the labour required for the actual analysis. Whether overhead costs such as those for internal and external communications (in general, costs arising from the organization) are to be included in either of the constants a or b is to some extent arbitrary.

Haeckel (1976), in a book on the rationalization of a clinical laboratory, gave a much more extended treatment of the cost aspects. We can cite two equations that express these aspects more explicitly than eqn. 9.1. The fixed costs per day are given by

$$K_f = \frac{L(1+\frac{S}{100}\cdot T_1)}{T_2 \cdot T_3} + \frac{G}{T_4} + E_f + R_f \qquad (9.2)$$

where

L = catalogue price of apparatus ;

S = service cost per year as a percentage of L ;

T_1 = T_2 minus guarantee period in years ;

T_2 = expected number of years that the apparatus can be used ;

T_3 = number of work days per year ;

G = cost of glassware ;

T_4 = expected number of days that glassware can be used ;

E_f = fixed costs of materials per series of analyses ;

R_f = fixed costs of reagents per series of analyses.

The variable costs per series of analysis are expressed by

$$K_v = (E_p + R_p) \cdot n + P.t_n \qquad (9.3)$$

where

E_p = cost of material per sample ;

R_p = cost of reagents per sample ;

P = cost of labour per minute ;

t_n = time required for one series of analyses.

(4) From the above, it is clear that the method to be preferred will be different for different laboratories. In general, simple non-instrumental and non-automated methods will be most economic when only a few determinations are to be made. Instrumentation and automation should be considered and are justified only for large series of analyses. However, it is possible that an advanced and costly instrumentation and automation scheme may be attractive for reasons other than economic, for instance, in order to achieve better reproducibility.

Table 9.II

Cost summary for 12 determinations per specimen : glucose, alkaline phosphatase, S.G.O.T., total protein, hydrogen carbonate, albumin, bilirubin, phosphate, sodium, potassium and calcium (from Horne, 1970)

Item	300 samples		600 samples		1,200 samples		1,800 samples	
	£	p	£	p	£	p	£	p
Amortization, maintenance and servicing over 10 years	24	80	24	80	24	80	24	80
Salaries	3	12	4	37	6	87	9	37
Materials : vials, control sera, reagents, etc.	15	32	30	30	60	20	90	20
Total daily cost	43	25	59	47	91	92	124	37
Cost per sample	14.4 p		9.9 p		7.7 p		6.9 p	

The trends and factors discussed above are reflected in a cost analysis made by Horne (1970) for chemical analysis with the Vickers Multichannel 300 apparatus. As shown in Table 9.II, the cost per analysis decreases when the number of analyses is increased : on going from 300 to 600 analyses, there is a sharp decrease, on going from 600 to 1200 the decrease is much smaller and eventually a nearly constant cost per analysis is obtained.

9.3. TIME ASPECTS

In order to determine the cost of an analysis the time required for the analysis has to be known (compare eqn. 9.3), and is therefore an important parameter. In addition, the time aspect in itself is important when judging the usefulness of the analysis (see Part IV). However, the time aspect consists of different parameters and a distinction has to be made between two characteristics. The dead time, t_d (or time lag), of an analysis can be defined as the time that elapses between the sampling and the reporting of the results. The second parameter defines the number of analyses per unit time that can be carried out by an analyst and/or with an instrument, and will be referred to as the (average) sampling time, t_a. It equals the (average) time between two successive samplings and thus the (average) time that elapses between the reporting of two successive results. The reciprocal of the sampling time is identical with the frequency of analysis and hence is a measure of the capacity of the procedure (operator and equipment). This frequency is, as has been observed, an important parameter for calculating the cost per analysis. However, it is not the only time parameter that is required for the calculation of the cost. For this calculation, it is also necessary to know the time utilized by the analyst (labour) and the instrument required for the analysis. This time is not necessarily identical with the dead time (or time lag) as defined above. An example may serve as illustration (a possible use of the parameters t_d and t_a will be treated in Part IV). A gas chromatographic procedure has a time lag equal to the time that elapses between the injection of the sample and the end of the elution of the last peak. However, if the sample is to be pre-treated in some way before subjecting it to the chromatographic separation, the time required for the pre-treatment has to be included in the time lag. Similarly, the time required for processing the chromatogram (calculation of the result) has to be included.

The time lag is essentially a parameter of importance when considered from the viewpoint of the person who requires the result, who is not interested in what has to be done in order to obtain that result but merely in the time he has to wait after he has supplied the sample.

The sampling time needs not be identical with the time lag. While the chromatograph is separating the sample into its components, the analyst often can prepare the next sample and calculate the results of the preceeding run. It is also clear that the time aspect required for calculating the cost of analysis is not necessarily equal to the time lag. It also need not be equal to the sampling time, as in many instances it is feasible that one analyst can operate several chromatographs simultaneously.

9.4. SOME RELATIONSHIPS BETWEEN CHARACTERISTICS

Some remarks should be made about the relationships between the parameters considered so far. Although in some instances these relationships can be stated as very clear and definite rules, they are often vague. Nevertheless, they have to be borne in mind when optimal procedures are to be established or when existing procedures are to be optimized. Optimization with respect to one parameter can easily make a procedure worse from another point of view. Fig. 9.2 shows schematically the relationships between some characteristics.

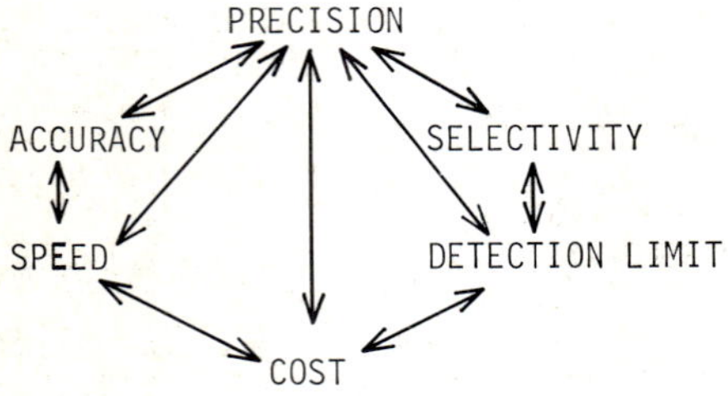

Fig. 9.2. Some relationships between the characteristics.

The following discussion is intended to clarify some of the relationships. A well described procedure has a certain precision, accuracy, etc. The precision of an insufficiently precise procedure can be improved in several ways, depending on the reason for the imprecision, one of the most common techniques being to carry out replicate analyses. Repeating a procedure n times results in a precision of the average result that is a factor $\sqrt{n}$ better than the precision of a result derived from a single analysis ; repeating a procedure also leads to an improvement in the signal-to-noise ratio. An application of this principle can be found in, for instance, nuclear magnetic resonance spectroscopy and is usually called signal averaging. This technique consists in measuring the spectrum 100 (or more) times and adding the spectra. Resonance peaks are simply added, whereas the (random) noise is magnified by only a factor $\sqrt{100}$ when 100 spectra are added. Obviously the time required for the measurement is increased 100-fold and the cost ascribed to the apparatus is increased by the same factor. It is clear that by the same technique the detection limit is also lowered. Another reason for a low precision can be a low selectivity of the procedure. An insufficiently selective procedure can result in inaccurate results. Inaccurate results for different samples in a number of instances can be regarded as a source of imprecision. These interactions can be circumvented by improving the selectivity of the procedure by, for instance, introducing a separation prior to the measurement or by introducing a more elaborate calibration procedure. However, such techniques will influence the speed and cost of analysis.

One is tempted to formulate some general rules for expressing the relationships between the characteristics. One of these rules might be that an increase in precision by a factor n will increase the cost by a factor n^2. However, it is doubtful whether the relationship between speed and precision will hold in general. Some procedures are more precise than others and some require less time than others because of the different principles on which they are based. Nevertheless, it can be considered as a general rule that a better precision and a reduction in the time of analysis will result in a higher cost of the analysis, due to either the use of more sophisticated instruments or more skilful labour. The development of more precise and rapid procedures may require more research and development.

REFERENCES

G. Bechtler, Organisation, Automatisation des Laboratoires - Biologie prospective, Colloque Pont à Mousson, G. Siest ed., Expansion scientifique française, Paris, 1972, p. 24.
A.W. Brown, Econ. Record., 45 (1969) 158.
J. Büttner, R. Borth, J.H. Boutwell, P.M.G. Broughton and R.C. Bowyer, Clin. Chim. Acta, 69 (1976) F1.
R. Haeckel, Rationalisierung des medizinischen Laboratoriums, G-I-T-Verlag Ernst Giebeler, Darmstadt, 1976.
T. Horne, Z. Anal. Chem., 252 (1970) 241.
F.A. Leemans, Anal. Chem., 43 (1971) 36A.

Chapter 10

CHARACTERIZATION OF CONTINUOUS PROCEDURES

10.1. CONTINUOUS VERSUS DISCONTINUOUS PROCEDURES

Most analytical procedures carried out in the laboratory can be considered as batch processes. Analysis by means of such procedures involves taking a certain amount of sample (for instance, by weighing), treating this sample in some way (heating, diluting, etc.) and performing one or more measurements on the pre-treated sample. Analysis of a discrete sample leads to one (set of) measurement(s) from which the identity or the composition of the sample can be derived. The analytical result is obtained with a certain precision and accuracy after a certain time, depending on the characteristics of the procedure used.

In contrast to these batch or discrete procedures, which for obvious reasons might also be called discontinuous procedures, a number of procedures are based on an entirely different way of handling the sample. These procedures involve continuous sampling, followed by continuous pre-treatment of and subsequent continuous measurements on the sample. As a result of continuous variations of the sample composition, the measurements will yield a continuously varying signal. The signal measured at a certain instant will not be related simply through the calibration constant to the sample compositon at that instant. The calibration function as used for the calculation of the sample composition from the measurement(s) has to be replaced by a time-dependent function. Such time-dependent functions are commonly used in, for instance, the electronics and process engineering and control fields. A concise discussion of the description of continuous procedures is given in this chapter ; for further details the reader is referred to textbooks on electronics (Connor, 1975), process engineering, control theory (Van der Grinten and Lenoir, 1973) or, in a more general sense, to (linear) systems theory (Zadeh and Desoer, 1963 ; Papoulis,

1965 ; Gabel and Roberts, 1973 ; Flagle et al., 1960).

In general, continuous methods can be applied only to the analysis of liquids, although in some instances application to solids is possible. Many analysers for control purposes were designed for the analysis of sample streams. Automated chemical analysis by using the continuous flow approach is nowadays also widely applied in the analytical laboratory, especially in the clinical laboratory. Some reviews on the application of continuous flow analysis were published by Blaedel and Laessig (1966), Skeggs (1966), Kies (1974), Snyder et al. (1976), and books by Siggia (1959) and Leithe (1964) also contain information on continuous analysis. Many applications and discussions of the principles, especially of the Technicon Auto Analyzer system,can be found in the series Advances in Automated Analysis (Technicon). The principles that will be described in the following sections apply not only to continuous procedures, but also to parts of procedures that operate on a continuous basis. For instance, detectors for gas chromatography, liquid chromatography, etc., can be described in the same way.

10.2. NOISE AND DRIFT

Noise and drift of (parts of) analytical instruments are familiar phenomena. These phenomena can pragmatically be defined as everything that contributes to the uncertainty of the measurement. It is, of course, easy to overcome them by using filters, but the quality of the analytical results is also influenced by these filters, in either a negative or a positive sense.

In order to characterize noise, or to quantify its magnitude, one can proceed in essentially the same manner as in estimating the precision for discontinuous analyses. When feeding a constant amount of sample to a continuous analyser, the continuous measurements will consist of a series of constant values if drift and noise are absent. In the presence of noise and drift, deviations from this constant value, or rather average value, of the signal will be observed. This effect is shown schematically in Fig. 10.1.

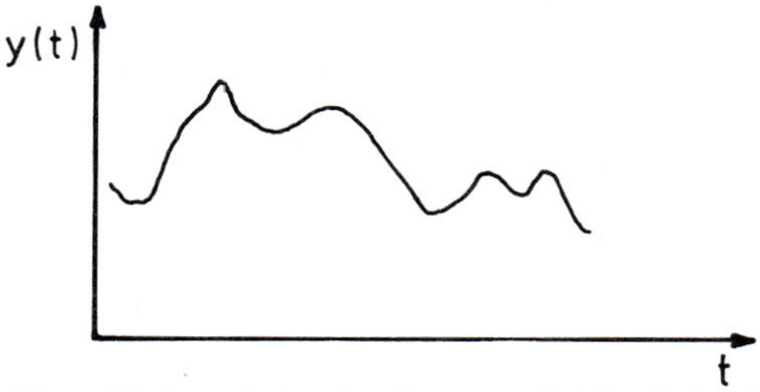

Fig. 10.1. Output of an analytical instrument for constant feed.

In agreement with the definition of precision in terms of the variance (or second moment), the magnitude of drift and noise is characterized by the variance, σ^2

$$\sigma^2 = \lim_{T\to\infty} \frac{1}{T} \int_0^T (\Delta y(t))^2 \, dt \cong \lim_{n\to\infty} \frac{1}{n} \sum_{i=1}^{n} (\Delta y(t_i))^2 \qquad (10.1)$$

where $\Delta y(t)$ is the difference between the signal at time t minus the average signal and n is the number of discrete measurements at times t_i.

Eqn. 10.1 can be used only for the characterization of stationary random fluctuations, where the term stationary implies that the expected values of the average signal and the variance are not affected by a shift in the time origin (Papoulis, 1965). In practice, the variance is estimated by taking the integral of eqn. 10.1 over a limited period of time. Then slow fluctuations (drift) certainly cannot be considered as random and stationary. In order to characterize the noise by its variance, the values of $\Delta y(t)$ must first be corrected for drift.

However, characterizing noise by its magnitude is not adequate for judging the practicability of continuous analytical procedures. A more elaborate characterization consists essentially of the quantitation of what is to be understood by quickly and slowly fluctuating signals. In a popularized way one can consider the fluctuating signal as being composed of a series of periodic sine or cosine functions, each with its own frequency and amplitude. (However, this picture can be misleading ; it certainly is incorrect in the mathematical sense, see Papoulis (1965)). Plotting the square of the amplitude versus the frequency yields the power spectrum of the noise. An example of a power spectrum, taken from a paper by Smit and Walg (1975), is shown in Fig. 10.2, where the power spectral density of a flame-ionization detector, expressed as the square of the output (noise) divided by the frequency is plotted versus the frequency, ν (cps, cycles per second). The value of this density function at

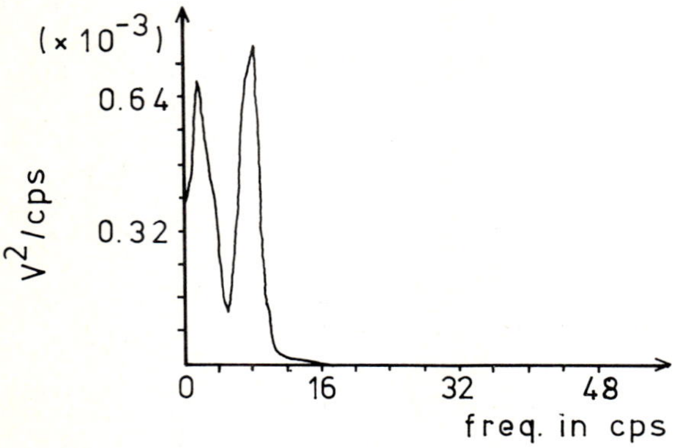

Fig. 10.2. Power spectrum of flame-ionization detector (Smit and Walg, 1975).

a frequency ν multiplied by dν represents the power of the noise with frequencies between ν and ν + dν. The density function might be determined by measuring the square of the output in the presence of filters with different frequencies cutting the noise above or below certain frequencies. From Fig. 10.2, it can be deduced that the application of a filter which cuts frequencies higher than 16 cps has virtually no effect on the magnitude of the noise. In this instance the power of the noise, which is equal to the squared amplitude or the variance, can be calculated by taking the integral of the density function over the interval 0-16 cps. Similarly, it can be seen that application of a filter which cuts frequencies above approximately 6 cps reduces the power of the noise by about half.

The information represented by the power spectrum can also be obtained by making use of auto-covariance functions or the related auto-correlation functions. When the nature of the fluctuations of the signal is random, and when one is dealing with stationary noise, the expected deviations (Δy) are determined largely by the variance of the noise and it is possible to give the probability of a certain deviation occurring at any instant. However, if at a certain time t the deviation $\Delta y(t)$ is known, the deviation $\Delta y(t+\Delta t)$ can only assume a value close to $\Delta y(t)$ unless Δt is large. This correlation can be expressed by the auto-covariance function, which is defined as

$$\Gamma\,(\Delta y(t),\ \Delta y(t+\Delta t)) = \Gamma\,(\Delta t) = \lim_{T\to\infty} \frac{1}{T} \int_{0}^{T} \Delta y(t).\Delta y(t+\Delta t).dt \qquad (10.2)$$

The auto-covariance is given by the average of the product of two deviations from the average value of the signal $[\Delta y(t)$ and $\Delta y(t+\Delta t)]$ separated by a time interval Δt. The auto-covariance for $\Delta t = 0$ is equal to the variance as defined by eqn. 10.1. The auto-covariance function describes the degree of coherence or the correlation between the magnitude of the signals at different time intervals. The correlation of the signal with itself is clearly larger than the correlation with the signal observed at some later stage. For large time intervals (and randomly fluctuating signals), there is no relationship between the two values observed. Then a combination of two positive deviations is equally probable as the combination of a positive with a negative deviation. Consequently the average of the product $\Delta y(t).\Delta y(t+\Delta t)$ for large Δt will be zero and the resulting value of the covariance is zero. An example of a covariance function is shown in Fig. 10.3.

Figs. 10.2 and 10.3 represent the same information about the flame-ionization detector, i.e., information about the magnitude and speed of the fluctuations. Mathematically, the relationship between the two functions is a Fourier transform (Wiener-Khinchin relations). The Fourier transform of the power spectrum yields the auto-covariance function (Papoulis, 1965). It is not our intention to consider the mathematical details, but rather to show the use of both functions. As Figs. 10.2 and 10.3 represent rather complicated functions, this use can be demonstrated better with some idealized functions. If the

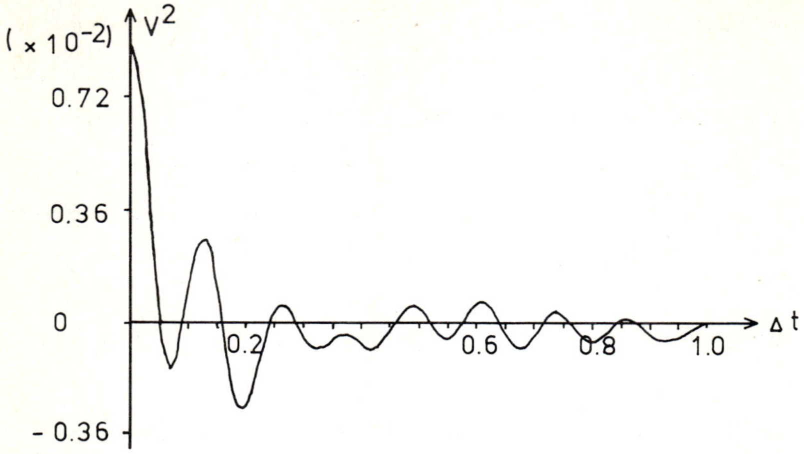

Fig. 10.3. Auto-covariance function of flame-ionization detector noise. (Smit and Walg, 1975).

auto-covariance function is exponential, the corresponding power spectrum is flat up to a certain frequency ν_o (see Fig. 10.4). The (time) constant or correlation time, τ_y, of the exponential auto-covariance function

$$R(\Delta t) = s^2 \, e^{-\Delta t/\tau_y} \tag{10.3}$$

is related to this frequency (in cps if τ_y and Δt are expressed in seconds ; it also can be expressed as the angular frequency ω_o in radials per second, rps) by the simple relationship $\omega_o = 1/\tau_y$ shown in Fig. 10.4. Clearly τ_y and ν_o are the same measure for the speed of the fluctuating process, whereas s^2 characterizes the magnitude.

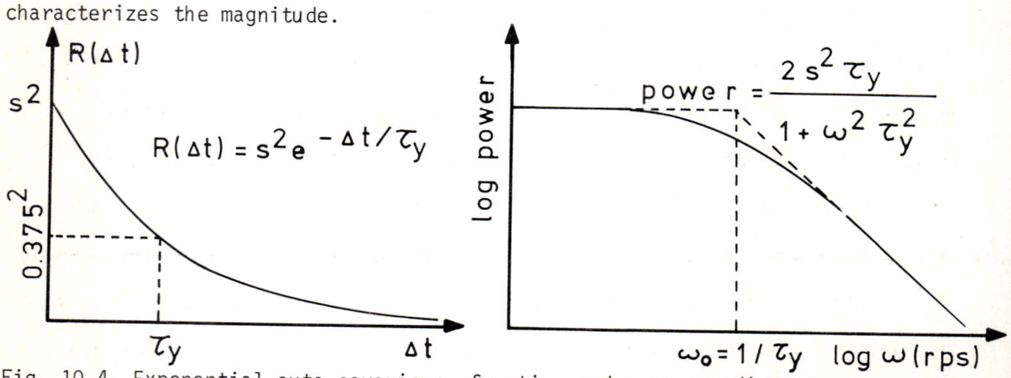

Fig. 10.4. Exponential auto-covariance function and corresponding power spectrum.

Applications of the auto-covariance function are discussed in more detail in Chapter 26 (Part IV).

10.3. RESPONSE AS A FUNCTION OF TIME

Continuous procedures are designed to measure (changes in) the composition of a continuous sample stream, and noise adds a fluctuation to the signal that cannot be ascribed to this composition. Hence translation of the (total) signal into sample composition by means of the calibration function leads to imprecise results. In this respect there is no essential difference with discontinuous procedures. However, when using continuous procedures the presence of noise is not the only reason for erroneous results being produced.

In general, instruments that operate continuously require some time to reach an equilibrium that corresponds to the sample composition. Continuous procedures cannot adequately measure the composition of a sample stream when the composition is changing rapidly. The magnitude of the fluctuations as shown by the instrument is in general smaller than the real fluctuations that would be obtained when applying the calibration function. Measuring with a slow, continuously working instrument is essentially the same as applying a filter in order to remove the noise. It clearly is desirable not to lose the information that has to be gathered by the continuous instrument. The response of the instrument has to be fast with respect to the fluctuations that have to be measured, which requires a quantification of what is to be understood by fast. As has been shown, the time aspect of the fluctuations can be quantified by the covariance function or the power spectrum (analogous to the description of the noise). The response of a continuous instrument can be characterized in essentially two ways : description by a time parameter or by a frequency parameter.

As an example to illustrate the response of a continuous procedure we can consider a flow cell as used in many instruments that are operated continuously. In some instances such a flow cell acts as a so called ideal dilutor (i.e., with very good mixing characteristics). Here the sample (or rather part of the sample stream) entering the cell is mixed with the entire contents as soon as it enters the cell. It will be clear that a sudden (stepwise) change of concentration in the sample stream from, say, 0 to a causes a gradual change from concentration 0 to a in the sample cell. If the volume of the flow cell is represented by V (ml) and the flow by v (ml/sec), the differential equation describing the concentration x(t) in the flow cell as a function of time is

$$\frac{dx}{dt} = \frac{v}{V}\,(a-x(t)) \tag{10.4}$$

At time t = 0 the concentration of the solution entering the cell changes from 0 to a and the concentration in the cell is given by

$$x(t) = a\,(1-e^{-\frac{v}{V}t}) \tag{10.5}$$

The dynamic behaviour of this flow cell is first order. The constant characterizing this first-order process is v/V and has the dimensions of reciprocal time ; this time is called the time constant of the cell ($\tau_a = \frac{V}{v}$). A small time constant and thus a fast response is obtained for large flow-rates and small volumes. If the flow cell, for instance, serves as a cell for measuring the absorbance A and if the measurement of the absorbance is very fast, the total response of the flow cell and transducer for measuring this absorbance is given by

$$y(t) = A(t) = Sx\,(1-e^{-t/\tau_a}) \tag{10.6}$$

where S is the sensitivity as introduced in Chapter 6. Fig. 10.5 serves as an illustration of such a response. From eqn. 10.6 and Fig. 10.4, it is clear that after a time equal to the time constant τ_a the response is 0.63 Sx. After a time $5\tau_a$ the value of y(t) is within 1% of its final value.

In practice, the behaviour of many (parts of) continuous procedures can be approximated by first-order processes, although the physical model may be very different from the one described. These procedures are adequately characterized by a first-order time constant τ_a. However, not all procedures

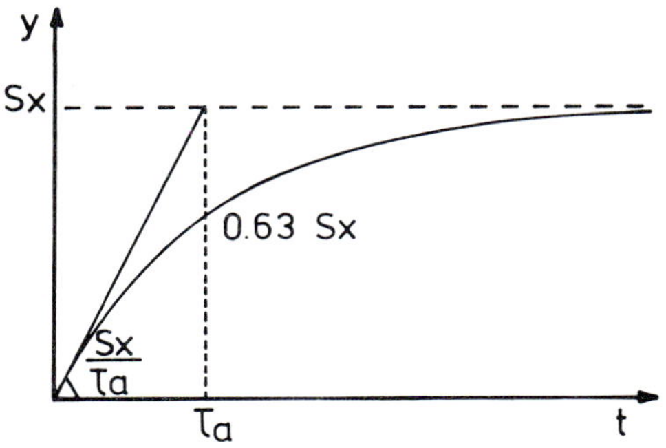

Fig. 10.5. First-order response of continuous procedure.

behave as first-order processes and therefore another characteristic parameter has to be introduced. In some instances a sudden change in the composition of the feed does not cause a response until a time t_d after the disturbance in concentration has been imposed. A physical model that is approximately valid for this type of response is a tube of certain length through which the sample stream is moving. A disturbance at one end at the tube will manifest itself at the other end of the tube at a time dependent on the length of the tube and the linear velocity of the moving liquid or gas. This time, t_d, is usually referred to as the dead time or time lag of the system.

It also is possible, and in many instances advantageous to describe the response of continuous procedures in another way. Instead of describing the response to stepwise changes in the concentration, the response to concentrations that fluctuate sinusoidally can be observed. Such a periodically changing feed will cause a periodically changing signal, a sine wave with the same frequency. This is only true for systems that can be described by a linear differential equation with constant coefficients. If the frequency is low the instrument will follow the changes in concentration such that the response at any time is given by the calibration function, $y(t) = Sx(t)$. For higher frequencies, the instrument cannot follow the changes and there will be a phase difference between the input and the output. Moreover, the amplitude of the signal fluctuations is smaller than the amplitude that might be expected from the calibration function. Before the measurement has reached its maximum possible response, the concentration is already decreasing, and before the minimum is reached the concentration is increasing again.

The phase difference and amplitudes are functions of the frequency. For a first-order process these dependences are as shown in Fig. 10.5 ; such diagrams usually are called Bode-diagrams. The frequency at which the response sharply decreases is here equal to the band width ; its relationship with the time constant is shown in Fig. 10.6. An instrument characterized by a time lag t_d in principle has an infinite band width. Sine waves of all frequencies pass undisturbed through the instrument, although the pass is characterized by a time delay (phase shift).

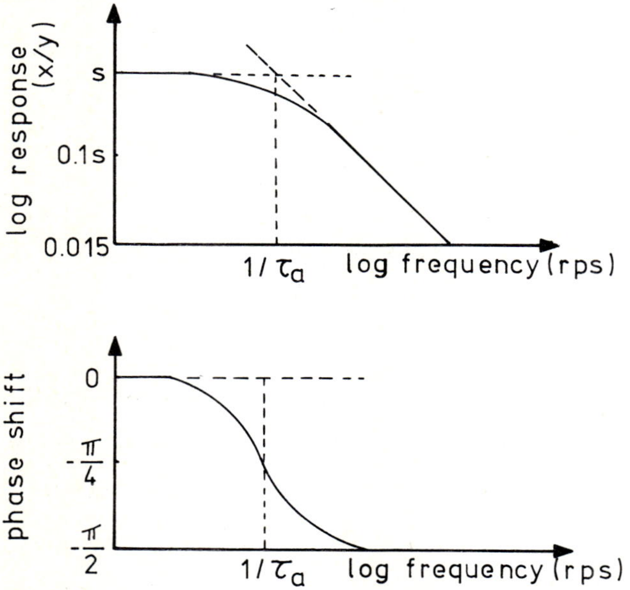

Fig. 10.6. Frequency response of first-order process.

We shall not consider the mathematical details, nor shall we discuss higher order processes. We shall remark only that the response as expressed by Fig. 10.6, is related to the step response of Fig. 10.5 (through Fourier and Laplace transforms).

10.4. DISCUSSION

Some aspects of continuous flow systems for automated chemical analysis may serve as an illustration of the use of the characteristics introduced in this chapter. Many systems of this type are in use in analytical and clinical laboratories where discrete samples have to be analysed. In order to be able to use a continuous flow system these discrete samples have to be converted into sample streams, which in practice is effected by creating a stream of sample 1 during a limited period of time followed by a stream of blank (wash fluid), a stream of sample 2, a stream of blank, and so on. The result is a series of stepwise changes in concentration, from 0 to x_1, from x_1 to 0, from 0 to x_2, etc., with a time interval Δt between the steps. If the continuous flow system shows a first-order response, the signals y measured as a function of time will show a pattern as represented by Fig. 10.7. Here the distance between the steps is about $5\tau_a$, with the result that a small plateau with a virtually constant response corresponding to y = Sx is obtained. If the distance is smaller than $5\tau_a$ the maximum response is not obtained. It is clear that the sampling time has to be at least twice $5\tau_a$ and the sampling frequency will be the reciprocal of that value. Hence a decrease in τ_a will permit a larger number of samples to be analysed within a certain time span with the same instrument.

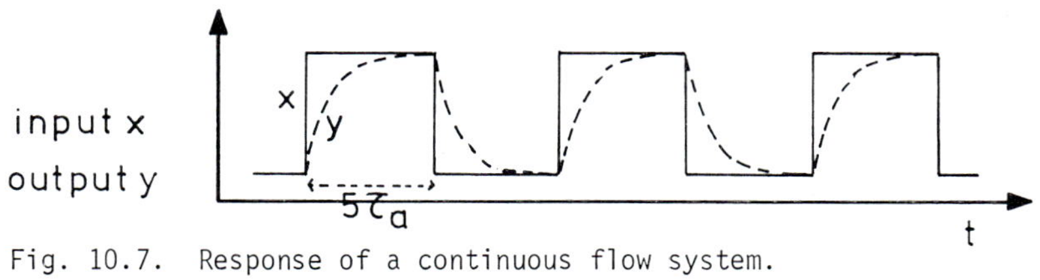

Fig. 10.7. Response of a continuous flow system.

In reality the response is more complicated owing to a more complicated pattern of the stream in the instrument. A more detailed analysis of the dynamic behaviour as reflected by the response can serve as an aid in improving the instrument. A discussion of these aspects is beyond the scope of this chapter, and the reader is referred to the paper by Snyder et al. (1976).

Although some aspects of the use of analytical procedures for monitoring purposes are treated in Chapter 26, a brief discussion on the characteristics introduced in relation to this monitoring is useful here. Variance and covariance functions have been introduced to characterize the output of the continuous procedure. If a stream with a constant concentration is fed into the instrument over a long period of time (long with respect to the time constant of the covariance function), the noise can be decreased by the process of averaging or filtering. The precision therefore can be increased at the expense of time. However, if the concentration is constantly varying, such a filtering procedure cannot be applied. There is always a risk that not only the noise is disappearing but also the (unknown) variations in the concentration. Filtering without losing essential information is possible only when the power density function of the variations to be measured does not appreciably overlap with the power density function of the noise.

The discussion of special correlation techniques in chromatography and spectroscopy falls outside the scope of this book. However, it is interesting to note that instead of multiplying the signal by itself as measured Δt earlier, which leads to the auto-covariance or auto-correlation, the signal (output) can also be multiplied by the feed (input) at a time Δt. In that case a cross-correlation between the input and output is obtained. This can be applied in chromatography when a special sample introduction technique is used ; it is called correlation chromatography and results in the enhancement of the signal-to-noise ratio (or a decrease in the limit of detection) (Annino, 1976). Similar applications of correlation techniques are used in spectroscopy (Horlick, 1973). Also outside the scope of this book falls the study of the characteristics of the continuous flow systems in order to improve the performance of such systems (see for instance, Snyder and Adler, 1976 a and b).

10.5. STOCHASTIC PROCESSES (MATHEMATICAL)

A stochastic process y is defined as a function which associates a random variable y(t) to each instant t

$$y : t \rightarrow y(t) \qquad (10.7)$$

The process is called stationary if the following three conditions are fulfilled

$$E\,(y(t)) = \mu(t) = \mu \qquad (10.8)$$

$$E\left[(y(t) - \mu)^2\right] = \sigma^2(t) = \sigma^2 \qquad (10.9)$$

$$E\left[(y(t) - \mu)\;(y(t+\Delta t) - \mu)\right] = \lambda(\Delta t) \qquad (10.10)$$

Conditions 10.8 and 10.9 indicate that the mean value and variance do not change with time. Condition 10.10 indicates that the covariance between two random variables y(t) and y(t+Δt) taken from the process depends only on the time interval Δt.

When the probability function of the random variables y(t) are unknown, these parameters must be estimated by using a realization of the process during a sufficiently long period of time. Such a realization is called a time series. In this way, the mean value $\overline{y}$ can be found from

$$\overline{y} = \frac{1}{T}\int_0^T y(t)\,dt \cong \frac{1}{n}\sum_{i=1}^{n} y(t_i) \qquad (10.11)$$

It is an estimate of the true mean value given by

$$\mu = \lim_{T\to\infty} \frac{1}{T}\int_0^T y(t)\,dt \qquad (10.12)$$

The variance σ^2 is given by

$$\sigma^2 = \lim_{T\to\infty} \frac{1}{T}\int_0^T (y(t)-\overline{y})^2\,dt = \frac{\int_0^\infty \Delta y(t)^2\,dt}{\int_0^\infty dt} \qquad (10.13)$$

where $\Delta y(t)$ is the deviation from the mean value $\overline{y}$

$$\Delta y(t) = y(t) - \overline{y} \tag{10.14}$$

The auto-covariance function given by eqn. 10.10 can be written as

$$\Gamma(\Delta y(t),\Delta y(t+\Delta t)) = \Gamma(\Delta t) = \lim_{T\to\infty} \frac{1}{T} \int_0^T \Delta y(t)\, \Delta y(t+\Delta t)\, dt \tag{10.15}$$

$$= \frac{\int_0^\infty \Delta y(t)\, \Delta y(t+\Delta t)\, dt}{\int_0^\infty dt}$$

It can be estimated by means of $C(\Delta t)$, given by

$$C(\Delta t) = \frac{1}{T} \int_0^T \Delta y(t)\, \Delta y(t+\Delta t)\, dt \cong \frac{1}{n} \sum_{i=1}^{n} \Delta y(t_i)\, \Delta y(t_{i+1}) \tag{10.16}$$

The auto-correlation function which expresses the auto-covariance in terms of the variance can then be estimated from

$$R(\Delta t) = \frac{C(\Delta t)}{s^2} \tag{10.17}$$

where s^2 is an estimate of σ^2 given by

$$s^2 = \frac{1}{T} \int_0^T (y(t)-\overline{y})^2\, dt \cong \frac{1}{n} \sum_{i=1}^{n} (y(t_i)-\overline{y})^2 \tag{10.18}$$

This of course far from exhausts the subject of time series analysis. In recent years the subject has witnessed a wide variety of developments. These can be divided into several important types of approaches among which spectral analysis and time domain analysis are the most important. In the latter type the regression methods, the smoothing methods and the Box-Jenkins technique are especially worth mentioning.

Readers interested in these developments can find details in the books by Box and Jenkins (1970), Nelson (1973), Koopmans (1974) and Papoulis (1965).

REFERENCES

R. Annino, J. Chromatogr. Sci., 14 (1976) 265.

W.J. Blaedel and R.H. Laessig, in N. Reilly and F.W. McLafferty (Editors), Advances in Analytical Chemistry, Vol. 5, Interscience, New York, 1966, p. 69.

G.E.P. Box and G.M. Jenkins, Time series analysis. Forecasting and Control, Holden Day, 1970.

F.R. Connor, Signals, Arnold, London, 1975.

C.D. Flagle, W.H. Huggins and R.H. Roy, Operations Research and Systems Engineering, John Hopkins, Baltimore, 1960.

R.A. Gabel and R.A. Roberts, Signals and Linear Systems, Wiley, New York, 1973.

G. Horlick, Anal. Chem., 45 (1973) 319.

H.L. Kies, in J. Penciner (Editor), Reviews in Analytical Chemistry, Vol. 2, Scientific Publications Division, Freund Publishing House, Tel Aviv, 1974, pp. 9, 167, 229.

L.H. Koopmans, The Spectral Analysis of Time Series, Academic Press, 1974.

W. Leithe, Analytische Chemie in der Industriellen Praxis, Akademie Verlag, Frankfurt, 1964.

C.R. Nelson, Applied Time Series Analysis for Managerial Forecasting, Holden Day, 1973.

A. Papoulis, Probability, Random Variables and Stochastic Processes, McGraw-Hill, New York, 1965.

S. Siggia, Continuous Analysis of Chemical Process Systems, Wiley, New York, 1959.

L.J. Skeggs, Anal. Chem., 38 (1966) 31A.

H.C. Smit and H.L. Walg, Chromatographia, 8 (1975) 311.

L.R. Snyder and H.J. Adler, Anal. Chem., 48 (1976 a) 1017.

L.R. Snyder and H.J. Adler, Anal. Chem., 48 (1976 b) 1022.

L. Snyder, J. Levine, R. Stoy and A. Conetta, Anal. Chem., 48 (1976) 942A.

P.M.E.M. van der Grinten and J.M.H. Lenoir, Statistische Procesbeheersing, Het Spectrum, Utrecht, 1973.

L.A. Zadeh and C.A. Desoer, Linear System Theory, McGraw-Hill, New York, 1963.

PART II

EXPERIMENTAL OPTIMIZATION

Chapter 11

SURVEY OF EXPERIMENTAL OPTIMIZATION METHODS

11.1. INTRODUCTION - THE NEED FOR FORMAL METHODS

One of the most general problems in analytical chemistry is the optimization of an existing procedure, the object being to optimize (maximize or minimize) the response of this procedure. Response in this context is to be understood as the quantity used to evaluate the method, i.e., the optimization criterion. In many instances this will be (or should be) one of the evaluation criteria (precision, sensitivity, etc.) discussed in Part I or a physical quantity related to one of these criteria (for example, optical absorbance, related to sensitivity). The optimization consists in the selection of the values of the parameters (continuous and discrete) such that the best possible response is obtained. As an example, consider the colorimetric determination of phosphate and suppose that the optimization criterion is the optical absorbance. One then wants to know the optimal concentrations of the molybdate and reducing agent, the optimal time for colour development, whether one should use ascorbic acid or tin (II) chloride for the reduction, whether an extraction step should be introduced, etc.

In general, there are two kinds of questions, namely :

(a) what is the optimal value of a parameter (such as a reagent concentration or a reaction time) ;

(b) what are the best experimental circumstances or the best attributes (should one add an extraction step, what is the best reagent, etc.).

It is sometimes necessary to ask the two kinds of questions together. Category (a) questions can be solved with the methods described in all the chapters of this part of the book. Questions of category (b) can be solved only with the factorial methods of Chapters 12 and 13.

Before describing the formal methods that have been developed for optimization in an optimal way, let us consider first how an optimization such as the determination of phosphate would usually be carried out in practice. It is assumed that only two parameters are of importance, namely the concentrations of ammoniummolybdate and the reducing agent. Probably the investigator will keep one of these factors constant and determine the optimal value of the other, then at this value he will determine the optimal value of the former. This is shown in Fig. 11.1, where A is the start (initial conditions) of the optimization procedure or search and O the optimum to be attained.

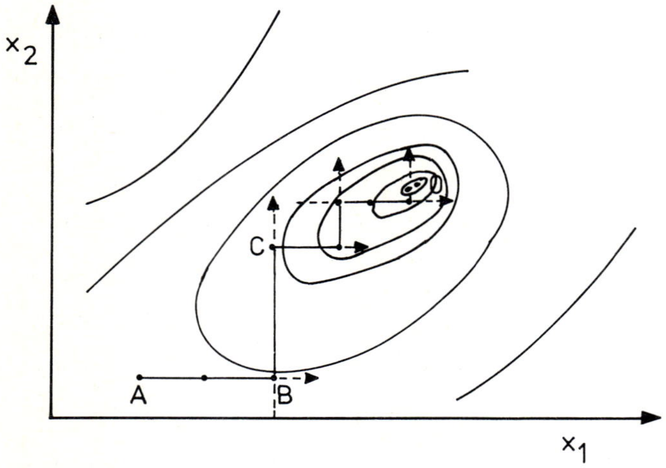

Fig. 11.1. Univariate search procedure . A = Starting point ; O = optimum (based on D.L. Massart et al., 1977)

Handling the optimization in this way has several disadvantages, the most important of which are the following

(1) Other and more important factors may influence the result ;

(2) The optimal value found is not the real optimal value because the factors interact. In Fig. 11.1, B would be obtained by optimizing factor x_1 at constant x_2 and C by subsequent optimization of x_2 at the value of B for x_1. Clearly, C is not optimal at all. The optimum could have been obtained by repetition of this "one factor at a time", single-step or univariate search procedure.

However, this necessitates a large number of experiments, particularly when more than two factors have to be considered. This leads us to the third main disadvantage :

(3) The univariate optimization strategy wastes labour because it is inefficient in the sense that a larger number of experiments than necessary are carried out.

Formal optimization methods usually allow one to avoid these difficulties ; they yield more information and require less work.

Nalimov (1972) remarked that one of the more noticeable trends in modern science is to pass on from the study of well organized systems to badly organized or diffuse systems. From the beginning of modern science in the 17th century to the middle of the 20th century, workers in the physical sciences tried to study well organized systems with as small a number of variables as possible. During many centuries it was impressed on new investigators that the one-factor experiment is the only acceptable one. This belief still holds and one has to consult only a few analytical chemistry journals to encounter many examples. Mathematical statistics and systems theory initiated a change in these beliefs from around 1935. In analytical chemistry, this has resulted in several different approaches to optimization.

One can carry out the kind of optimization discussed in this part of the book in two general ways :

(i) The analytical approach, where the word "analytical" is used in its mathematical sense. In analytical chemistry, this means that one has to identify the underlying physico-chemical principles and to develop an exact equation that describes the process.

(ii) The black box approach : one considers the method from the purely experimental side, i.e., one observes the effects of changing the factors on the response, and one does this by changing all of the factors more or less at the same time.

These are the methods we wish to discuss in this book and we call them formal optimization methods. It should be noted here that the black box is a notion stemming from systems theory and that it will be discussed in more detail in Part V.

The first method can be called the semi-analytical approach, and consists in describing approximately the process studied using a mathematical equation obtained by regression techniques. Most so-called simultaneous optimization procedures (see section 11.2) belong to this category. In the second approach one does not try to understand the procedure, not even in an approximate way. The sequential optimization procedures described in Chapter 14 are examples of this pure black-box approach.

11.2. OPTIMIZATION STRATEGIES

An optimization method consists of three stages. In the first stage, one decides on the objective function (or response) according to which the method (the output of the black box) will be judged. As stated above, this will usually consist of one of the evaluation criteria discussed in Part I. Often, one criterion will be insufficient and composite criteria, such as cost per unit of time, will be employed.

We have already discussed the difficulty of choosing a criterion in Part I, and we have tried to show how some of the criteria can be formulated in a formal way. However, no general methods are available at present for optimizing methods subject to two or more criteria. Some possible means of doing this are discussed in the chapter on multicriteria analysis in Part III. For the optimization procedures described in Part II, one will always have to choose a single criterion that will constitute the objective function. In the second stage, one decides the factors that have an influence on the objective function. In many instances the processes underlying the method are sufficiently well known to be able to decide this without experimentation. When this is not so, one investigates the method using techniques such as the analysis of variance or factorial experimentation. The latter not only allows one to identify those parameters which have an effect but also whether these factors are independent or, on the contrary, interact. Factorial experimentation is discussed in Chapter 12. The final stage is the optimization itself.

When the response is plotted as a function of two factors, a response surface results and the optimization consists in finding a maximum or minimum on this surface. It can be represented as in Fig. 11.2 or else using contour lines, such as in Fig. 11.1. Response-surface concepts can be generalized (but not represented visually) for more than two factors.

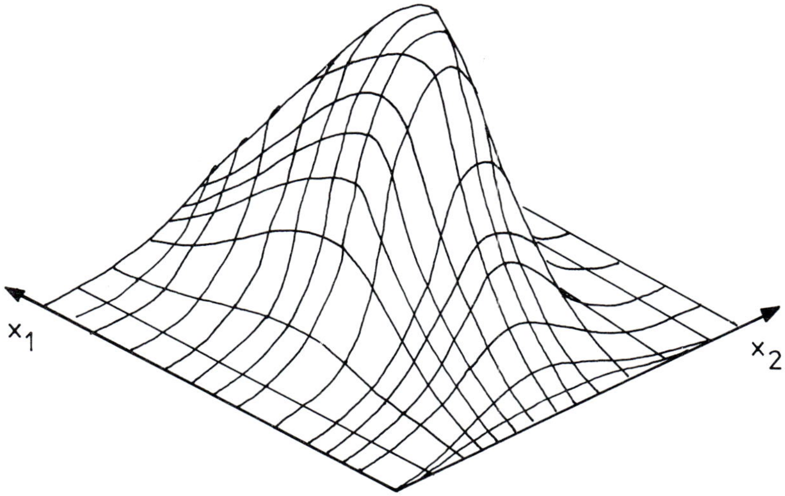

Fig. 11.2. Response surface for two factors.

One can choose among many different strategies or designs. Usually, one makes a distinction between simultaneous (or pre-planned) and sequential designs (see also section 11.1). The former entails carrying out a rather large number of experiments according to a pre-arranged plan ; factorial design is a typical example. A sequential design consists in carrying out only a few, often only one, experiments at a time and using these to determine the experiment to be carried out next ; the Simplex method is an example. Mixed approaches also exist. Simultaneous optimization strategies are discussed in Chapter 13 and sequential strategies in Chapter 14. In some instances, one makes use of the data assembled in the manner described in Chapters 12 - 14 to try and describe the response surface or the gradient along it, for example by regression analysis (see the semi-analytical approach, section 11.1). This is described in Chapter 15.

REFERENCES

D.L. Massart, H. De Clercq and R. Smits, Reviews on Analytical Chemistry, W. Fresenius ed., Akademiai Kiado, Budapest, 1977.
V.V. Nalimov, Reinstoffe in Wissenschaft und Technik, Akademie-Verlag, Berlin, 1972.

Chapter 12

FACTORIAL ANALYSIS

12.1. DESCRIPTION OF THE METHOD

In Chapter 4, we have seen that ANOVA can be used to investigate which parameters of a procedure have an influence on the result. A linear model such as

$$y_{hijk} = \mu + \alpha_h + \beta_i + \gamma_k + e_{hijk} \tag{12.1}$$

can then be written. The quantity y_{hijk} is thought to be composed of a mean value, effects α_h, β_i and γ_k and an error e_{hijk}. If the ANOVA confirms that the effects are significant, one knows that one should optimize the values of parameters A, B and C responsible for effects α_h, β_i and γ_k. Several ways of doing this are described in the following sections. Let us note here that eqn. 12.1 gives a descriptive model, which allows one to test the significance of the parameters (factors) but which cannot be used directly for optimization purposes. Regression equations such as

$$y = b_o + b_A x_A + b_B x_B + b_C x_C \tag{12.2}$$

on the contrary can be used directly for optimization purposes. In eqn. 12.2 y is again the signal or measurement value to be optimized, x_A, x_B and x_C are values of parameters A, B and C and b_o, b_A, b_B and b_C are the estimates of β_o, β_A, β_B, and β_C in the model

$$y = \beta_o + \beta_A x_A + \beta_B x_B + \beta_C x_C \tag{12.3}$$

The optimal value (i.e., the highest or lowest value) of y can be obtained from eqn. 12.2 and yields the optimal values of the parameters A, B and C. This is described in detail in Chapter 15. It must be noted here that in many practical instances this model is too simple, mainly because it assumes that the parameters are independent. It is found that it is usually not correct for optimization experiments. In fact, we have already seen an example of interactions in Chapter 4 (laboratory-sample interaction). In this chapter we shall investigate how to take this into account and how to construct a more realistic ANOVA model than that given in eqn. 12.1. The means of completing regression eqn. 12.2 is described in Chapter 15. ANOVA can be carried out in such a way that not only the effects of single parameters but also the interactions among them are detected. In Chapter 4 we have seen that this kind of ANOVA can be called factorial analysis. Before investigating how to carry out a factorial analysis, let us consider an example of interacting parameters in an extraction procedure such as the extraction of a metal ion with a chelate-forming agent such as dithizone in the presence of a sequestering agent such as EDTA. Both the pH and the concentration of the sequestering agent determine the extent of extraction. The effect of the pH is, however, not the same at high and low concentrations of EDTA and they are therefore interacting parameters.

Let us assume now that a measurement may depend on three factors. To investigate if these are significant, one carries out a so-called factorial experiment for three factors at two levels, which means that the effects of three factors are investigated at two values (levels) of each. This then constitutes a 2^3 design. In Table 12.1, A, B and C are the factors and + and - indicate a measure at the higher or the lower level, respectively. The numbers in the table represent the eight experiments. In the example of the phosphate determination given in the previous chapter, A could be the concentration of ammonium molybdate and the + and - levels would represent 1.0 and 0.5 M, respectively.

Let us consider, for example, the effect of factor A. If one compares experiments 1 and 5, one observes that in both experiments the levels at which B and C are measured are the same but that for A two different values are used. The difference between the results obtained with these experiments is therefore

Table 12.I

A 2^3 factorial design

Experiment *	A	B	C	Result
1	+	+	+	y_1
2	+	+	-	y_2
3	+	-	+	y_3
4	+	-	-	y_4
5	-	+	+	y_5
6	-	+	-	y_6
7	-	-	+	y_7
8	-	-	-	y_8

* In the terminology of the literature on factorial analysis, the experiments are often called treatments.

an estimate of the effect of A when B and C are at the + level. The difference between the results obtained in experiments 2 and 6 constitutes another estimate of the effect of A, this time at the + level for B and the - level for C. In total, four estimates for the effect of A can be obtained and an average effect of A can be calculated using all eight experiments. The effect of A can be estimated from

$$\frac{1}{4}\left[(y_1 + y_2 + y_3 + y_4) - (y_5 + y_6 + y_7 + y_8)\right] \qquad (12.4)$$

In the same way, the other main effects can be investigated. The next question that should be asked is whether the effect of, for instance, A depends on B (i.e., do A and B interact ?).

One can re-state the conclusion about the difference $y_1 - y_5$ obtained in experiments 1 and 5 in the following way : $y_1 - y_5$ is an estimate of the effect of A at the higher level of B for constant C. The difference between y_3 and y_7 is then an estimate of the effect of A at the lower level of B for the same constant value of C, and

$$\frac{1}{2}\left[(y_1 - y_5) - (y_3 - y_7)\right] = \frac{1}{2}\left[(y_1 + y_7) - (y_3 + y_5)\right] \qquad (12.5)$$

can then be used to evaluate whether the effect of A is the same at both levels of B.

One can also interchange the letters A and B and estimate whether the effect of B is the same at both levels of A. This too is an estimate of the interaction between A and B (written as A x B or AB) and, without going into detail, it appears that this is also given by

$$\frac{1}{2}\left[(y_1 - y_3) - (y_5 - y_7)\right] = \frac{1}{2}\left[(y_1 + y_7) - (y_3 + y_5)\right]$$

A second estimate of the effect can be obtained at the low level of C, and is given by

$$\frac{1}{2}\left[(y_2 + y_8) - (y_6 + y_4)\right]$$

The interaction (A x B) can therefore be obtained from

$$\frac{1}{4}\left[(y_1 + y_2 + y_7 + y_8) - (y_3 + y_4 + y_5 + y_6)\right] \qquad (12.6)$$

Again, all of the results are used to evaluate the interaction A x B, and the other two-factor interactions (AC, BC) can be handled in the same way. One can then proceed to investigate whether the interaction of A and B is the same at low and high levels of C. If this is not so, a three-factor interaction ABC exists.

It becomes tedious to write down the equations that are necessary in order to obtain all of these effects, particularly if four or more factors are used. For the four-factor model, there are four main effects (A, B, C, D), six two-factor interactions (AB, AC, AD, BC, BD, CD), four three-factor interactions (ABD, ABC, ACD, BCD) and one four-factor interaction (ABCD). An easier method of obtaining the estimates is therefore necessary, and to do this we write the factorial experiment of Table 12.I. in another way (Table 12.II).

The levels for the interactions in Table 12.II are obtained by multiplication according to the usual algebraical rules. For example, experiment 4, with - levels for B and C and a + level for A, yields a (-) x (-) = + level for the BC interaction and a (+) x (-) x (-) = + level for the ABC interaction. The effects are then obtained (apart from the 1/4 factors) by subtracting the results obtained at the - level from the results obtained at the + level. This can be verified by writing down the estimates for A or AB directly from Table 12.II and comparing them with the equation given earlier. The calculation of the estimates of the effects is therefore relatively simple.

Table 12.II

A 2^3 factorial design with interactions

Effect	1	2	3	4	5	6	7	8
A	+	+	+	+	-	-	-	-
B	+	+	-	-	+	+	-	-
C	+	-	+	-	+	-	+	-
AB	+	+	-	-	-	-	+	+
AC	+	-	+	-	-	+	-	+
BC	+	-	-	+	+	-	-	+
ABC	+	-	-	+	-	+	+	-
Result	y_1	y_2	y_3	y_4	y_5	y_6	y_7	y_8

Several other schemes allow the easy calculation of the effects in two-level designs. One of these was applied by Kamenev et al. (1966 a) in a study on the effect of the composition of the supporting electrolyte on the anodic polarographic peaks of some cations. He distinguished the following four factors : cation radius of the supporting electrolyte ; valence of cation of the supporting electrolyte ; anion of the supporting electrolyte ; and test element. Three of these were studied in a full factorial experiment (in fact, a half replication, see the next chapter, for four factors, was used but for our purpose we may consider this here as a full factorial for three factors). For the experimental design given in Table 12.III, the calculations can be carried out conveniently as shown in Table 12.IV.

Starting with the column "Result", one successively adds together the results two by two and writes these results in column 1. For example 998 is the sum of 568 and 430, 918 the sum of 394 and 524, etc.

Table 12.III

Experimental design and results obtained by Kamenev et al. (1966 a)

Experiment	Factor A	B	C	Result (sum of six replicates)
1	-	-	-	568
2	+	-	-	430
3	-	+	-	394
4	+	+	-	524
5	-	-	+	399
6	+	-	+	581
7	-	+	+	588
8	+	+	+	434

Table 12.IV

Calculation of the effects from the experimental design in Table 12.III (based on Kamenev, 1966 a)

Experiment	Result	1	2	3	Effect
1	568	998	1916	3918	Sum
2	430	918	2002	20	A
3	394	980	-8	-38	B
4	524	1022	28	-68	AB
5	399	-138	-80	86	C
6	581	130	42	36	AC
7	588	182	268	122	BC
8	434	-154	-336	-604	ABC

Then one subtracts the first result from the second, the third from the fourth, etc. These results are also added to column 1. For example, -138 is obtained by subtracting 568 from 430. One now proceeds in the same way with the results of column 1, obtaining in this way the results in column 2, and eventually using the results from column 2 to obtain the results in column 3. These can then be identified with the effects given in the last column. One can easily verify that the result obtained, for example, in column 3, row 2 is the result of summing the results of experiments 2, 4, 6 and 8 and subtracting those of experiments 1, 3, 5 and 7. This is indeed a measure of the effect of factor A. A second important step in the interpretation of the results consists in testing the significance of the observed effects. In the two-level case two possibilities exist. One can apply a t-test to compare for each effect, interactions included, the results obtained at the + level with those obtained at the - level.

An example can be found in the paper by Kamenev et al. (1966 a). In the worked examples section, his data are used to illustrate this method of interpreting the results of a factorial design. The second possibility is to apply an analysis of variance, which we call here factorial analysis. Factorial analysis can be applied with equal ease to the multi-level case.

The ANOVA in Chapter 4 consists essentially in splitting up a total sum of squares into sums of squares for the factors (or main effects) considered and the residual error. If one has to carry out an ANOVA for the factorial experiment described in Table 12.I, one would have to divide the total sum of squares into three sums of squares (for A, B and C) and the residual error. In the present instance, i.e., when a factorial analysis is carried out, one must add to this four sums of squares for the interactions. How this is done is discussed in the mathematical section and shown in a worked example.

In the mathematical section of this chapter, the two-way multi-level case is investigated and a generalization is also given. In the worked example section, a practical calculation scheme is given.

The effects (factors, parameters) that have been found to be meaningful by factorial experimentation and analysis can then be optimized using either sequential (Chapter 13) or simultaneous (Chapter 14) strategies.

One difficulty in the application of factorial experiments about which the reader must be warned is the possibility of under-estimating or usually over-estimating the importance of the effects. Suppose that all determinations with factor A at the + level are carried out on one day (or by one analyst) and those with A at the - level the next day (or by another analyst). There is a significant between-days (or between analysts) component in the residual error but the day is not taken into account as a factor. Then all of the determinations made on the first day will, for example, be slightly higher than they would have been on the second day. When carrying out the analysis, this source of variation is considered as part of the A effect, thereby over-estimating it and, as the residual error is obtained by difference (see worked examples), under-estimating the latter. This can lead to the erroneous conclusion that a significant effect is present. This difficulty can be avoided by randomizing the sequence according to which the experiments are carried out.

12.2. MATHEMATICAL SECTION

In the section concerning two-way analysis of variance in Chapter 4, it was assumed that the effects of the two factors A and B are independent. The model obtained with this assumption was called the additive model.

We shall now again study a fixed-effects model with two factors A and B but assume that the effect of each factor may depend upon the level reached by the other. Again, we shall assume that the numbers of observations for each combination (i,j) of values taken by the two factors A and B are all equal to J.

By analogy with the notations of Chapter 4, we write

$$y_{hij} = \mu_{hi} + e_{hij} \qquad \begin{array}{l} h = 1, 2, \ldots, p_1 \\ i = 1, 2, \ldots, p_2 \\ j = 1, 2, \ldots, J \end{array} \tag{12.7}$$

where

$$\mu_{hi} = \mu + \alpha_h + \beta_i + \gamma_{hi} \tag{12.8}$$

and

$$\mu = \frac{1}{p_1 p_2} \sum_{h=1}^{p_1} \sum_{i=1}^{p_2} \mu_{hi} \tag{12.9}$$

$$\alpha_h = \frac{1}{p_2} \sum_{i=1}^{p_2} \mu_{hi} - \mu \quad \text{and} \quad \sum_{h=1}^{p_1} \alpha_h = 0 \tag{12.10}$$

$$\beta_i = \frac{1}{p_1} \sum_{h=1}^{p_1} \mu_{hi} - \mu \quad \text{and} \quad \sum_{i=1}^{p_2} \beta_i = 0 \tag{12.11}$$

$$\gamma_{hi} = \mu_{hi} - \frac{1}{p_2} \sum_{i=1}^{p_2} \mu_{hi} - \frac{1}{p_1} \sum_{h=1}^{p_1} \mu_{hi} + \mu \quad \text{and} \quad \sum_{h=1}^{p_1} \gamma_{hi} = \sum_{i=1}^{p_2} \gamma_{hi} = 0 \tag{12.12}$$

The decomposition of the total sum of squares of deviations with respect to the general average is given by

$$\sum_{h=1}^{p_1} \sum_{i=1}^{p_2} \sum_{j=1}^{J} (y_{hij} - \bar{y}_{...})^2 = J\, p_2 \sum_{h=1}^{p_1} (\bar{y}_{h..} - \bar{y}_{...})^2$$

$$+ J\, p_1 \sum_{i=1}^{p_2} (\bar{y}_{.i.} - \bar{y}_{...})^2 + J \sum_{h=1}^{p_1} \sum_{i=1}^{p_2} (\bar{y}_{hi.} - \bar{y}_{h..} - \bar{y}_{.i.} + \bar{y}_{...})^2$$

$$+ \sum_{h=1}^{p_1} \sum_{i=1}^{p_2} \sum_{j=1}^{J} (y_{hij} - \bar{y}_{hi.})^2$$

This sum of squares can also be written in the form

$$SS_t = SS_A + SS_B + SS_{AB} + SS_r \qquad (12.13)$$

The numbers of degrees of freedom of these sums of squares are given by

SS_t : $n - 1$

SS_A : $p_1 - 1$

SS_B : $p_2 - 1$

SS_{AB} : $(p_1 - 1)\ (p_2 - 1)$

SS_r : $p_1 p_2\ (J - 1)$

Three hypotheses can now be considered

$$H_1 : \alpha_h = 0 \qquad h = 1, 2, \ldots, p_1 \qquad (12.14)$$

$$H_2 : \beta_i = 0 \qquad i = 1, 2, \ldots, p_2 \qquad (12.15)$$

$$H_3 : \gamma_{hi} = 0 \qquad h = 1, 2, \ldots, p_1\ ;\ i = 1, 2, \ldots, p_2 \qquad (12.16)$$

Hypotheses H_1 and H_2

As in Chapter 4, it can be shown that the values

$$\frac{J\, p_2 \sum_{h=1}^{p_1} (\bar{y}_{h..} - \bar{y}_{...})^2 / (p_1 - 1)}{\sum_{h=1}^{p_1} \sum_{i=1}^{p_2} \sum_{j=1}^{J} (y_{hij} - \bar{y}_{hi.})^2 / p_1 p_2 (J - 1)} = \frac{SS_A / (p_1 - 1)}{SS_r / p_1 p_2 (J - 1)}$$

and

$$\frac{J\, p_1 \sum_{i=1}^{p_2} (\bar{y}_{.i.} - \bar{y}_{...})^2 / (p_2 - 1)}{\sum_{h=1}^{p_1} \sum_{i=1}^{p_2} \sum_{j=1}^{J} (y_{hij} - \bar{y}_{hi.})^2 / p_1 p_2 (J - 1)} = \frac{SS_B / (p_2 - 1)}{SS_r / p_1 p_2 (J - 1)}$$

have an F-distribution under hypotheses H_1 and H_2. The parameters of these F distributions are $(p_1 - 1, p_1p_2J-p_1p_2)$ and $(p_2 - 1, p_1p_2J-p_1p_2)$, respectively. In this way, hypotheses H_1 and H_2 can be tested.

Hypothesis H_3

Under hypothesis H_3, the value

$$\frac{J \sum_{h=1}^{p_1} \sum_{i=1}^{p_2} (\bar{y}_{hi.} - \bar{y}_{h..} - \bar{y}_{.i.} + \bar{y}_{...})^2 / (p_1 - 1)(p_2 - 1)}{\sum_{h=1}^{p_1} \sum_{i=1}^{p_2} \sum_{j=1}^{J} (y_{hij} - \bar{y}_{hi.})^2 / p_1 p_2 (J - 1)} = \frac{SS_{AB} / (p_1 - 1)(p_2 - 1)}{SS_r / p_1 p_2 (J - 1)}$$

has an F-distribution with parameters $(p_1 - 1)(p_2 - 1)$ and $p_1p_2 (J - 1)$. This makes it possible to test hypotheses H_3.

The equations given above are used when the calculations are carried out by computer. They are, however, not in a suitable form for manual calculations, in which case, one would prefer to introduce the correction factor for the mean (see section 4.1.7). The factorial analysis is then calculated as follows

$$\text{Correction factor } C = \frac{y_{...}^2}{p_1 p_2 J}$$

$$SS_A = \frac{1}{p_1 J} \sum_{i=1}^{p_2} y_{.i.}^2 - C$$

$$SS_B = \frac{1}{p_2 J} \sum_{h=1}^{p_1} y_{h..}^2 - C$$

$$SS_{AB} = \frac{1}{J} \sum_{h=1}^{p_1} \sum_{i=1}^{p_2} y_{hi.}^2 - C - SS_A - SS_B$$

$$SS_t = \sum_{h=1}^{p_1} \sum_{i=1}^{p_2} \sum_{j=1}^{J} y_{hij}^2 - C$$

$$SS_r = SS_t - SS_A - SS_B - SS_{AB}$$

The mean squares and F-values are then obtained from the sum of squares by division through the appropriate number of degrees of freedom.

Once the mean squares have been derived, the tests are easy to derive. For example, let us consider hypothesis $H_3 : \gamma_{hi} = 0 \quad \forall h,i$. After calculating the sum of squares, F_{AB} is given by

$$F_{AB} = \frac{SS_{AB} / (p_1 - 1)(p_2 - 1)}{SS_r / (J - 1) p_1 p_2}$$

This hypothesis is then accepted at a level α if

$$F_{AB} < F_{\alpha,((p_1-1)(p_2-1),(J-1)p_1p_2)}$$

meaning that the interaction terms are significantly different from zero.

This procedure can be generalized when there are f factors instead of two. The model used for splitting the sum of squares dependent upon f factors is as follows

$$y_{i_1 i_2 i_3 \ldots i_f j} = \alpha_{i_1} + \beta_{i_2} + \ldots + \xi_{i_f}$$
$$+ (\alpha\beta)_{i_1, i_2} + (\alpha\gamma)_{i_1, i_3} + \ldots$$
$$+ (\alpha\beta\gamma)_{i_1 i_2 i_3} + (\alpha\beta\delta)_{i_1 i_2 i_4} + \ldots$$
$$+$$
$$\vdots$$
$$+ (\alpha\beta\ldots\xi)_{i_1 i_2 i_3 \ldots i_f} + e_{i_1 i_2 i_3 \ldots i_f j}$$

with $i_1 = 1, 2, \ldots, p_1$

$i_2 = 1, 2, \ldots, p_2$

$\vdots$

$i_f = 1, 2, \ldots, p_f$

$j = 1, 2, \ldots, J$

This gives the following general relationship

$$y_{i_1, i_2, \ldots, i_f j} - \bar{y}_{\cdot\cdot\ \cdot\cdot} = (y_{i_1 i_2 \ldots i_f j} - \bar{y}_{i_1 i_2 \ldots i_f \cdot})$$

Single factor

$$+(\bar{y}_{i_1\cdot\ \cdot\cdot} - \bar{y}_{\cdot\cdot\ \cdot\cdot})$$
$$+(\bar{y}_{\cdot\ i_2 \cdot\cdot} - \bar{y}_{\cdot\cdot\ \cdot\cdot})$$
$$\vdots$$
$$+(\bar{y}_{\cdot\cdot\ i_f \cdot} - \bar{y}_{\cdot\cdot\ \cdot\cdot})$$

Interaction of two factors

$$+(\bar{y}_{i_1 i_2\ \cdot\ \cdot\cdot} - \bar{y}_{i_1\cdot\ \cdot\cdot} - \bar{y}_{\cdot i_2\cdot\ \cdot\cdot} + \bar{y}_{\cdot\cdot\ \cdot\cdot})$$
$$+(\bar{y}_{i_1 \cdot i_3 \cdot\ \cdot\cdot} - \bar{y}_{i_1\cdot\ \cdot\cdot} - \bar{y}_{\cdot\cdot i_3\ \cdot\cdot} + \bar{y}_{\cdot\cdot\ \cdot\cdot})$$
$$+ \ldots$$
$$\vdots$$
$$+(\bar{y}_{\cdot\cdot\ i_{f-1} i_f \cdot} - \bar{y}_{\cdot\cdot\ i_{f-1}\cdot\cdot} - \bar{y}_{\cdot\cdot\ \cdot\ i_f\ \cdot} + \bar{y}_{\cdot\cdot\ \cdot\cdot})$$

Interaction of three factors

$$+(\bar{y}_{i_1 i_2 i_3 \cdot\ \cdot} - \ldots + \ldots -$$
$$+$$
$$\vdots$$
$$+$$

Interaction of f factors

$$\vdots$$
$$(\bar{y}_{i_1 i_2 \ldots i_f \cdot} - \bar{y}_{\cdot i_2 i_3 \ldots i_f \cdot} - \bar{y}_{i_1 \cdot i_3 i_4 \ldots i_f \cdot} - \bar{y}_{\cdot i_2 \cdot i_f \cdot} + \bar{y}_{\cdot\cdot i_3 \ldots i_f \cdot} + \quad - \quad + \quad - \ldots\ldots$$

This leads to a sum-of-squares relationship in the classical way.

When the hypotheses of significance of the different interactions have been made, the sums-of-squares expression can be simplified. If, for instance, all interactions of g or more factors are not significant, all of these sums of squares are added to the residual sum of squares SS_r. The total sum of squares SS_t is then divided into terms corresponding to the factors, terms corresponding to all interactions of up to g-1 factors and the residual sum of squares SS_r.

12.3. EXAMPLES

Although factorial experimentation has been known for many years, it seems that it has been applied in analytical chemistry only since about 1960. It has been used most extensively by Russian workers. Alimarin et al. (1971) cited several examples in connection with optimization in analytical chemistry. Most of the factorial designs used were incomplete designs (see the next chapter), but a few complete factorial experiments were cited. Apart from the already cited work by Kamenev et al. (1966 a) the following applications were considered : optimization of peak height in the amalgam polarography of lead (Kamenev et al., 1966 b) ; optimization of the accuracy of a differential photometric method of determining antipyrine (Belikov et al., 1967) ; optimization of the absorbance (maximization) in a photometric method for the determination of phenol (Barskii and Noskov, 1965) ; and optimization of spot area in a paper chromatographic separation of fatty acid salts (Luk'yanov and Kosinskaya, 1964). A general discussion of the application of factorial designs in analytical chemistry was also given by Wernimont (1969).

Although there are a few early examples in the western literature (for example, the optimization of barium sulphate precipitation described by Moris and Bozolek, 1959), applications of factorial analysis are so infrequent that in 1974 an international journal for general analytical chemistry still accepted an article (Davies, 1975) written to introduce the technique in analytical chemistry ! Davies (1975) used a 2^3 factorial experiment with triplication [meaning that each set of variables (each treatment) is used three times] to study the titration of ascorbic acid with iron (III) ions and *vice versa*. The criterion was the method bias and the three factors were the temperature, the initial mineral acidity in the flask and the reagent concentration in the flask.

Wu and Suffet (1977) described the optimization of a helix continuous liquid-liquid extraction apparatus. In an initial 2^5 experiment they investigated the effects of the following parameters : helix winding diameter, coil length, flow-rate, water to solvent ratio and the use of a pre-mixer. This first factorial design permitted the elimination of two of the five parameters and gave a rough idea about the optimal levels of the remaining parameters. A 2^3 factorial experiment was then run in order to obtain a more precise idea of the optimal levels.

All of the applications cited are examples of two-level designs and these therefore constitute the largest percentage of applications. Multi-level designs have also been used. Vanroelen et al. (1976), for instance, measured the absorbance at three levels of three variables in the optimization of the photometric determination of phosphate. The design is therefore a 3^3 design with triplication. Their results showed, for example, that the three parameters are significant and that there is also a significant interaction between the concentration of ammonium molybdate and that of perchloric acid. This is considered reasonable from the chemical point of view as the pH can be expected to have an influence on the formation of the ammonium phosphomolybdate complex, depending on the molybdate concentration.

A very complex factorial design was used by Van Eenaeme et al. (1974). This is an excellent example of a case where errors of the second type are also considered (see section 3.2.1.1). Most applications cited so far are concerned with the comparison of two levels of a continuous variable (concentration, temperature) but Van Eenaeme et al. also compared different attributes such as the use of two different injectors in a gas-liquid chromatographic procedure.

Van Eenaeme et al. studied the existence of ghost peaks. Ghost peaks resulting from surface phenomena (adsorption or surface reactions) are an important source of error in the quantitative determination of low-molecular-weight fatty acids by gas-liquid chromatography. Van Eenaeme et al. compared two column packings, four fatty acids, two injectors, three ghost eluting substances and two carrier gases. This is therefore a very complex application, as it constitutes a $2^3 \times 3 \times 4$ design. The experiment allowed the conclusion to be drawn, for example, that a carrier gas containing formic acid was much better than one without it and that one kind of injector (Pyrex) was much better than another (metal).

We have chosen to present two examples here. The first (Kamenev et al., 1966 a), is a two-level design and the significance testing is carried out with a t-test. The second (Vanroelen et al., 1976), is a three-level design with significance testing by analysis of variance (factorial analysis).

Table 12.V gives the results obtained by Kamenev et al. (1966 a), simplified by us by the elimination of one parameter. Each experiment consisted in three polarographic determinations carried out on different days, each determination being carried out twice. This means, for example, that experiment 1 consists of six measurements of the polarographic peak height with a supporting electrolyte consisting of lithium nitrate (levels A, B and C at the - level).

Kamenev et al. took precautions to prevent an under-estimation of the residual error (see section 12.1) by non-random variations during the day by randomizing the sequence of the experiments on the different days. The randomized order is given in Table 12.V.

The first question asked was whether the days should be considered as a factor or not. An analysis of variance (according to our terminology, in fact a factorial analysis) was therefore carried out first. Kamenev et al. used

Table 12.V

Simplified experimental design and results obtained by Kamenev et al. (1966 a)

Factor and variation level				Order in which experiments were carried out			Results obtained, mm					
Experiment	A (electrolyte cation radius)	B (valence of supporting electrolyte cation)	C (supporting electrolyte anion)	Day 1	Day 2	Day 3	Day 1		Day 2		Day 3	
1	-	-	-	2	16	17	90	100	94	97	94	93
2	+	-	-	5	9	20	73	74	72	75	64	72
3	-	+	-	3	14	23	66	70	66	64	64	64
4	+	+	-	8	15	22	86	89	82	90	88	89
5	-	-	+	1	12	24	67	67	67	66	66	66
6	+	-	+	7	13	19	102	95	90	94	100	100
7	-	+	+	4	10	21	100	100	98	95	98	97
8	+	+	+	6	11	18	72	70	73	76	69	74

Levels A : + 1.40 Å (corresponding to K^+, Ba^{2+})

- 0.78 Å (corresponding to Li^+, Mg^{2+})

B : + : 2+   - : 1+

C : + : Br^- - : NO_3^-

Table 12.VI

Analysis of variance of the results of Table 12.V

Source of Variation	Sum of squares	Degrees of Freedom	Mean	F
Conditions	8226.25	7	1175.18	
Days	21.12	2	10.56	1.29
Interaction (Conditions days)	147.88	14	10.56	1.29
Residual	196.00	24	8.17	-
Total	8595.25	47		

a two-way layout with interaction. The factors were "days" and "conditions", the latter including the effects A, B and C.

The effect of the conditions is highly significant, but not the effect of the days or the interaction. As the days can be eliminated as a factor, the six determinations constituting one experiment can be summed. This now yields Table 12.III (section 12.1).

In section 12.1, we saw how the extent of the effect was determined. It should be remembered that the results in column 3 in Table 12.IV are the result of a subtraction of the results at the + level from those at the - level. To obtain the effect for one determination one should divide by 6 (one experiment is the sum of six determinations) and again by 4 (the effect in column 3 is the result of the subtraction of four results from four others, i.e., of four comparisons).

Let us consider, for example, the effect of factor A. If one uses a t-test to do this, this means that one compares the results obtained at the A - level with those at the A + level. This is done using eqn. 3.2 or its simplified form. The difference $\bar{y}_1 - \bar{y}_2$ is the effect of A and can therefore be obtained from Table 12.IV. It follows that $\bar{y}_1 - \bar{y}_2 = 20/24$, $\sqrt{1/n_1 + 1/n_2} = \sqrt{2/24}$. The standard deviation is an estimate of σ, the true standard deviation common to both populations. It is therefore calculated from all of the data with 40 degrees of freedom as there are 48 observations, 40 of which can be considered to be independent (the data are gathered into eight sums). According to Kamenev et al., $s^2 = 9.2$. At a level of significance of $\alpha = 0.05$, 40 degrees of freedom, $t_{\alpha/2} = 2.02$. When

$$t_{\alpha/2} \cdot s \cdot \sqrt{\frac{1}{n_1} + \frac{1}{n_2}} > \bar{y}_1 - \bar{y}_2$$

the A effect is therefore considered to be not significant.

$$2.02 \cdot \sqrt{9.2} \cdot \sqrt{\frac{2}{24}} > 20/24$$

or, to arrive at Kamenev et al.'s way of presenting the results

$$2.02 \cdot \sqrt{9.2 \quad . \quad 48} > 20$$

$$42.4 > 20$$

and the effect of A is considered not to be significant. In general, all effects with a value of less than 42.4 in Table 12.V will be considered not to be significant. This is the case for effects A, B and AC, A is significant at the 5% level and the other effects at the 0.1% level.

12.4. WORKED EXAMPLE

The interference of a calcium salt on the atomic-absorption signal of manganese using a graphite furnace was studied. The effects of three factors were investigated, two at two levels and one at three levels. The experimental values are the peak heights of the manganese signal in the presence of the interfering substance expressed as a percentage of the manganese peak height without interferent.

The following factors were chosen :

A : the ashing temperature at the levels, A_1 = 1050°C and A_2 = 1150°C

B : the ashing time at B_1 = 20 sec and B_2 = 40 sec

C : the argon flow-rate at C_1 = 1.0 l/min, C_2 = 1.5 l/min and C_3 = 2.0 l/min.

The experiment was carried out in triplicate. Therefore, 3 x (2 x 2 x 3) = 36 experimental results were obtained (Table 12.VII). In order to avoid under-estimation of the residual error, the experiments should be carried out in a random sequence.

Table 12.VII. Experimental results on interferences on a manganese atomic -absorption signal

A_1		A_2		
B_1	B_2	B_1	B_2	
73.95	84.81	85.35	99.49	
74.79	83.55	80.69	97.01	C_1
74.89	84.51	85.92	95.59	
63.86	101.25	113.33	98.48	
55.05	108.67	110.97	100.85	C_2
52.85	100.65	109.49	95.93	
58.78	70.16	62.70	64.60	
66.67	78.90	60.33	68.73	C_3
57.24	86.72	63.04	62.39	

To facilitate calculations, the results are simplified by subtracting 80 from each of the 36 values (Table 12.VIII).

Table 12.VIII

Data for the factorial analysis

A_1		A_2		
B_1	B_2	B_1	B_2	
-6.05	4.81	5.35	19.49	
-5.21	3.55	0.69	17.01	C_1
-5.11	4.51	5.92	15.59	
-16.14	21.25	33.33	18.48	
-24.95	28.67	30.97	20.85	C_2
-27.15	20.65	29.49	15.93	
-21.22	-9.84	-17.30	-15.40	
-13.33	-1.10	-19.67	-11.37	C_3
-22.76	6.72	-16.96	-17.61	

The total sum of squares can be broken down as follows

$$SS_t = SS_A + SS_B + SS_C + SS_{AB} + SS_{AC} + SS_{BC} + SS_{ABC} + SS_r \quad (12.17)$$

Tables 12.IX-12.XVI were constructed in order to permit the calculation of the different terms of eqn. 12.17.

Table 12.IX

The three replicate results in Table 12.VIII are added. Each number is the sum of three values

A_1		A_2		
B_1	B_2	B_1	B_2	
-16.37	12.87	11.96	52.09	C_1
-68.24	70.57	93.79	55.26	C_2
-57.31	-4.22	-53.93	-44.38	C_3

Table 12.X

The C_i values of Table 12.IX are added. Each entry obtained is the sum of nine values

A_1		A_2	
B_1	B_2	B_1	B_2
-141.92	79.22	51.82	62.97

Table 12.XI

The B_i values of Table 12.IX are added. Each entry is the sum of six values

A_1	A_2	
-3.50	64.05	C_1
+2.33	149.05	C_2
-61.53	-98.31	C_3

Table 12.XII

The A_i values of Table 12.IX are added. Each entry is the sum of six values

B_1	B_2	
-4.41	64.96	C_1
25.55	125.83	C_2
-111.24	-48.60	C_3

Table 12.XIII

The C_i values of Table 12.XI or the B_i values of Table 12.X are added. Each entry is the sum of eighteen values

A_1	A_2
-62.70	114.79

Table 12.XIV

The C_i values of Table 12.XII or the A_i values of Table 12.X are added. Each entry is the sum of eighteen values

B_1	B_2
-90.10	142.19

Table 12.XV

The A_i values of Table 12.XI or the B_i values of Table 12.XII are added. Each entry is the sum of twelve values

C_1	C_2	C_3
60.55	151.38	-159.84

Calculation of the correction factor

$$C = \frac{y^2\ldots}{n} = \frac{(52.09)^2}{36} = 75.37$$

Calculation of the sums of squares :

Effect of factor A (Table 12.XIII)

$$SS_A = \frac{1}{18}((-62.70)^2 + (114.79)^2) - 75.37 = 875.08$$

Degrees of freedom : 1

Effect of factor B (Table 12.XIV)

$$SS_B = \frac{1}{18}((-90.10)^2 + (142.19)^2) - 75.37 = 1498.85$$

Degrees of freedom : 1

Effect of factor C (Table 12.XV)

$$SS_C = \frac{1}{12}((60.55)^2 + (151.38)^2 + (-159.84)^2) - C = 4268.88$$

Degrees of freedom : 2

Effect of the interaction A x B (Table 12.X)

$$SS_{AB} = \frac{1}{9}((-141.82)^2 + (79.22)^2 + (51.82)^2 + (62.97)^2) - SS_A - SS_B - C = 1224.78$$

Degrees of freedom : 1 x 1 = 1

Effect of the interaction A x C (Table 12.XI)

$$SS_{AC} = \frac{1}{6}((-3.50)^2 + (2.33)^2 + (-61.53)^2 + (64.05)^2 + (149.05)^2 + (-98.31)^2)$$
$$- SS_A - SS_C - C = 1411.80$$

Degrees of freedom : 1 x 2 = 2

Effect of the interaction B x C (Table 12.XII)

$$S_{BC} = \frac{1}{6}((-4.41)^2 + (25.55)^2 + (-111.24)^2 + (64.96)^2 + (125.83)^2 + (-48.60)^2)$$
$$- SS_B - SS_C - C = 67.16$$

Degrees of freedom : 1 x 2 = 2

Effect of the interaction A x B x C (Table 12.IX)

$$SS_{ABC} = \frac{1}{3}\left((-16.37)^2 + (-68.24)^2 + \ldots + (-44.38)^2\right) - SS_{AB} - SS_{AC} - SS_{BC}$$
$$- SS_A - SS_B - SS_C - C = 1563.86$$

Degrees of freedom : 1 x 1 x 2 = 2

Calculation of the residual error

$$SS_r = SS_t - SS_A - SS_B - SS_C - SS_{AB} - SS_{AC} - SS_{BC} - SS_{ABC}$$

SS_t is calculated from Table 12.VIII

$$SS_t = (-6.05)^2 + (-5.21)^2 + \ldots + (-17.61)^2 - C$$
$$SS_r = 365.99$$

The total number of degrees of freedom is 35 (n-1). The number of degrees of freedom of the residual sum of squares is 35 - 11 = 24. The sums of squares and degrees of freedom are summarized in Table 12.XVI.

Table 12.XVI

ANOVA table

Effect		Sum of squares	Degrees of freedom	Variance
main factors	A	875.08	1	875.08
	B	1498.85	1	1498.85
	C	4268.88	2	2134.44
interaction between two factors :	AB	1224.78	1	1224.78
	AC	1411.80	2	705.90
	BC	67.16	2	33.58
interaction between all factors :	ABC	1563.86	2	781.93
residual		365.99	24	15.25

As null hypotheses, it is stated that the observed effects do not differ significantly from the residual, so all variances are part of the experimental error. To test the null hypotheses, the residual variance should be compared with the variances due to the three factors and their combinations.

First, the highest level of significance has to be tested, i.e., the effect of the ABC interaction. If that effect is found to be non-significant, the variance due to it should be incorporated into the residual, which enables one to obtain an improved residual error. If the factorial experiment is carried out without replicates, the variance of the highest level of interaction is an estimate of the residual.

The comparison of two variances is carried out with an F-test. In the case of the ABC interaction, F = 781.93/15.25 = 51.27. In the F-table at the 1% significance level, F = 5.61 with 2 and 24 degrees of freedom. A very high significance is found (P << 0.001). Except for the interaction BC, it can be observed that all factors and their interactions contribute significantly to the total variance. The regression eqn. 12.2 becomes

$$y = b_o + b_A x_A + b_B x_B + b_C x_C + b_D x_A x_B + b_E x_A x_C + b_F x_A x_B x_C$$

From the results in Table 12.VII, it is obvious that $A_1B_2C_2$, $A_2B_2C_1$ and $A_2B_2C_2$ are good combinations. Further optimization can then be carried out by performing, e.g., another factorial experiment in the provisional optimal area and choosing narrower levels. Levels between A_1 and A_2, between C_1 and C_2 and values higher than B_2 could be tried.

REFERENCES

I.P. Alimarin, L.M. Petrukhin and G.I. Malofleva, Zh. Anal. Khim., 26 (1971) 2019.
V.D. Barskii and V.V. Noskov, Zavod. Lab., 31 (1965) 349.
V.G. Belikov, N.I. Kokovkin-Shcherbak and S.Kh. Matsueva, Zavod. Lab., 33 (1967) 1049.
L. Davies, Talanta, 22 (1975) 371.
A.I. Kamenev, V.B. Luk'yanov, V.N. Figurovskaya and E.N. Vinogradova, Zh. Anal. Khim., 21 (1966 a) 535.
A.I. Kamenev, V.B. Luk'yanov and E.N. Vinogradova, Vestn. Mosk. Un-ta. Sr. II, Khimya, No 3 (1966 b) 115, as cited by Alimarin et al. (1971).
V.B. Luk'yanov and E.A. Kosinskaya, Zavod. Lab., 30 (1964) 869.
A.G.C. Moris and S.J. Bozolek, Anal. Chim. Acta, 21 (1959) 215.
C. Van Eenaeme, J.M. Bienfait, O. Lambot and A. Pondant, J. Chromatogr. Sci., 12 (1974) 398.

C. Vanroelen, R. Smits, P. Van den Winkel and D.L. Massart, Z. anal. Chem., 280 (1976) 21.
G. Wernimont, Materials Research and Standards, 9 (9) (1969) 8.
C. Wu and I.H. Suffet, Anal. Chem., 49 (1977) 231.

Chapter 13

SIMULTANEOUS EXPERIMENTAL DESIGNS

13.1. COMPLETE FACTORIAL DESIGNS

In a simultaneous (or pre-planned) optimization design, the measurements are carried out according to a fixed plan. After the results have been obtained, the optimum can be determined.

If one carries out a complete factorial two-level experiment for four factors, one obtains 16 (2^4) experimental values and the selection of the one with the best response constitutes an optimization. If the optimum is to be determined with more precision, one can plan a second factorial experiment around the provisional optimum or obtain an estimate in a mathematical way. The latter possibility is considered in Chapter 15. Let us consider here the first possibility and let us suppose, for example, that a factorial experiment is carried out for the optimization of the molybdenum blue colorimetric method for phosphate. As factors we chose ammonium molybdate reagent (at concentrations 0.5 and 1 M), tin (II) chloride reagent (at concentrations of 0.5 and 5%), the reduction time (5 and 30 min) and the acidity of the reaction medium (0.5 and 1 M).

The factorial experiment indicated that the reduction time and the acidity of the reaction medium have no significant effect and that the best result is found at the concentrations of 1 M ammonium molybdate and 5% tin (II) chloride. The result can be considered to be satisfactory and in subsequent phosphate determination procedures these concentrations will be used. On the other hand, it may be suspected that even better results could be obtained and a second factorial experiment is planned. The exact plan which is decided upon will depend, of course, on the results obtained but it could consist, for example, of a three-level, two-factor design (i.e., a 3^2 experiment), e.g., 0.8, 1 and 1.5 M ammonium molybdate and 3, 5 and 10% tin (II) chloride.

The main advantage of using factorial designs for optimization purposes is that in the region selected for the search, one arrives not only at a (provisional) optimum, but also at a thorough understanding of the importance of the effects studied. There are, however, also several disadvantages, the most important being the large number of experiments to be carried out. The minimal number of experiments for a p-factor, q-level experiment is q^p, which becomes rapidly prohibitive when the number of factors or levels increases. Moreover, this minimal number of experiments does not allow for complete statistical significance testing. If all interactions are considered, for a 2^4 experiment, this necessitates 15 sums of squares (4 main effects, 11 interactions). As one disposes of only 15 degrees of freedom (16 measurements - 1) (see Chapters 4 and 12), this leaves no degrees of freedom available for the estimation of the residual error. These extra degrees of freedom can only be obtained by replication of the determinations. If each determination is carried out twice, 31 degrees of freedom are obtained and 16 are available for estimating the residual error. This means, however, that 32 experiments have been carried out instead of 16. One can conclude, therefore, that the amount of experimentation required when carrying out complete factorial experiments is large. In the following sections some more economical factorial methods are described.

13.2. INCOMPLETE FACTORIAL DESIGNS

13.2.1. Neglection of higher order terms

The complete p-factorial design considers main effects and interactions up to p-order interactions. However, one can usually assume that third and higher order interactions are not important and can be neglected. In the four-factor, two-level case, the model then reduces to

$$y_{hijkl} = \mu + \alpha_h + \beta_i + \gamma_k + \delta_l + \varepsilon_{hi} + \eta_{hk} + \theta_{hl} + \iota_{ik} + \kappa_{il} + \lambda_{kl} + e_{hijkl}$$

For statistical testing, the total sum of squares is split into the following component sums of squares : four main effects, six two-factor interactions and the residual error. For the latter, one disposes of 15 - 10 = 5 degrees of freedom. One observes that enough degrees of freedom are now available for the estimation of the residual error without having to replicate the measurements. One should understand that, in fact, the higher order interactions have been incorporated in the residual error. If one of these interactions is important, this will lead to an over-estimation of the residual error and, as the significance of the main effects and two-factor interactions is determined with reference to the residual error, the significance of the effects studied may be under-estimated.

13.2.2. Partial factorials

Partial factorials according to Plackett and Burman (1946) have been discussed in section 5.2. They allow the estimation only of main effects, and if there are interactions, which as we have said earlier is frequently the case, the conclusions obtained may be in error. Partial factorials have therefore been used only rarely in the experimental optimization of procedures. Arpadjan et al. (1974) used them in the initial stage of an optimization procedure to obtain a rough idea of the location of the optimum. After having approximately located the optimum in this way they proceeded with an experimental design permitting the estimation of interactions to locate it with more precision.

13.2.3. Latin squares

The Latin square arrangement has been used in only a very few instances in analytical chemistry and, as far as we know, never for the purpose of optimization in the sense used in this part of the book. It is essentially a three-way design for factorial experimentation and can therefore be used for the selection of meaningful factors. For this reason, we shall explain the arrangement but not discuss its mathematics. A good account of the latter was given by Scheffé

(1959). Latin square arrangements have been used for a few applications which may perhaps be considered as within the scope of this book (although not completely within the scope of Part II since, as we said, there appear to be no such applications in the literature). Before giving an analytical example, it is convenient to introduce the Latin square method using an example originating from agricultural experimentation. Suppose that five varieties of some economically valuable plant are to be compared in terms of their yields. Planting the five varieties in five plots next to each other may lead to error because the location of the plot may influence the result. To avoid this effect, the field is divided into 25 plots arranged in five rows and five columns. The varieties are planted so that they appear once in each row and once in each column. If the varieties are called a, b, c, d and e, this could lead to the following distribution

d	c	e	b	a
c	a	d	e	b
a	e	b	c	d
e	b	a	d	c
b	d	c	a	e

The rows and the columns are two factors and the varieties the third. Mottet and Bontemps (1973) used this arrangement for the densitometric analysis of TLC spots. It is known that the results may depend on the plate and the location on the plate. They studied a set of seven plates, on each of which they applied spots of seven concentrations in seven locations from right to left. The factors in this instance are the plate, the location on the plate and the concentration. Mottet and Bontemps (1973) used this method not for a study of the factors, but to increase the precision of the determination. This application lies outside the scope of this book, but the experimental setup could have been used to determine the influence of the three factors (location, plate and concentration). A second example is an application of a so-called Greco-Latin square to the quantitative microscopy of urine. The Greco-Latin square is an expanded version of the Latin square that allows the investigation of four factors. Winkel et al.

(1974) carried out this investigation to examine the relative contributions of the technician preparing the urine specimen, the technician reading the urine slide, the time elapsed since the receipt of the urine specimen and the effect of the microscope used.

It can be observed that Latin squares are incomplete three-factorial experiments. As only m^2 experiments are carried out, Latin square arrangements permit a reduction with a factor m of the number of experiments. This means that only incomplete information is obtained and it appears, therefore, that for optimization purposes this method has the disadvantage of being misleading when interactions are present.

13.2.4. Fractional factorials

In a 2^4 experiment, the effect of each main factor is obtained from eight comparisons, in which all 16 experiments are used. This can constitute too large a degree of replication for the purpose considered and one can wonder whether it is not possible to obtain estimates of the main factors and principal interactions with a smaller amount of work. An answer resides in the use of "fractional factorials" (or "fractional factorial designs" or "fractional replication"). A clear paper on this subject was written by Davies and Hay (1950) and we shall follow their arguments to a large extent. Their paper is strongly recommended to those who may consider applying this method. Another important paper on this subject was written by Finney (1946).

Suppose that it is required to investigate four factors (A, B, C and D) and that one does not wish to carry out 2^4 experiments, but considers 2^3 acceptable. This is half the number of experiments normally required and the design will therefore be called a half-replication. It is also equal to the number of experiments required for three factors and, therefore, one way of arriving at the half-design is to start with a complete factorial experiment for three factors (A, B and C) and to see how one can incorporate factor D without requiring additional experimentation.

Let us therefore write again the table from Chapter 12 for a complete 2^3 design (Table 13.I).

Table 13.I

A complete factorial experiment for three factors

Effect	1	2	3	4	5	6	7	8
A	+	+	+	+	-	-	-	-
B	+	+	-	-	+	+	-	-
C	+	-	+	-	+	-	+	-
AB	+	+	-	-	-	-	+	+
AC	+	-	+	-	-	+	-	+
BC	+	-	-	+	+	-	-	+
ABC	+	-	-	+	-	+	+	-
Result	y_1	y_2	y_3	y_4	y_5	y_6	y_7	y_8

In section 13.2.1 it was concluded that often higher order interactions are of little importance. In that section, therefore, this led us to incorporate the higher interactions in the residual error. In the plan here we shall replace the third-order interaction with the additional effect D. The row for ABC in Table 13.I therefore now represents factor D and the table is re-written as a consequence (Table 13.II).

Table 13.II

Table derived from Table 13.I by equating ABC to D

Effect	1	2	3	4	5	6	7	8
A	+	+	+	+	-	-	-	-
B	+	+	-	-	+	+	-	-
C	+	-	+	-	+	-	+	-
D	+	-	-	+	-	+	+	-
AB	+	+	-	-	-	-	+	+
AC	+	-	+	-	-	+	-	+
BC	+	-	-	+	+	-	-	+

The levels for the interactions with D can be calculated according to the multiplication method described in the preceding chapter. Some of them are given in Table 13.III.

Table 13.III

Some interactions with factor D calculated from Table 13.II

Effect	1	2	3	4	5	6	7	8
AD	+	-	-	+	+	-	-	+
BD	+	-	+	-	-	+	-	+
CD	+	+	-	-	-	-	+	+
ABD	+	-	+	-	+	-	+	-

When the Tables 13.II and 13.III are observed closely, it can be seen that the levels for AB and CD, AC and BD and AD and BC are identical. One could also have given the levels for the three-factor interactions. It is found that ABD is equal to C. In the same way, the levels for ACD are equal to those for B, those for BCD are equal to those for A, while ABC is equated to D. One can say that A and BCD or AC and BD are aliases, and Table 13.IV can now be written to show this more clearly.

Table 13.IV

Half-replication of a two-level, four-factor experiment

Effects	1	2	3	4	5	6	7	8
A(=BCD)	+	+	+	+	-	-	-	-
B(=ACD)	+	+	-	-	+	+	-	-
C(=ABD)	+	-	+	-	+	-	+	-
D(=ABC)	+	-	-	+	-	+	+	-
AB(=CD)	+	+	-	-	-	-	+	+
AC(=BD)	+	-	+	-	-	+	-	+
BC(=AD)	+	-	-	+	+	-	-	+

By equating D to ABC, one is therefore able to construct a design with half the experiments. The disadvantages to be offset against this are a lower precision in the estimation of the magnitude of the effects and the impossibility of obtaining estimates of effects free from other effects. The main effects are confounded with third-order interactions and, as one of the premises was that third-order interactions can be neglected, this is of little importance. More serious is that the second-order effects are paired. However, in practice, by chemical reasoning one usually is able to conclude that one of the pair is more important than the other, and the latter is then neglected. It is also possible that one knows that interactions with one of the factors are negligible. If this factor is D, for example, this would permit the determination of AB, AC and BC free from interference. When this is not the case, additional experimentation could be necessary in order to make a distinction between the two.

The three-factor scheme can also be expanded to accommodate even larger numbers of factors, with of course even less precision and more aliases. To incorporate a fifth factor E, one must assume that one of the two-factor interactions is negligible so that it can be equated to E (for example, BC = E in Table 13.V). This now corresponds to a quarter of the full design for five factors and is therefore called a quarter-replication.

It becomes rather complicated to define all of the aliases by writing all of the combinations in tabular form and therefore an easier way of deriving them is necessary. Let us return, therefore, to the half-design of Table 13.IV, which was obtained by equating ABC to D. These are multiplied and the result is called the defining contrast

$$I = ABCD$$

and the aliases are obtained by multiplying the defining contrast with each of the effects. The rules for these multiplications are the usual rules of algebra with the additional condition that $A^2 = B^2 = \ldots = 1$. For example, the alias for A is obtained by multiplying A with ABCD. The result is $A^2BCD = BCD$, while for AC it is $A^2BC^2D = BD$.

When there is more than one additional factor, one obtains two defining contrasts in the way described above. When ABC is equated to D and BC to E, this yields

$$I = ABCD$$

and

$$I = BCE$$

A third defining contrast is obtained by multiplication of the two which have already been obtained. It is equal to ABCD x BCE = ADE, so that

I = ABCD = BCE = ADE

The aliases for each effect are now obtained in the usual way. For A this yields, for example

$A = A^2BCD = ABCE = A^2DE$

or

A = BCD = ABCE = DE

The table for the quarter-replication of a five-factor design is now given by Table 13.V.

Table 13.V

A quarter-replication of a design for five factors

Effects	1	2	3	4	5	6	7	8
A(=BCD=ABCE=DE)	+	+	+	+	-	-	-	-
B(=ACD=CE=ABDE)	+	+	-	-	+	+	-	-
C(=ABD=BE=ACDE)	+	-	+	-	+	-	+	-
D(=ABC=BCDE=AE)	+	-	-	+	-	+	+	-
E(=ABCDE=BC=AD)	+	-	-	+	+	-	-	+
AB(=CD=ACE=BDE)	+	+	-	-	-	-	+	+
AC(=BD=ABE=CDE)	+	-	+	-	-	+	-	+

In summary, with eight experiments one has the following possibilities (Davies and Hay, 1950)

seven factors, if all interactions are negligible ;

six factors and one two-factor interaction if all other interactions are negligible ;

five factors and the interaction of one factor with each of two others if all other interactions are negligible ;

four factors and the interactions between three of these factors if the interactions with the fourth are negligible ;

three factors and all the interactions, including the third order interaction.

If one applies one of these possibilities with eight unreplicated experiments, the design itself does not allow statistical testing, because in each instance seven degrees of freedom are used up and only seven are available. This means that no degrees of freedom remain for the residual error. One should be reminded, however, that the residual error in fact estimates the within-group variation, i.e., the precision on a replicated measurement when no other causes of variation are present. In some instances, one may have prior knowledge about the value of the precision. This can then be used instead of the residual error. This is also true for unreplicated complete factorial designs. If one does not want or is not able to use prior information on the precision, but still needs a statistical test of the significance of the factors and/or interactions, one has no alternative but to carry out a replication of the results.

A fractional factorial analysis is carried out in exactly the same way as explained in Chapter 12. To show this we can use again Kamenev et al.'s (1966) example. In the preceding chapter, we showed how these authors used a full three-factorial design for the study of a polarographic method. At that time, we wrote that we simplified Kamenev et al.'s application by the elimination of one factor. Indeed, they in reality used a half-replica of a four-factor experiment. Leaving out the sequences according to which the experiments were carried out, Table 13.V can therefore be rewritten and completed as shown in Table 13.VI.

Levels A : + : 1.40 Å (corresponding to K^+, Ba^{2+})

- : 0.78 Å (corresponding to Li^+, Mg^{2+})

B : + : 2+

- : 1+

C : + : Br^-

- : NO_3^-

D : + : Cd^{2+}

- : Tl^+

Table 13.VI

Experimental design and results obtained by Kamenev et al. (1966)

Experiment	A (electrolyte cation radius)	B(valence of supporting electrolyte cation)	C (supporting electrolyte anion)	D (test element)	Results obtained, mm: Day 1		Day 2		Day 3	
1	-	-	-	-	90	100	94	97	94	93
2	+	-	-	+	73	74	72	75	64	72
3	-	+	-	+	66	70	66	64	64	64
4	+	+	-	-	86	89	82	90	88	89
5	-	-	+	+	67	67	67	66	66	66
6	+	-	+	-	102	95	90	94	100	100
7	-	+	+	-	100	100	98	95	98	95
8	+	+	+	+	72	70	73	76	69	74

The results in Table 13.VI are now treated in the same way as those in Table 12.V, to yield Table 12.IV. The results cannot be assigned to single effects as was the case in this table, but to a combination of two effects. Instead of concluding that A, B and AC are not significant, we now conclude that A and its alias BCD, B and ACD and AC and BD are not significant, while C and ABD, BC and AD and ABC and D are significant at the 0.1% level. Because in a fractional factorial experiment it is supposed that third-order effects are not significant, this would mean that C and D are significant and either BC and AD or both.

As is also the case for complete factorial designs, fractional factorial designs have not been used very often in analytical chemical work except in the U.S.S.R., where they appear to be used more or less routinely. Alimarin et al. (1971) gave about 30 examples.

In the western literature, there have been only a few applications up to 1974, which is surprising in view of the existence of several good books on the subject (for example, Cochran and Cox, 1957) and the fact that the few existing papers are very convincing. Rubin et al. (1971), for example, described how they established optimal conditions for the determination of arginine-, glutamic acid- and lysine-accepting transfer ribonucleic acid, starting with 10 variables. The initial fractional factorial allowed a reduction of the variables to five and a first adjustment of the levels of these variables to more optimal values. A half-replica of the full factorial design for five variables allowed the elimination of one more variable. This was followed by several central composite designs [a second-order design which is discussed very briefly in Chapter 15 and in more detail by Cochran and Cox (1957)]. The complete study was carried out in 360 - 540 individual trials, compared with more than 1000 when conventional single-factor procedures are used. Owing to the revival of interest in planned optimization experiments noted in the last few years, some more applications have been published recently, for example by Morgan and Deming (1974) and by Van Eenaeme et al. (1974).

13.3. USE FOR OPTIMIZATION PURPOSES

As stated before, the best result in a factorial experiment can be selected and the levels of the factors for this experiment are then considered as optimal. If one is not satisfied with this simple (and in many instances very effective) procedure, one can proceed in two different ways :

(a) A new experiment can be carried out with closer spaced levels around the provisional optimum. This will lead to good results if the true optimum is situated within the bounds given by the levels of the original designs. If the true optimum is situated beyond these original levels, no real amelioration will be obtained from the second experiment.

(b) The results are used to calculate the coefficients of an equation that describes the response surface. This mathematical model is then optimized algebraically. The latter procedure is described in more detail in Chapter 15.

REFERENCES

I.P. Alimarin, O.M. Metrukhin and G.I. Malofeeva, Zh. Anal. Khim., 26 (1971) 2019.
S. Arpadjan, K. Doerffel, K. Holland-Letz, H. Muck and M. Paunach, Z. anal. Chem., 270 (1974) 257.
W.G. Cochran and G.M. Cox, Experimental Designs, Wiley, New York, 1957.
O.L. Davies and W.A. Hay, Biometrics, 6 (1950) 233.
D.J. Finney, J. Agric. Sci., 36 (1946) 184.
A.I. Kamenev, V.B. Luk'yanov, V.N. Figurovskaya and E.N. Vonogradova, Zh. Anal. Khim., 21 (1966) 535.
S.L. Morgan and S.N. Deming, Anal. Chem., 46 (1974) 1170.
J. Mottet and R. Bontemps, VIIth Symposium on Chromatography and Electrophoresis, Presses Académiques Européennes, Brussels, 1973.
R.L. Plackett and J.P. Burman, Biometrika, 23 (1946) 305.
I.B. Rubin, T.J. Mitchell and G. Goldstein, Anal. Chem., 43 (1971) 717.
H. Scheffé, The Analysis of Variance, Wiley, New York, 1959.
C. Van Eenaeme, J.M. Bienfait, O. Lambot and A. Pondant, J. Chromatogr. Sci., 12 (1974) 404.
P. Winkel, B.E. Statland and K. Jørgenson, Clin. Chem., 20 (1974) 436.

Chapter 14

SEQUENTIAL EXPERIMENTAL DESIGNS

14.1. ONE-PARAMETER METHODS

The optimization of a single variable can be carried out according to both sequential and simultaneous (pre-planned) designs and there are many such methods available. The designs can be used with regularly or irregularly sized intervals, they can be accelerated or not, etc. It is not our purpose to give a complete account of all these methods, which are discussed in detail in Beveridge and Schechter's (1970) book. We shall confine our discussion to the only method which, to our knowledge, has been proposed in the analytical chemical literature, namely the uniplex method, and to one other method, the mathematics of which are very appealing. Both are sequential unequal interval methods, which means that the experiments are carried out one at a time and with variable step sizes. Both methods are confined to unimodal functions (i.e., functions with only one optimum). If the function is multimodal it must be divided into unimodal regions. The unimodal single-variable function is very common in analytical chemistry. Very typical situations are the Van Deemter equation for the dependence of the plate height on flow velocity in chromatography for a fixed solvent- stationary phase combination and the distribution coefficients in the anion exchange of metal ion complexes as a function of the complexing agent. These curves are often obtained when two factors are competing. In the Van Deemter equation, for example, the longitudinal diffusion and the mass transfer terms have opposing effects. The former causes a decrease in plate height when the flow velocity increases, while the latter has an increasing effect. The result is a minimal (and optimal) plate height at intermediate flow values.

Many analytical chemists probably feel that it is not necessary or advantageous to use these methods because one is rarely interested in the very precise location of an optimum. Therefore, one will be content to measure the variable, the optimal value of which is sought, at regular intervals and to declare that the value at which the highest result was obtained is the optimum. In explaining the formal methods of this section, it is not our purpose to declare the other methods invalid in all instances. We think that situations may arise (for example, small amounts of sample available for the optimization study, or very costly reagents) where the experimental design methods should be of value. Further, they allow one to gain a better understanding of the philosophy of sequential search methods in general.

14.1.1. The use of Fibonacci numbers

Fibonacci numbers are due to the 13th century mathematician Leonardo of Pisa, who was also called Fibonacci. Fibonacci numbers (cf., the Fibonacci series) are defined by the recursive relationship

$$t_{n+2} = t_n + t_{n+1} \tag{14.1}$$

with $t_0 = 1$, $t_1 = 1$. In other words, each number of the series is the sum of the two preceding numbers. The Fibonacci series therefore begins as follows : 1, 1, 2, 3, 5, 8, 13, 21, 34, 55, 89, 144, 233, ...

If we call $\alpha = \frac{1}{2}(1 + \sqrt{5})$, then the general term is

$$t_{n-1} = \frac{\alpha^n - (-\alpha)^{-n}}{\alpha + \alpha^{-1}} \tag{14.2}$$

which can also be written as

$$t_{n-1} = \frac{1}{\sqrt{5}} \left[\left(\frac{1+\sqrt{5}}{2}\right) n - \left(\frac{1-\sqrt{5}}{2}\right) n \right] \tag{14.3}$$

These numbers can be used to direct a restricted region search, meaning that the limits of the region to be searched are known. This often happens in analytical chemistry. In the metal complex anion-exchange example given in the introduction and using hydrochloric acid as the complexing agent, the region of search for the optimal concentration of hydrochloric acid is naturally restricted to 0 - 12 N. The situation in which no such restrictions are given is called an open-ended search. The Van Deemter equation is an example, as no *a priori* upper limits for the flow velocity are given (although, of course, practical limitations may exist).

The philosophy of this search method is to eliminate parts of the region to be searched from consideration, thereby narrowing at each cycle the region in which the optimum can be situated.

Consider the case in which the maximum of a function y(x) must be found in a region x_A - x_B. This function, which is unknown to the experimenter, is depicted in Fig. 14.1.

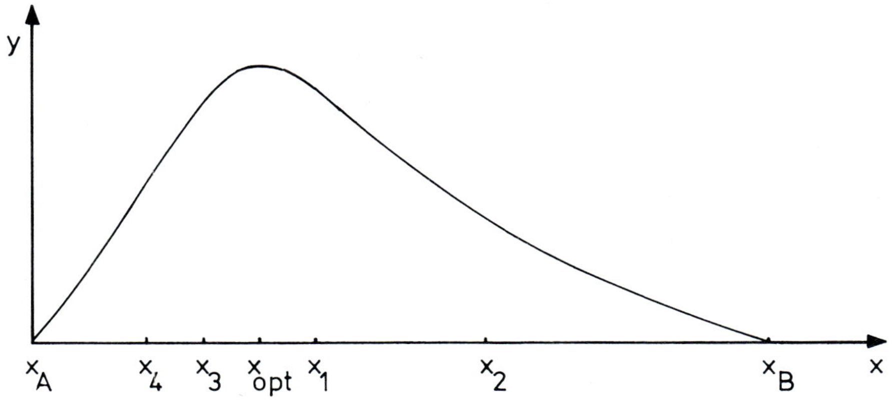

Fig. 14.1. Example of the first stages in a Fibonacci search.

The value to be found is x_{OPT}. Two experiments are carried out with the parameter values x_1 and x_2, chosen in such a way that the distance $x_A - x_1$ is equal to $x_2 - x_B$. The resulting $y(x_1)$ and $y(x_2)$ values are recorded and it is observed that $y(x_1) > y(x_2)$.

The experimenter assumes that the function is unimodal. He is therefore able to conclude that the maximum is not situated in the $x_2 - x_B$ region, to eliminate this region from further consideration and to concentrate on the $x_A - x_2$ region. In this region he has already one experimental result $[y(x_1)]$ at his disposal while $x_A - x_2$ can be considered in its turn as a restricted region in which a search has to be carried out. One can then repeat the strategy of the first cycle by selecting x_3 so that the distance $x_A - x_3$ is equal to $x_1 - x_2$. In the present instance, this leads to the elimination of the region $x_1 - x_2$ and the selection of x_4, so that $x_A - x_4 = x_3 - x_1$, etc.

The Fibonacci search can be shown to be very effective, meaning that a very small number of experiments is necessary. This is particularly true when the optimal value must be known very precisely. This can be shown by comparing the Fibonacci method with the simplest possible search method, the pre-planned regular interval design. This design is very common in analytical chemistry and consists in determining the result of the experiment at regular intervals (for example, with 0, 1, 2, ..., 12 N hydrochloric acid). To delineate a region in which the optimum falls, equal to one tenth of the original region, one needs 19 experiments, while the Fibonacci method demands only 6.

When the optimal region must be one thousandth of the original region, 1999 and 16 experiments are necessary with the two methods, respectively. In fact, the Fibonacci search method is the best available at this moment as far as effectiveness is concerned (Beveridge and Schechter, 1970). It suffers from one disadvantage, namely that it only works optimally in the absence of experimental error. Random errors can lead to the exclusion of the wrong region. Steps should therefore be taken to be sure that, when it is decided that $y(x_N) > y(x_{N+1})$, this corresponds to the truth. This may necessitate duplication (or multiplication) of the experiments so that in practice and certainly in analytical chemistry the Fibonacci search is not necessarily as effective as predicted. Until now, we have discussed the principle of the Fibonacci search procedure and its disadvantages and advantages but we have not explained how Fibonacci numbers are used in this procedure. This application is described below.

Phase 1

Selection of the number of experiments.

The experimenter must decide on the width, a, of the optimal region which he will accept compared with the original search region, A. The Fibonacci series instantly indicates which number is the one immediately larger than A/a. If this is the n+1th number of the series, then n experiments will be needed. For example, if A/a is 50, then the smallest Fibonacci number which is higher than A/a is 55. This is the tenth number in the series, t_9, and therefore nine experiments will be necessary.

Phase 2

(a) Step 1. The first two experiments.

Let us call the length of the original search region L_1 and the distances between the experiments x_1 and x_2 and the boundaries, l_1

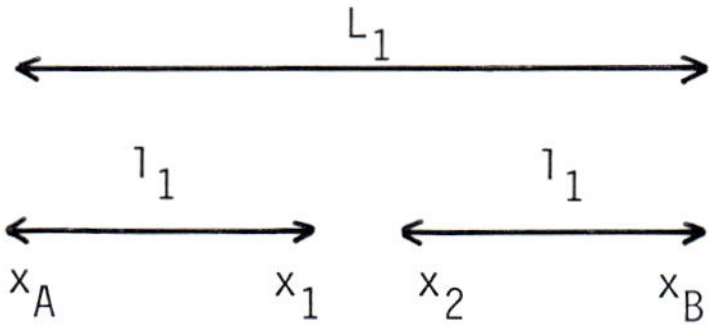

It has been shown that taking

$$l_1 = \frac{t_{n-2}}{t_n} \cdot L_1 \qquad (14.4)$$

yields an optimal number of experiments.

For the example described under Phase 1

$$l_1 = \frac{21}{55} L_1$$

and therefore $x_1 = x_A + l_1$ and $x_2 = x_B - l_1$.

One of the intervals $x_A - x_1$ or $x_2 - x_B$ is eliminated according to whether $y(x_1) < y(x_2)$ or vice versa, as explained earlier. The length of the remaining region is

$$L_2 = L_1 - \frac{t_{n-2}}{t_n} L_1 = \frac{t_{n-1}}{t_n} L_1 \tag{14.5}$$

(indeed, $t_{n-1} + t_{n-2} = t_n$, see eqn. 14.1).

(b) Step 2. Let us suppose that $x_2 - x_B$ was eliminated ; x_1 is retained and we have to determine l_2, so that this is equal to the distances between x_1 and x_2 and between a new experiment x_3 and x_A.

The following general equation is then applied

$$l_k = \frac{t_{n-(k+1)}}{t_{n-(k-1)}} L_k \tag{14.6}$$

so that

$$l_2 = \frac{t_{n-3}}{t_{n-1}} \cdot L_2$$

also,

$$L_k = \frac{t_{n-(k-1)}}{t_{n-(k-2)}} L_{k-1} \tag{14.7}$$

so that

$$L_3 = \frac{t_{n-2}}{t_{n-1}} L_2$$

(c) Step 3, etc.

One proceeds in the same way until n-1 experiments have been performed.

<u>Phase 3</u>

For the last experiment

$$\frac{l_{n-1}}{L_{n-1}} = \frac{t_{n-[(n-1)+1]}}{t_{n-[(n-1)-1]}} = \frac{t_0}{t_2} = \frac{1}{2} \tag{14.8}$$

which means that the distance between the last-but-one experiment and the boundary of the remaining region is half of the length of this region. In other words, the last-but-one experiment was situated at the centre of the remaining search region. The last or nth experiment should also be placed at this point. If the two experiments are carried out with the same x value no new information is gained. Therefore, the last experiment is placed at the smallest distance which is thought to give a measurable difference in response. If $y(x_n) > y(x_{n-1})$, the optimum is situated in the interval $x_{n-1} - x_{n-2}$; if $y(x_{n-1}) > y(x_n)$, it is to be found in $x_n - x_{n-3}$.

14.1.2. The uniplex method

King and Deming (1974) proposed another sequential unequal interval method called uniplex. In contrast with the Fibonacci search procedure it is open-ended, which means that no *a priori* boundaries for the variables need to be given. If, for chemical reasons, these boundaries must be introduced this can be done by allocating an artificial very low response to points falling outside the boundaries. In this way, the search will automatically be restricted to the chosen region. The method is, in fact, based on the modified simplex method which we shall discuss in section 14.2.2. An initial interval, which is called simplex for reasons explained in section 14.2.1, is made to move by reflections, contractions and expansions to the optimum. At no time, however, is any part of the search region excluded so that the method can work in the presence of experimental error. This is its principal advantage compared with the Fibonacci procedure. One starts with the selection of two points (vertices) (see Fig. 14.2), making up the first simplex. $y(x_W)$ and $y(x_B)$ are measured and $y(x_B)$ is found to be the best response. It is clear that the next move will be to explore the region $x > x_B$. This is done by reflecting the line segment $x_W - x_B$ along the x-axis

$$x_R = x_B + (x_B - x_W)$$

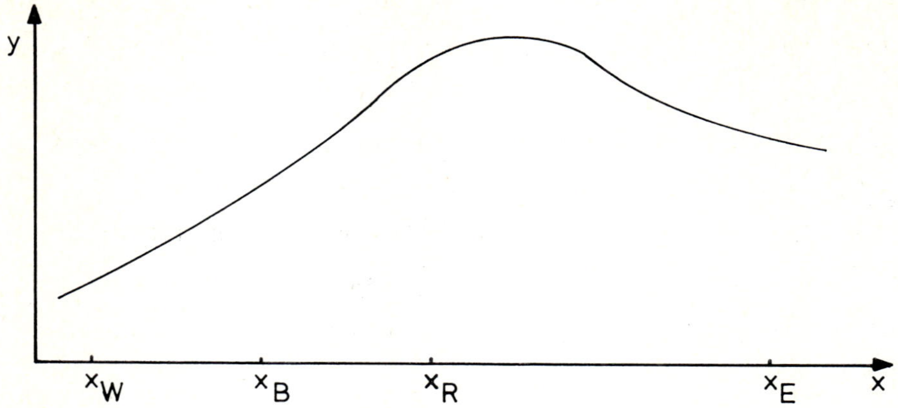

Fig. 14.2. Initial stages in a uniplex procedure.

The measurement of $y(x_R)$ will lead to one of two possible conclusions

(1) $y(x_R) > y(x_B)$. The simplex is moving in the right direction and its movement should therefore be accelerated. A new point is calculated according to $x_E = x_B + \gamma\ (x_B - x_W)$, where $\gamma > 1$ and is usually arbitrarily fixed at 2. If $y(x_E) > y(x_R)$, the new simplex is $x_B - x_E$, if $y(x_E) < y(x_R)$, the simplex has moved too far and the new simplex is $x_B - x_R$. This is the case in Fig. 14.2.

(2) $y(x_R) < y(x_B)$. In moving to x_R the simplex moves too far. There are two new possibilities :

(a) $y(x_R) > y(x_W)$: it is probable (although not certain) that the optimum is situated nearer to x_R than to x_W. Therefore, a new vertex x_{CR} is calculated so that

$$x_{CR} = x_B + \beta(x_B - x_W)$$

with $\beta < 1$ and usually = 0.5. The new simplex is $x_B - x_{CR}$.

(b) $y(x_R) < y(x_W)$: the optimum is situated presumably in the interval $x_B - x_W$. A new vertex x_{CW} is calculated :

$$x_{CW} = x_B - \beta(x_B - x_W)$$

The new simplex is $x_B - x_W$.
The simplex $x_B - x_E$ or $x_B - x_R$ or $x_B - x_{CW}$ or $x_B - x_{CR}$ obtained is considered as a new simplex $x_B - x_W$ and a new cycle can be started. The simplex will move to the optimum and will eventually contract. When the length of the interval has reached a pre-determined small value, the procedure is stopped.

14.1.3. An application

The only known application of uniplex at this moment is the application with which King and Deming (1974) introduce this technique. It consists in a search for the optimal ratio of chromate and hydrogen ion for the production of dichromate. The dichromate yield is derived from its optical absorbance. The single variable here is the speed at which chromate is added by a pump. The maximum speed is 1000 steps sec^{-1} and the minimum 0. As hydrochloric acid was added at the rate (1000 - chromate rate) steps sec^{-1} after the chromate, this is another way of expressing the ratio between both reagents. Table 14.I gives the first 12 points evaluated and Table 14.II the sequence of simplexes.

The starting simplex is 100. - 200.. As 100. yields the worst response, it is eliminated and its reflection 300. is selected. The new point yields a better result than the other two, so that an expansion is carried out. This yields vertex 4 = 400. The new simplex is 200. - 400. and the worst response in it is obtained with a value of 200. Therefore, 200. is rejected and a new vertex, 600., is obtained by reflection. This yields a better result than 200. and 400.. Therefore an expansion is attempted, which fails as 800. leads to a lower result than both 400. and 600.. The new simplex is 400. - 600.. As 400. has the lowest response, this is eliminated. The new vertex would be 800. This has already been evaluated. As the absorbance is lower at 800. than at 400. and 600., rule 2(b) of the preceding section is applied. The new simplex is 500. - 600.. From now on the simplex is contracted at each step until step 12 is reached. As the absorbance has changed in the last three steps by only 0.005 and the simplex has narrowed down to 3 units, one could stop there. If one wants the optimum with still greater precision, one can continue. This was done by King and Deming and after the evaluation of 26 vertices they situated the optimum at 533.

Table 14.I

Results of an optimization study by King and Deming (1974)

Vertex	Chromate pump speed	Net absorbance
1	100.	0.1918
2	200.	0.3666
3	300.	0.5254
4	400.	0.6742
5	600.	0.7625
6	800.	0.4299
7	500.	0.8078
8	550.	0.8370
9	525.	0.8402
10	537.5	0.8519
11	531.25	0.8524
12	534.375	0.8521

Table 14.II

Sequence of simplexes in the optimization study of Table 14.I

Sequence Number	Uniplex
1	100.-200.
2	200.-400.
3	400.-600.
4	500.-600.
5	500.-550.
6	525.-550.
7	525.-537.5
8	531.25-537.5
9	531.25-534.375

14.2. MULTIPLE PARAMETER METHODS

In this section, methods in which several factors at a time are varied according to a sequential design will be discussed. These methods have been called evolutionary operations methods (EVOP), a term which seems to have been introduced into analytical chemistry by Deming and Morgan (1973). There are several such methods, but only one seems to be used systematically, namely the simplex method.

14.2.1. The simplex method

A simplex is a geometric figure defined by a number of points equal to one more than the number of parameters considered in the optimization or, to put it another way, to one more than the number of dimensions of the factor space. For the simplest multi-factor problem, namely an optimization of two parameters, the simplex is therefore a triangle. This example will be used to introduce the technique. Consider the isoresponse surface given in Fig. 14.3. This figure is adapted from Long (1969) and describes the optimization of a colorimetric determination of sulphur dioxide. The numbers along the isoresponse lines are absorbances and the highest absorbance is considered to be the optimum.

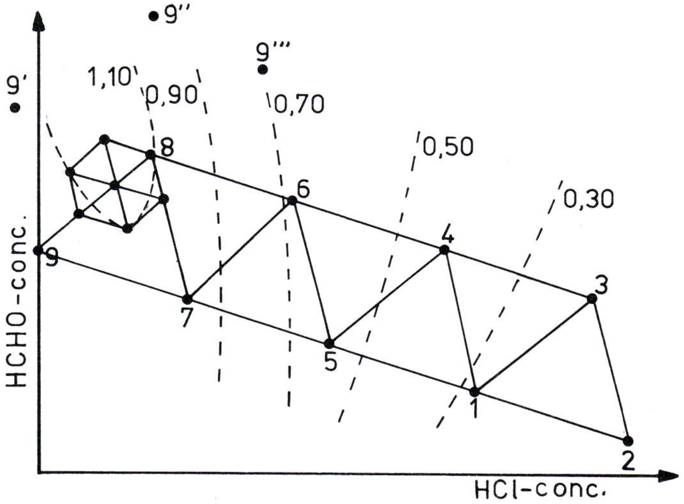

Fig. 14.3. Example of simplex optimization (adapted from Long, 1969).

The optimization starts with points 1, 2 and 3. These points form an equilateral triangle and point 2 shows the worst response of the three. It is logical to conclude that the response will probably be higher in the direction opposite to this point. Therefore, the triangle is reflected so that point 4 opposite to point 2 is obtained. An experiment is now run with the parameter values of point 4. Points 1, 3 and 4 are considered to form together a new simplex. The procedure is now repeated.

It appears that point 3 yields the lowest absorbance. Point 3 is therefore rejected and point 5 is obtained. In this way, using successive simplexes, one moves rapidly along the response surface. This procedure is described by the following rules.

Rule 1 : the new simplex is formed by rejecting the point with the worst result in the preceding simplex and replacing it with its mirror image across the line defined by the two remaining points.

In the initial stages of an optimization, the new point in a simplex will usually yield a better result than at least one of the two remaining points, because the simplexes will tend to move towards the optimum. When the new point does not cause a move in this general direction a change in the progression axis is necessary. When the new point has the worst response of the simplex, it is impossible to apply rule 1, as this would lead to reflection back to the point which was itself the worst one in the preceding simplex. The repetition of rule 1 would then lead to an oscillation between two simplexes. For example, consider simplex 6, 7 and 8 in Fig. 14.3. Point 6 has the lowest absorbance and is replaced by 9, its mirror image across the line 7-8. Point 9 has the least desirable response in the new simplex. Rule 1 would lead back to point 6, then again to point 9, etc. Therefore, one now applies rule 2.

Rule 2 : if the newly obtained point in a simplex has the worst response, do not apply rule 1 but instead eliminate the point with the second lowest response and obtain its mirror image to form the new simplex.

The effect of this is to change the direction of progression towards the optimum. This will most often happen in the region of the optimum. If a point is obtained near to it, all of the other new points will overshoot the top of the response curve. A change in direction is then indicated. In the region of the optimum, the effect is that the simplexes circle around the provisional optimal point. For example, in Fig. 14.3 the application of rule 2 would lead to the rejection of the second lowest point, 7. Its reflection yields 9', a point with a negative hydrochloric acid concentration. Let us suppose for the moment that this is possible and that 9' would yield the lowest response.

Rule 2 leads to 9". The response of this point is lower than the response of 8 but better than that of point 9'. Rule 1 leads back to 9. If on the contrary 9" has a response lower than 9', 9"' is selected as the new point (rule 2).

In both instances, point 8 is retained in consecutive simplexes, which is interpreted as indicating that this point is situated as near to the optimum as one can get with the initially chosen simplex. The situation could also result from an erroneously high response from point 8. To make sure that this is not the case, one applies rule 3.

Rule 3 : if one point is retained in three successive simplexes, determine again the response at this point. If it is the highest in the last three simplexes it is considered as the optimum which can be attained with simplexes of the chosen size. If not, the simplex had become fastened to a false maximum and one starts again. One difficulty which has still to be resolved is what to do in practice when one encounters a situation such as that exemplified by point 9'. To avoid it one identifies the constraints or the boundaries between which the simplex may move. For example, when the parameters are concentrations, these are bracketed between a low limit, usually set at 0, and a high limit, usually the highest concentration which may be present but sometimes the highest concentration which is considered to be practical. Once this has been done, one applies rule 4.

Rule 4 : if a point falls outside one of the boundaries, assign an artificially low response to it and proceed with rules 1-3. The effect of applying rule 4 is that the outlying point is automatically rejected without bringing the succession of simplexes to an end.

The two-factor case can be generalized to the n-factor case. When the two-dimensional simplex can be obtained geometrically, this is no longer possible for three or more dimensions although the principle is exactly the same. When the vertex to be rejected has been determined, the coordinates of n retained vertices are summed for each factor and divided by n/2. From the resultant values one subtracts the coordinates of the rejected point. The result yields the coordinates of the new vertex. This is done best by using Table 14.III.

Table 14.III

Calculation of vertex for a simplex with n dimensions (adapted from Long, 1969, and Spendley et al., 1962)

	Vertex no.	Factor				
		x_1	x_2	x_3	...	x_4
(n retained vertices)		(Coordinates of retained vertices)				
Sums of retained coordinates						
2/n (Sums)						
Coordinates of discarded vertex ..						
Coordinates of new vertex						

14.2.2. The modified simplex method

In the original simplex method the step size is fixed. When it is too small, it takes many experiments to find the optimum ; when it is too large, the optimum is determined with insufficient precision. In the latter instance, one can start a new simplex around the provisional optimum with a smaller step size. This was the method used by Long (see Fig. 14.3). However, a modified simplex method in which the step size is variable throughout the whole procedure offers a more elegant (and efficient) solution. The principal disadvantage is that the simplicity of the calculations in the original simplex method no longer exists. The principles of the method are retained but additionally provision is made for the expansion or contraction of simplexes.

The uniplex design which was discussed in section 14.1.2 is in fact derived from the modified simplex procedure. As we have explained the uniplex in some detail, it is not necessary to do this again here for the modified simplex method. Let us simply recall that the philosophy is to expand or accelerate the simplex in the directions which seem favourable and to contract it in the directions that are unfavourable. This method, which is due to Nelder and Mead (1965), was introduced into analytical chemistry by Morgan and Deming (1974). It is explained here, using the notation of the latter workers for the two-dimensional case. This again yields a triangle (which is now no longer equilateral) as the simplex. The initial simplex is called BNW. In this simplex, the best response is obtained for vertex B and the worst for vertex W. The latter is therefore rejected and reflected. If $\overline{P}$ is the centroid of the line segment BN, then the reflected vertex R is obtained by

$$R = \overline{P} + (\overline{P} - W)$$

The response in R can be higher than in B, lower than in B but higher than in N or lower than in N. According to which of the three possibilities is found to be true, the following steps are undertaken.

(a) Response at R > response at B.

The simplex seems to move fast in a favourable direction. An expansion is therefore attempted by generating vertex E :

$$E = \overline{P} + \gamma(\overline{P} - W)$$

where γ is usually 2. If the response at E is also better than at B, the E is retained and the new simplex is BNE. If not, the expansion is considered to have failed and the new simplex is BNR. One proceeds in the usual manner (rejection of the worst vertex and reflection).

(b) Response at B > response at R > response at N.

The new simplex is BNR. No expansion or contraction is envisaged.

(c) Response at N > response at R.

The simplex has moved too far and it should be contracted. If the response at R is not worse than at W, the new vertex C_R is best situated nearer to R than to W

$$C_R = \overline{P} + \beta(\overline{P} - W)$$

where β is usually 0.5. If the response at R is also worse than that at W, the new vertex C_W should be situated nearer to W

$$C_W = \overline{P} - \beta(\overline{P} - W)$$

The new simplex is BNC_R or BNC_W and one proceeds in the usual manner.

14.2.3. An example

Krause and Lott (1974) described applications of the simplex technique to the optimization of automatic analysers. Their simplest example consists in the minimization of interaction (a measure of carry-over) in a continuous Technicon-type analyser. The variables were the sample-to-wash ratio (% sample) and the flow cell pull-through rate (% pull-through). Their results are given in Table 14.IV. In a first attempt (vertices 1 - 13), they located the optimum at vertex 9. To be certain that this is indeed the optimal vertex, they started the optimization again with an initial simplex (14, 15, 16) with very different conditions compared with those used in the first attempt. The sequence of simplexes leads to 24, 26, 27. The parameter values of 26 and 27 are equal to those of vertices 10 and 8, respectively. By rejection of the worst value (24), one therefore arrives through reflection at simplex 8, 10, 9 and the same optimum 9 is obtained.

A good example of a modified simplex can be found in an optimization of a colorimetric method for cholesterol in blood or serum (Morgan and Deming, 1974). This paper contains a detailed description of optimisation using either the

Table 14.IV

Optimization of a continuous flow system (Krause and Lott, 1974)

Note : Values in parentheses are those predicted by the simplex but which could not always be obtained experimentally

Vertex No.	% sample	% pull-through	% response (interaction)	Vertices retained from previous simplex
1	25.0 -	24.0 -	37.3	-
2	25.0 -	34.0 -	22.4	-
3	33.3 -	29.0 -	27.2	-
4	33.3 (33.3)	40.1 (39.0)	19.2	2.3
5	25.0 (25.0)	45.7 (45.1)	13.5	2.4
6	33.3 (33.3)	51.2 (51.8)	12.0	4.5
7	25.0 (25.0)	56.8 (56.8)	5.4	5.6
8	33.3 (33.3)	63.0 (62.3)	5.1	6.7
9	25.0 (25.0)	69.1 (68.6)	4.0	7.8
10	33.3 (33.3)	75.3 (75.3)	4.2	8.9
11	25.0 (25.0)	82.7 (82.4)	4.2	9.10
12	14.3 (16.7)	77.2 (76.5)	6.9	9.11
13	14.3 (14.3)	63.0 (63.6)	5.4	9.12
14	85.7 -	40.1 -	36.8	-
15	85.7 -	51.2 -	29.0	-
16	75.0 -	45.7 -	28.1	-
17	75.0 (75.0)	56.8 (56.8)	27.3	15.16
18	66.6 (64.3)	51.2 (51.3)	24.8	16.17
19	66.6 (66.6)	63.0 (62.3)	20.8	17.18
20	60.0 (58.2)	56.8 (57.4)	20.3	18.19
21	60.0 (60.0)	69.1 (68.6)	18.6	19.20
22	54.5 (53.4)	63.0 (62.9)	15.5	20.21
23	54.5 (54.5)	75.3 (75.3)	14.2	21.22
24	45.5 (49.0)	69.1 (69.2)	8.4	22.23
25	45.5 (45.5)	82.7 (81.4)	9.2	23.24
26	33.3 (36.5)	75.3 (76.5)	4.4	24.25
27	33.3 (33.3)	63.0 (61.7)	5.0	25.26

modified simplex or a fractional factorial experiment. It seems a pity to give a shortened version of such an excellent article here ; we shall not do this but, on the contrary, we urge the reader to read the complete, original version.

14.2.4. Other applications

The introduction of the method goes back to the early 1960 s(Spendley et al., 1962). Not surprisingly, Spendley et al.'s example is a chemical (but not an analytical chemical) one. The modified simplex was first proposed by Nelder and Mead (1965). The first application in analytical chemistry dates from 1969 (see also section 14.2.1). It consists of an optimization of the absorbance in a colorimetric method for sulphur dioxide (Long, 1969). The optimization of absorbance or related quantities has remained one of the most important applications. Morgan and Deming's work in this context has already been cited. Other applications of this kind are due to Vanroelen et al. (1976), who optimized a colorimetric method for phosphate and to Parker et al. (1975). The latter carried out a very detailed study on the optimization of experimental factors in atomic-absorption spectrophotometry. Other applications are due to Houle et al. (1970) and Czech (1973a, b).

Several chromatographic applications were also proposed. Rainey et al. (1974) used resolution as the criterion for the optimization of a phospholipid separation, while Smits et al. (1975) used a theoretical information criterion for the optimization of the composition of the eluent used in a cation-exchange separation of five metal ions. The theoretical background for this criterion was given by Massart and Smits (1974). This criterion is, as we now recognize, more complicated than necessary.

Morgan and Deming (1975) optimized a GLC separation of isomeric octanes using the so-called peak separation function. The same criterion was used by Holderith et al. (1976) for a GLC separation of methylbenzenes and by Detaevernier et al. (1976) in the liquid chromatography of tricyclic antidepressants (uniplex application).

A few other applications have also been proposed, for example, the optimization of a titration (Meus et al., 1975) and of a gravimetric determination (Parczewski et al., 1974).

14.2.5. Difficulties in the application of simplex optimization methods. Comparison with factorial designs

The simplex method is conceptually so simple that one can expect it to find more widespread use than factorial designs. This simplicity carries in itself the danger that some of its users would be led to think that there are no difficulties in its use and that it should always lead straight to an optimum.

In fact, the principal promoters of the technique, Deming and Morgan, wrote an article with King (King et al., 1975) to delineate "the propagation of mistakes and misunderstandings" in the application of simplex methods. Let us therefore investigate some of the problems that can arise.

These difficulties can be classified in three categories, namely occasions where simplex methods are of no use or where their application fails, mistakes to be avoided, and difficulties in deciding how to apply the method.

Simplex methods cannot be used when one of the variables is not continuous. For example, if a factor such as a choice between a Pyrex and a metal GLC injector is included, one has to use a factorial design. Simplex methods may fail if there is more than one optimum (this is also true for factorial experiments). The simplex method will certainly lead to a local optimum, but this may not be the "overall" optimum. In many instances, one may be sure that there will be only one optimum, but in a few situations many do exist. An example occurs in the optimization of chromatographic separations. If in the factor interval searched the order of elution or the sequence of migration rates of components A and B changes, there will be a local optimum for the sequence AB and one for the sequence BA. One way of making sure that there is only one optimum is to carry out the optimization twice, from two very different starting simplexes. If one arrives at the same optimum, as was the case in the example from Krause and Lott (1974) (see section 14.2.3), one can be reasonably sure that only one optimum exists. Among applications where errors have been made, King et al. (1975) cited studies by Houle et al. (1970) and Czech (1973a, b). We shall not go into each of the errors noted by King, because these errors seem to be inter-related. The authors cited used the sample volume as a factor in a colorimetric system where the absorbance was the response to be optimzed. Clearly, the choice of this factor is unfortunate, because when the amount of sample is included as a factor, the simplex naturally moves to a higher level of sample. It would have been better to determine the absorbance at a constant sample volume. Of course, this kind of error can be made also in factorial designs. The simplicity of the simplex method may lead the user, however, to apply it without exercising sufficient intellectual effort. The mathematical difficulty of factorial designs prohibits this. It should be noted, however, that some chemists reject simplex methods precisely on these grounds. They claim that contrary to factorial designs, simplex methods do not offer any information except about the location of the optimum, and not about the influence of the factors involved. It is true that for statistical significance testing, factorial designs are more appropriate. However, the results from a simplex optimization can be used to obtain least-squares mathematical models valid for portions of the factor space in the same way as those obtained with factorial designs. The resulting equations should permit an evaluation of the importance of the factors.

The last difficulty resides in the selection of the parameters determining the initial simplex, i.e., which factors should be selected, what should be the size of the initial step and which should be the vertices of the initial simplex. According to Yarbro and Deming (1974), who discussed these three questions, it is preferable to include as many factors as possible. This does not increase the number of experiments to a large extent, as it would for factorial designs. Simplex methods applied in this way are less apt to miss a significant factor. The authors point out that pre-selection of factors by statistical significance testing is not fool-proof. If the levels are too close a significant effect may be found to be statistically insignificant, while if the levels are too distant, they might be situated around the optimum in such a way that an effect is erroneously found to be non-significant. A question which can then be asked is what happens during the optimization when an insignificant factor has been included. An answer to this question was given by Parker et al. (1975).

In their work on the optimization of absorbance for calcium in AAS, they included one factor which they were sure could not influence the process (the volume of water in a 100 ml graduated cylinder on a laboratory bench some distance from the spectrometer). It was found that when one uses the modified simplex technique, such a non-significant factor also converges. In the interpretation of simplex results, one should therefore avoid considering a factor significant simply because it converges.

REFERENCES

G.S.G. Beveridge and R.S. Schechter, Optimization : Theory and Practice, McGraw Hill, New York, 1970.
F.P. Czech, J. Assoc. Offic. Anal. Chem., 56 (1973a) 1489.
F.P. Czech, J. Assoc. Offic. Anal. Chem., 56 (1973b) 1496.
S.N. Deming and S.L. Morgan, Anal. Chem., 45 (1973) 278A.
M.R. Detaevernier, L. Dryon and D.L. Massart, J. Chromatogr., 128 (1976) 204.
J. Holderith, T. Toth and A. Våradi, J. Chromatogr., 119 (1976) 215.
M.J. Houle, D.E. Long and D. Smette, Anal. Lett., 3 (1970) 401.
P.G. King and S.N. Deming, Anal. Chem., 46 (1974) 1476.
P.G. King, S.N. Deming and S.L. Morgan, Anal. Lett., 8 (1975) 369.
R.D. Krause and J.A. Lott, Clin. Chem., 20 (1974) 775.
D.E. Long, Anal. Chim. Acta, 46 (1969) 193.
D.L. Massart and R. Smits, Anal. Chem., 46 (1974) 283.
M. Meus, A. Parczewski and A. Rokosz, Chem. Anal. (Warsaw), 20 (1975) 247.
S.L. Morgan and S.N. Deming, Anal. Chem., 46 (1974) 1170.
S.L. Morgan and S.N. Deming, J. Chromatogr., 112 (1975) 267.
J.A. Nelder and R. Mead, Computer J., 7 (1965) 308.
A. Parczewski, A. Rokozs and M. Kasprzycka, Chem. Anal. (Warsaw), 19 (1974) 107.
L.R. Parker, S.L. Morgan and S.N. Deming, Appl. Spectrosc., 29 (1975) 429.
M.L. Rainey, S.W. McClear and W.C. Purdy, Abstracts "Biochemische Analytik 74 combined with 1st European Congress of Clinical Chemistry", München, April 1974.
R. Smits, C. Vanroelen and D.L. Massart, Z. anal. Chem., 273 (1975) 1.
W. Spendley, G.R. Hext and F.R. Himsworth, Technometrics, 4 (1962) 441.
C. Vanroelen, R. Smits, P. Van den Winkel and D.L. Massart, Z. anal. Chem., 280 (1976) 21.
L.A. Yarbro and S.N. Deming, Anal. Chim. Acta, 73 (1974) 391.

Chapter 15

STEEPEST ASCENT AND REGRESSION METHODS

15.1. INTRODUCTION

The effect of variables that have an influence upon the response of a procedure is described by the response surface. In the methods described in the preceding chapters, the optimum of this surface is sought (often in a very efficient manner) without trying to describe it. It is evident that if one succeeds in describing the surface accurately in a mathematical way, one should be able to establish the optimum with great precision.

We shall distinguish between two classes of methods. In the methods of the first class, the surface as such is not described but only the gradient along it. One determines the direction of the gradient and carries out experiments in a sequential way along this line of steepest ascent. In the methods of the second class, the surface is described in a more or less approximate way, usually with regression methods and using the results of factorial or related experiments. In the following sections, we shall describe these techniques using the simplest possible example, namely the dependence of the response on two factors. It should be understood, however, that the application of these methods can be extended to more than two factors and that they become more interesting when there are more factors.

15.2. STEEPEST ASCENT METHODS

Consider four experiments, constituting a 2^2 factorial experiment (see Fig. 15.1). The slope in the x_1 direction is proportional to the sum of the responses of experiments 4 and 2 minus the sum of the responses of experiments 3 and 1, i.e., proportional to $r_1 = (y_2 + y_4) - (y_3 + y_1)$. In fact, this slope

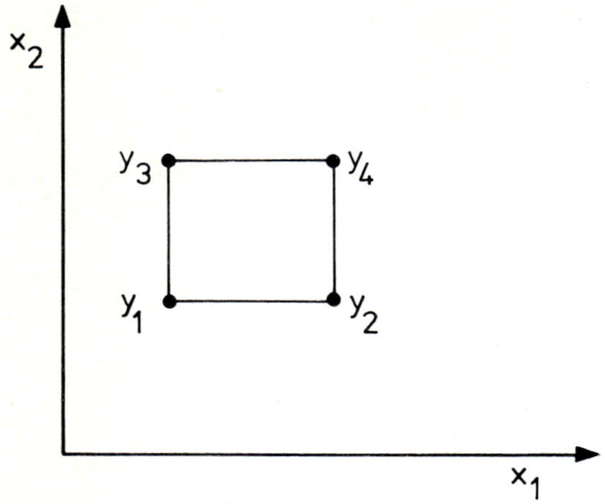

Fig. 15.1. A 2^2 factorial experiment.

is equal to twice (two observations at each level) the difference between the two levels. In this example, x_1 and x_2 are scaled units (see the next section) and the difference between the levels is the same in the two directions.

In the same way, the slope in the x_2 direction is proportional to $r_2 = (y_3 + y_4) - (y_1 + y_2)$. Let us suppose that in a given application both r_1 and r_2 are negative. Clearly, the optimum must be situated lower and more to the left. One will then carry out an experiment C with coordinates x_1 and x_2 lower than and more to the left than experiment 0, a new 2^2 factorial experiment around this point C or a new factorial experiment in which C is one of the experiments. One determines point C by moving from 0 in the best possible direction with a step size S. However, one still has to determine the best possible direction. It is logical that the displacement in directions x_1 and x_2 should be proportional to r_1 and r_2. Indeed, if the slope changes faster in the x_2 than in the x_1 direction, one should move a longer distance in the first direction.

This situation is depicted in Fig. 15.2. The direction of movement is given by a line segment g, the length of which is $\sqrt{r_1^2 + r_2^2}$. As the real length of the step must be S, the displacement in the x_1 direction should be (r_1/g) S and in the x_2 direction (r_2/g) S.

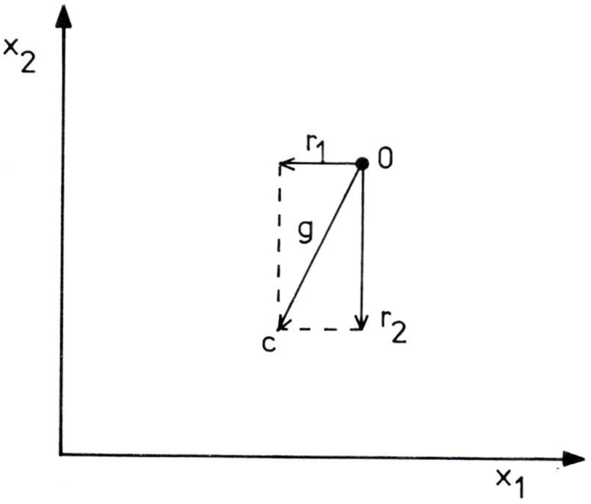

Fig. 15.2. Direction of gradient in a steepest ascent procedure.

The way of doing this in practice is illustrated with an example taken from Brooks (1959). The experimental region is shown in Fig. 15.3 and the allowed number of experiments is 16. The optimum (unknown to the experimenter) is point P and the starting point is the centre or base point 0 of the experimental region.

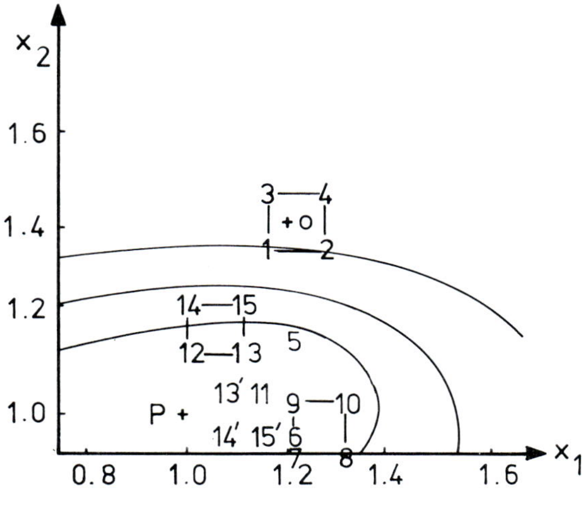

Fig. 15.3. Example of application of steepest ascent (adapted from Brooks, 1959).

Around 0 a 2^2 factorial experiment is carried out with a difference in levels of 0.125. The experiments are denoted by the numbers 1 - 4 and the responses are given in Table 15.I. From the results one calculates

$$r_1 = (y_2 + y_4) - (y_3 + y_1) = -0.0152$$

$$r_2 = (y_3 + y_4) - (y_1 + y_2) = -0.3956$$

$$g = \sqrt{r_1^2 + r_2^2} = 0.39589$$

The initial step size was fixed at 0.25 so that the displacement from 0 in the x_1 direction is given by $(r_1/g) \cdot 0.25 = -0.0095$ and in the x_2 direction by $(r_2/g) \cdot 0.25 = -0.2498$. The coordinates of the new experiment (5) are therefore $x_{15} = 1.2214 - 0.0095 = 1.2119$ and $x_{25} = 1.4033 - 0.2498 = 1.1535$. The result obtained, $y_5 = 0.7204$, is much higher than the average result, 0.2166, obtained in the first four trials. One concludes that one is moving in the right direction and therefore one decides to take another step in the same direction and with the same length (trial No. 6). Again, an improvement is obtained. Another step of the same length is impossible, because this would lead to a point beyond the limits of the experimental region. Therefore, the step size is shortened to such an extent (S = 0.004) that the new point 7 still falls in the experimental region.

Because one has reached the limits of the experimental region, further movement must be made in another direction. This direction is determined from a new 2^2 design, of which point 7 is one of the experiments together with 8, 9 and 10. From y_7, y_8, y_9 and y_{10}, one obtains new r_1, r_2, g and displacements in the x_1 and x_2 direction :

$$r_1 = y_{10} + y_8 - (y_9 + y_7) = -0.1494$$

$$r_2 = y_9 + y_{10} - (y_7 + y_8) = 0.1038$$

$$g = 0.1819$$

and, shortening the step size to 0.15, the displacement in the x_1 direction is equal to (-0.1494 / 0.1819) . 0.15 = -0.1231 and in the x_2 direction 0.0856. The new point obtained in this way (11) yields a response y_{11} of 0.9392, which is a clear improvement over the four experiments in the second 2^2 design.

Table 15.I

Example of steepest ascent optimisation (from Brooks, 1959)

Experiment j	x_{1j}	x_{2j}	x_j
0	1.2214	1.4033	not determined
1	1.1589	1.3408	0.3362
2	1.2839	1.3408	0.2947
3	1.1589	1.4658	0.1045
4	1.2839	1.4658	0.1308
5	1.2119	1.1535	0.7204
6	1.2024	0.9037	0.8018
7	1.2024	0.9033	0.8237
8	1.3274	0.9033	0.7738
9	1.2024	1.0283	0.9004
10	1.3274	1.0283	0.8009
11	1.1418	1.0513	0.9392
12	1.0187	1.1368	0.8449
13	1.1087	1.1368	0.7498
14	1.0187	1.2268	0.6120
15	1.1087	1.2268	0.6210
16	1.0429	1.0943	0.9361

One therefore continues in the same direction. Point 12 now gives a lower result, so that one concludes that a change in direction is necessary. A new factorial experiment with point 12 as one of the corner points and reduced distances between levels is carried out. This results in a gradient g leading to the final, sixteenth measurement with a response y_{16} of 0.9361 (the response at the optimum, P, being 1). It should be noted that the experimenter did not try to use the experience accumulated in all the experiments, but only the last one. If he had, he would, for example, have returned to point 11 to carry out a factorial experiment instead of using point 12. This experiment would have involved points 13', 14' and 15' and would doubtlessly have led to a still better estimate of the optimum. He could also have used point 12 and, reasoning that 11 yields a better response, chosen points 13 and two points opposite to 14 and 15. This too would have led to a better final result.

In this section we have tried to explain the principle of the method using as little mathematics as possible. In practice, at least in the few existing analytical chemical applications, the response is usually described in an approximate way by a first-order polynomial obtained by regression analysis (see section 15.3). The maximal gradient on this (hyper-)surface is then determined. One example of such a procedure can be found in a study by Arpadjan et al. (1974), who optimized emission optical methods and gel-chromatographic separation procedures. A mathematical development for this and related methods can be found in Beveridge and Schechter (1970).

15.3. REGRESSION METHODS

15.3.1. Location of the optimum

In these methods, one tries to describe the response surface in the region where the optimum is to be found using a mathematical equation. In most instances, one uses generalized polynomials to approximate the real (and unknown) surface.

A generalized polynomial can be written as

$$y = b_0 + b_1x_1 + \ldots + b_nx_n + b_{11}x_1^2 + b_{12}x_1x_2 + \ldots + b_{nn}x_n^2 + b_{111}x_1^3 + b_{112}x_1^2x_2 + \ldots + b_{nnn}x_n^3 + \ldots \qquad (15.1)$$

where y is the approximate response, $x_1, \ldots, x_n$ are the factors and $b_0, \ldots, b_{nnn}$ are the coefficients estimating the true and unknown coefficients $\beta_0, \ldots, \beta_{nnn}$. The degree of a term in such a polynomial is defined as the number of variables multiplied together and the degree of the polynomial of eqn. 15.1, truncated after term $b_{nnn}x_n^3$ is therefore 3. If the same equation were truncated after any one of the terms $b_{11}x_1^2$ to $b_{nn}x_n^2$, a second-degree polynomial would result.

In experimental optimization, one usually applies only first- or second-degree polynomials. The latter is, of course, a better approximation than the former. Also, a second-degree polynomial can contain a minimum or a maximum. Therefore, first-degree polynomials are usually applied in the first stages of an optimization, while second-degree polynomials are used in the vicinity of the optimum. Often the determination of the maximum or minimum in a second-degree polynomial constitutes the last step in an optimization procedure.

The coefficients $b_0, \ldots, b_{nnn}$ are determined from a number of equations at least equal to the number of coefficients. If this number is r, this means that r experimental responses y_j must be determined. The method of obtaining the coefficients $b_0, \ldots, b_{nnn}$ from these experimental data is shown below and explained in the mathematical section (section 15.5). The description of regression methods given here follows closely Beveridge and Schechter's (1970) book.

The coefficients can be determined from any set of experimental responses containing r or more data, but their calculation is much easier when the experiments are organized in a factorial experiment. For this purpose, the factorial experiment must be planned more carefully than when it is used only for statistical tests (see Chapters 12 and 13). In particular, it is preferable to use orthogonal factorial plans. To begin with, it is convenient first to scale the factors and to express the factor levels in scaled units. This is done by using an equation such as

$$\zeta_i = \frac{x_i - x_{i,o}}{R_i} \qquad (15.2)$$

where R_i is the range of variation of interest for the ith factor, x_i the value of the ith factor, ζ_i its scaled value and $x_{i,o}$ the value of the ith factor for the so-called base point. Suppose that the concentration of a reagent is one of the factors investigated and that a two-level factorial experiment must be carried out around the value 2 M (the base point value for this factor). For practical reasons, the concentration can vary only between 0 and 4 M. If the levels are situated at 1 and 3 M, the scaled values are

$$\frac{1 - 2}{2} = -1/2$$

and

$$\frac{3 - 2}{2} = 1/2$$

as the range of variation around the base point is 2. The factorial experiment is then carried out symmetrically around the base point. For a 2^2 design, this means that the base point is the centre of a square in the x_1 - x_2 graph, the factors x_1 and x_2 being expressed in scaled units. For a 2^3 design it is the centre of a cube in the (scaled) x_1 - x_2 - x_3 graph (see Fig. 15.4). The responses (y_i values) are recorded not only at the levels prescribed by the factorial plan, as was the case for statistical testing in Chapters 12 and 13, but also at the base point and the responses at the factor levels (for example, at the corners of the square in Fig. 15.4) are expressed as deviations from y_o, the experimental result at the base point (centre of the square).

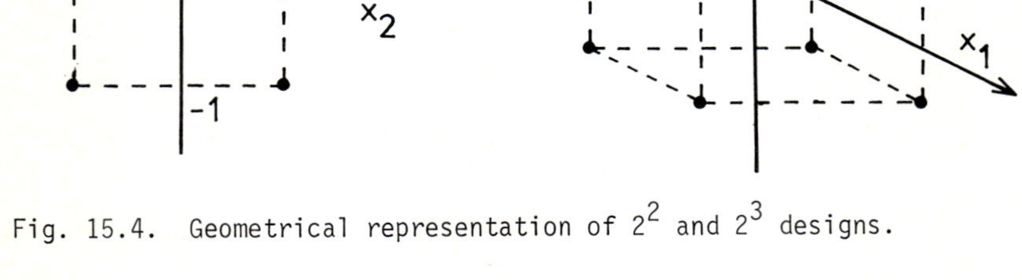

Fig. 15.4. Geometrical representation of 2^2 and 2^3 designs.

The determination of the experimental result at the base point and the expression of the other results as deviations permits the elimination of one of the unknown b values. It can be shown that eqn. 15.1 is now transformed into

$$y' = b_1\zeta_1 + \ldots + b_n\zeta_n + b_{11}\zeta_1^2 + b_{12}\zeta_1\zeta_2 + \ldots + b_{nn}\zeta_n^2 + b_{111}\zeta_1^3 + b_{112}\zeta_1^2\zeta_2 + \ldots + b_{nnn}\zeta_n^3 + \ldots \quad (15.3)$$

The b coefficients used in eqn. 15.3 are estimates of the regression coefficients β for the scaled variables and not, as in eqn. 15.1, estimates of the regression coefficients for the x variables.

Let us first consider the simplest possible case, i.e., a first-degree polynomial for two factors. Eqn. 15.3 then reduces to

$$y' = b_1\zeta_1 + b_2\zeta_2 \quad (15.4)$$

It can be shown that the coefficients b_1 and b_2 can be obtained from

$$b_i = \frac{D_i}{C_{ii}} \quad (15.5)$$

where

$$D_i = \sum_{j=1}^{r} \zeta_{ij}\, y'_j \quad (15.6)$$

and

$$C_{ii} = \sum_{j=1}^{r} \zeta_{ij}^2 \quad (15.7)$$

Let us apply this equation to the results of the factorial experiment described in Table 15.II.

Table 15.II

A 2^2 factorial experiment

Experiment	Location of original values		Location of scaled values		Original result	Reduced result
j	x_{1j}	x_{2j}	ζ_{1j}	ζ_{2j}	y	y'
0	2	3	0	0	6	0
1	3	5	1	1	8.5	2.5
2	3	1	1	-1	7.5	1.5
3	1	5	-1	1	5.5	-0.5
4	1	1	-1	-1	4	-2.0

$$D_1 = 1.(2.5) + 1.(1.5) - 1.(-0.5) - 1.(-2.0) = 6.5$$

$$D_2 = 1.(2.5) - 1.(1.5) + 1.(-0.5) - 1.(-2.0) = 2.5$$

$$C_{11} = 1^2 + 1^2 + (-1)^2 + (-1)^2 = 4 = C_{22}$$

$$b_1 = \frac{6.5}{4} = 1.625$$

$$b_2 = \frac{2.5}{4} = 0.625$$

$$y' = 1.625 \cdot \zeta_1 + 0.625 \cdot \zeta_2$$

or in the original scale

$$(y-6) = 1.625\ (x_1 - 2) + 0.312\ (x_2 - 3)$$

$$y = 1.814 + 1.625\ x_1 + 0.312\ x_2$$

The two-level factorial plan is not sufficient for the determination of all the coefficients when one needs to fit a quadratic surface. The approximating equation for the two-factor case is then given by

$$y' = b_1\zeta_1 + b_2\zeta_2 + b_{12}\zeta_1\zeta_2 + b_{11}\zeta_1^2 + b_{22}\zeta_2^2 \qquad (15.8)$$

and contains five coefficients. The 2^2 factorial experiment yields only four y_i values and therefore four equations. It can be shown that the four equations permit one to calculate b_1, b_2, b_{12} and $(b_{11} + b_{22})$ in the following way

$$b_1 = \frac{D_1}{C_{11}} \tag{15.9}$$

$$b_2 = \frac{D_2}{C_{22}} \tag{15.10}$$

$$b_{12} = \frac{D_{12}}{C_{1122}} \tag{15.11}$$

where

$$D_{ik} = \sum_{j=1}^{r} \zeta_{ij} \zeta_{kj} y'_j \tag{15.12}$$

$$C_{iikk} = \sum_{j=1}^{r} \zeta_{ij}^2 \zeta_{kj}^2 \tag{15.13}$$

For the two-level design of Table 15.II

$$C_{11} = C_{22} = C_{1122} = C_{2211} = 4$$

$$D_{12} = 2.5 - 1.5 + 0.5 - 2.0 = -0.5$$

$$b_{12} = \frac{D_{12}}{C_{1122}} = \frac{-0.5}{4} = -0.125$$

$$b_{11}C_{1111} + b_{22}C_{2211} = D_{11} \tag{15.14}$$

$$b_{11}C_{1122} + b_{22}C_{2222} = D_{22} \tag{15.15}$$

As all of the C coefficients have the same value (4), these two equations reduce to a single equation :

$$4\ b_{11} + 4\ b_{22} = \sum_{j=1}^{r} \zeta_{ij}^2\ y'_j \tag{15.16}$$

If one wants to obtain separate values of b_{11} and b_{22}, additional experiments have to be carried out. One means of doing this is to add a third level, so that ζ_{ij} can take values of -1, 0 and +1. The base point experiment serves as one of the nine experiments and so do the four experiments of the two-level factorial. Four additional experiments have to be added, namely experiments 5 - 8 in Table 15.III.

Table 15.III

A 2^3 factorial design

Experiment	Location of scaled values	
j	ζ_{1j}	ζ_{2j}
0	0	0
1	1	1
2	1	-1
3	-1	1
4	-1	-1
5	0	-1
6	0	1
7	-1	0
8	1	0

The C values are now

$$C_{1111} = C_{2222} = 6$$

$$C_{1122} = C_{2211} = 4$$

so that eqns 15.14 and 15.15 become

$$6\ b_{11} + 4\ b_{22} = D_{11} \qquad (15.17)$$

$$4\ b_{11} + 6\ b_{22} = D_{22} \qquad (15.18)$$

Two distinct equations are therefore obtained, so that b_{11} and b_{22} can be determined separately. The most important disadvantage of using the three-level design for this purpose is that many more experiments are carried out than necessary. For the three-factor case, one carries out 27 experiments to obtain the values of 10 coefficients. A more efficient procedure is the use of the central composite design introduced by Box and Wilson (1951). One adds 2n + 1 experiments to the 2^n design, one of the additional experiments being performed at the base point and the other along the coordinate axes (see Fig. 15.5) at a distance α.

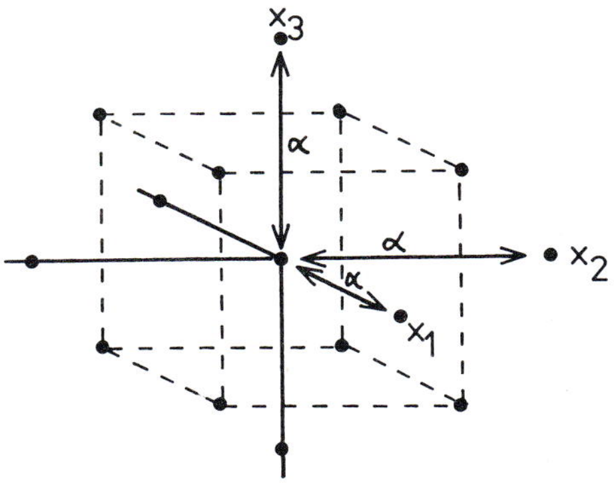

Fig. 15.5. Central composite design for three factors.

For the two-factor case, this procedure requires nine experiments and is therefore not more economical than the three-level design. If $\alpha = 1$, the design is converted into the 3^2 design. For a three-factor problem, however, one needs only 15 experiments while the three-level design needs 27.

As the two-factor central composite design reduces to the 3^2 design, we shall now consider a three-factor problem. The design is given in Table 15.IV. The value of α is now preferably 1.215. In this instance, the design is again orthogonal, which simplifies the calculations (see also section 15.5). The b coefficients are obtained from

$$b_i = \frac{D_i}{C_{ii}} \tag{15.19}$$

$$b_{ik} = \frac{D_{ik}}{C_{ikik}} \tag{15.20}$$

and

$$b_{11}C_{1111} + b_{22}C_{2211} + b_{33}C_{3311} = D_{11}$$

$$b_{11}C_{1122} + b_{22}C_{2222} + b_{33}C_{3322} = D_{22}$$

$$b_{11}C_{1133} + b_{22}C_{2233} + b_{33}C_{3333} = D_{33}$$

Table 15.IV

A three-factor central composite design

	Experiment j	Scaled value level ζ_{1j}	ζ_{2j}	ζ_{3j}
Base point	0	0	0	0
Two-level design	1	1	1	1
	2	1	1	-1
	3	1	-1	1
	4	1	-1	-1
	5	-1	1	1
	6	-1	1	-1
	7	-1	-1	1
	8	-1	-1	-1
Additional 2n points along coordinate axes	9	α	0	0
	10	$-\alpha$	0	0
	11	0	α	0
	12	0	$-\alpha$	0
	13	0	0	α
	14	0	0	$-\alpha$

For example

$$C_{2222} = 0^4 + 1^4 + 1^4 + (-1)^4 + (-1)^4 + 1^4 + 1^4 + (-1)^4 + (-1)^4 + 0^4$$
$$+ 0^4 + \alpha^4 + (-\alpha)^4 + 0^4 + 0^4 = 12.35 = C_{iiii}$$

(there is a small calculation error in Beveridge's book, as there the value 17.64 is given).

An application of the three-level central composite design for the optimization of the assay of transfer ribonucleic acid is due to Rubin et al. (1971). They used five successive central composite designs because the final and optimal parameter values turned out to be a considerable distance away from the initial values. This shows that one should not think that it suffices to have a simple equation (here a quadratic equation) for a response surface in order to be able to calculate the optimum with absolute certainty. Very often even quadratic equations are only gross approximations of the response surface, valid only in the experimental region. A four-factor example concerning the optimization of atomic-absorption spectrophotometric experimental conditions was given by Cellier and Stace (1966).

Instead of polynomials, one can use other models, such as the Gaussian model. Olansky and Deming (1976) located the optimum of a colorimetric procedure using the simplex procedure and, in order to understand the response in the neighbourhood of the optimum, they carried out a factorial experiment. Assuming an approximate Gaussian behaviour, they fitted the data to the following model :

$$z = k_1 \exp \left[-(((x_2/x_1 + x_2 + 2.0)) - k_2)^2) / (2 k_3^2) \right] \times$$
$$(2.0/x_1 + x_2 + 2.0)) (1 - \exp (-k_4 x_1))$$

where k_1, ..., k_4 are the coefficients obtained by least-squares fitting and x_1 and x_2 are the two factors.

It is also not necessary to use orthogonal designs. Calculations are much easier when such a design is used but, on the other hand, computer programs for multiple regression are readily available, so that ease of computation is often an unimportant factor in the choice of levels and designs. An example of an application of non-orthogonal designs to least-squares fitting was given by Morgan and Deming (1974). They used the results of 16 experiments of a factorial design, 26 results obtained at the vertices of a simplex optimization procedure and 25 other observations in order to arrive at a quadratic equation.

Other designs exist and the reader who wishes to be acquainted with them should consult the books by Davies (1956) and Cochran and Cox (1957). Good preliminary introductions can be found in articles by Leuenberger et al. (1976) and Koehler (1960) for pharmaceutical and industrial chemists, respectively.

15.3.2. Determination of the optimum from regression equations

15.3.2.1. Linear programming

If the equation is of the form

$$y = b_0 + b_1x_1 + b_2x_2 + \ldots + b_nx_n$$

and additional constraints are given, such as boundaries between which the parameters may be considered, the optimum can be found by the method of linear programming (see Chapter 21).

15.3.2.2. Quadratic surfaces

In the simplest possible case, when there is only one variable

$$y = b_0 + b_1x_1 + b_{11}x_1^2$$

the optimum is found for the value of x_1 where

$$\frac{dy}{dx_1} = 0$$

For two factors, one finds the optimum by partial differentiation with respect to x_1 and x_2

$$\frac{\partial y}{\partial x_1} = 0$$

and

$$\frac{\partial y}{\partial x_2} = 0$$

For the equation

$$y = b_0 + b_1x_1 + b_{11}x_1^2 + b_2x_2 + b_{22}x_2^2 + b_{12}x_1x_2$$

this yields the system

$$b_1 + 2b_{11}x_1 + b_{12}x_2 = 0$$

$$b_2 + 2b_{22}x_2 + b_{12}x_1 = 0$$

15.4. A COMPARISON OF METHODS

In section 14.2.5, we have already made a comparison of the simplex and factorial methods. Brooks (1959) compared the factorial design, the univariate and the steepest ascent method, and his conclusion is that the steepest ascent method is better than the univariate method, which is itself better than the factorial method. These surprisingly poor achievements of the factorial method are due to the fact that very often the quadratic functions used do not adequately describe the surface. It should be noted, however, that Brooks limited his research to two-factor problems and that one might expect that the univariate method becomes less efficient when there are more factors.

Beveridge and Schechter (1970) concluded that there is no criterion for defining the effectiveness of a particular optimization design, so that no method can be singled out as the best one. In general, it seems that sequential methods such as the simplex or steepest ascent method are more effective in the sense that they approach the optimum more rapidly. These methods are also indicated for the optimization of procedures that take a considerable time and cannot be carried out simultaneously, because one can stop the optimization after a few experiments without having found the optimum, but still with a sufficient increase in response. Factorial methods are to be preferred for lengthy procedures that can be carried out simultaneously. They usually permit a better understanding of the surface, they allow time trends to be eliminated and they are indicated for the optimization of procedures in which a number of observations can be obtained simultaneously or in rapid succession.

15.5. MATHEMATICAL SECTION : MULTIPLE REGRESSION

15.5.1. Surface fitting

The representation of the influence of a set of n variables $x_1, x_2, \ldots, x_n$ on a response variable y (sometimes called the dependent variable) can be described by a mathematical function. Geometrically, this problem can be solved by the construction of a surface in an (n+1) dimensional space. In this space, the first n dimensions correspond to the variables $x_1, x_2, \ldots, x_n$ and the (n+1)th dimension to the response variable y. The general algebraic form of the response surface is given by

$$y = f(x_1, x_2, \ldots, x_n) \qquad (15.21)$$

The problem that remains to be solved is the nature of the function f. In general, a hypothesis of the following type is made : f belongs to a class of functions for which one or several parameters are unknown. This hypothesis can be written in the following way :

$$y = f(x_1, x_2, \ldots, x_n \; ; \; \beta_0, \beta_1, \ldots, \beta_m) \qquad (15.22)$$

To determine estimates $b_0, b_1, \ldots, b_m$ of the unknown parameters $\beta_0, \beta_1, \ldots, \beta_m$, the r experimentally measured values of the variable y $(y_1, y_2, \ldots, y_r)$ are compared with the values given by the function 15.22. In the same way as in section 3.2.8 (see eqn. 3.33), the measured value y_j is expressed as the sum of function f and an error e_j

$$y_j = f(x_{1j}, x_{2j}, \ldots, x_{nj} \; ; \beta_0, \beta_1, \ldots, \beta_m) + e_j \qquad j=1,2,\ldots,r \qquad (15.23)$$

The parameters are estimated by minimizing the sum of the squares of the differences between the measured values (y_j) and the values given by the function f

$$\text{Min} \sum_{j=1}^{r} (y_j - f(x_{1j}, x_{2j}, \ldots, x_{nj} ; \beta_0, \beta_1, \ldots, \beta_m))^2 \qquad (15.24)$$

This technique for estimating the β parameters is called the least-squares fitting method. The estimates are obtained by setting equal to zero the partial derivatives of eqn. 15.24 with respect to all β parameters. In the following sections, some special types of functions are examined.

15.5.2. Fitting of plane surfaces

The general linear function of the variables $x_1, x_2, \ldots, x_n$ can be written as

$$y = \beta_0 + \beta_1 x_1 + \beta_2 x_2 + \ldots + \beta_n x_n \qquad (15.25)$$

In this particular case of eqn. 15.22, the number of parameters is n+1. Expression 15.24 for obtaining the least-squares estimates now becomes

$$\underset{\beta_0, \beta_1, \ldots, \beta_n}{\text{Min}} \sum_{j=1}^{r} (y_j - \beta_0 - \beta_1 x_{1j} - \ldots - \beta_n x_{nj})^2 \qquad (15.26)$$

The estimates are found by setting to zero the partial derivatives of this expression with respect to the parameters. This yields n+1 linear equations with n+1 unknown values.

Very often it will be convenient to introduce deviations from a given base point for both the x and y variables and in some instances to scale the x variables. This leads to new variables, y' and ζ, given by

$$y' = y - y_0 \qquad (15.27)$$

and

$$\zeta_i = \frac{x_i - x_{i,o}}{R_i} \qquad (15.28)$$

where R_i is a scaling factor and $(x_{1,0}, x_{2,0}, \ldots, x_{n,0}, y_0)$ represents the base point. By these transformations, eqn. 15.25 now becomes

$$y' = \beta_1\zeta_1 + \beta_2\zeta_2 + \ldots + \beta_n\zeta_n \qquad (15.29)$$

It must be observed that the parameters β are different from those in eqn. 15.25 and that the presence of a base point makes it possible to disregard the parameter β_0.

In the same way as in the general case, it is now possible to obtain estimates of the β parameters by considering the following expression

$$\underset{\beta_1, \beta_2, \ldots, \beta_n}{\text{Min}} \quad \sum_{j=1}^{r} (y'_j - \beta_1\zeta_{1j} - \beta_2\zeta_{2j} - \ldots - \beta_n\zeta_{nj})^2 \qquad (15.30)$$

Setting to zero the partial derivatives of this expression with respect to the parameters β, the following n equations for determining the estimates b_1, b_2, ..., b_n are obtained

$$\sum_{j=1}^{r} \zeta_{1j} (y'_j - \sum_{k=1}^{n} b_k \zeta_{kj}) = 0$$

$$\vdots$$

$$\sum_{j=1}^{r} \zeta_{nj} (y'_j - \sum_{k=1}^{n} b_k \zeta_{kj}) = 0$$

and this in turn gives the following general equation

$$\sum_{j=1}^{r} \sum_{k=1}^{n} \zeta_{ij} \zeta_{kj} b_k = \sum_{j=1}^{r} \zeta_{ij} y'_j \qquad i = 1,2,\ldots,n \qquad (15.31)$$

By defining

$$C_{ik} = \sum_{j=1}^{r} \zeta_{ij}\, \zeta_{kj} \qquad (15.32)$$

and

$$D_i = \sum_{j=1}^{r} \zeta_{ij}\, y'_j \qquad (15.33)$$

the general equation becomes

$$\sum_{k=1}^{n} C_{ik}\, b_k = D_i \qquad (15.34)$$

Often when the influence of the variables $\zeta_1, \zeta_2, \ldots, \zeta_n$ on a dependent variable y' is being investigated, it is possible to choose the values of the ζ variables freely. Suppose the values ζ_{ij} are chosen in such a way that all C_{ik} are zero when i is different from k. Eqn. 15.34 then takes the form

$$C_{kk}\, b_k = D_k \qquad k = 1,2,\ldots,n \qquad (15.35)$$

a trivial solution of which is given by

$$b_k = \frac{D_k}{C_{kk}}$$

A set of experiments which has this property is called an orthogonal design. This term results from the property that the vectors $\vec{C}_i$ given by

$$\vec{C}_i = \begin{bmatrix} C_{i1} \\ C_{i2} \\ \vdots \\ C_{in} \end{bmatrix}$$

are orthogonal vectors in an n-dimensional space.

An example of an orthogonal design with three variables, x_1, x_2 and x_3, and for which $x_{1,0}$, $x_{2,0}$, $x_{3,0}$ and y_0 are all zero, is given in Table 15.IV.

It can easily be seen that

$$C_{ik} = 0 \qquad \text{if } i \neq k$$

and

$$C_{kk} = 8 \qquad k = 1,2,3$$

If only experiments 2, 3, 4 and 8 or experiments 1, 5, 6 and 7 are considered, C_{ik} (for $i \neq k$) are also all zero.

15.5.3. Fitting of a quadratic surface

When considering n variables, the general quadratic function contains all linear terms, all squares of variables and all cross products. It can be written as

$$y' = \sum_{k=1}^{n} \beta_k \zeta_k + \sum_{i=1}^{n} \sum_{k=i}^{n} \beta_{ik} \zeta_i \zeta_k \qquad (15.36)$$

In this equation, it has already been assumed that if all ζ variables are zero, then so is the response variable y. The number of coefficients β to be determined is $n + (n(n+1))/2$ and at least as many measurements must be made for each of the variables. The least-squares estimates of the parameters are given by

$$\underset{\substack{\beta_1 \cdots \beta_n \\ \beta_{11} \cdots \beta_{nn}}}{\text{Min}} \quad \sum_{j=1}^{n} (y'_j - \sum_{k=1}^{n} \beta_k \zeta_k - \sum_{i=1}^{n} \sum_{k=i}^{n} \beta_{ik} \zeta_{ij} \zeta_{kj})^2 \qquad (15.37)$$

Setting to zero the derivatives of this expression with respect to the parameters, one obtains two sets of linear equations that make it possible to calculate estimates b_k and b_{ik} of β_k and β_{ik}

$$\sum_{k=1}^{n} b_k C_{km} + \sum_{i=1}^{n} \sum_{k=i}^{n} b_{ik} C_{ikm} = D_m \qquad m = 1,2,\ldots,n \tag{15.38}$$

and

$$\sum_{k=1}^{n} b_k C_{kms} + \sum_{i=1}^{n} \sum_{k=i}^{n} b_{ik} C_{ikms} = D_{ms} \qquad \begin{matrix} m = 1,2,\ldots,n \\ s = r, \ldots, n \end{matrix} \tag{15.39}$$

where the following definitions are used

$$D_m = \sum_{j=1}^{r} y'_j \zeta_{mj} \tag{15.40}$$

$$D_{ms} = \sum_{j=1}^{r} y'_j \zeta_{mj} \zeta_{sj} \tag{15.41}$$

$$C_{km} = \sum_{j=1}^{r} \zeta_{kj} \zeta_{mj} \tag{15.42}$$

$$C_{kms} = \sum_{j=1}^{r} \zeta_{kj} \zeta_{mj} \zeta_{sj} \tag{15.43}$$

and

$$C_{ikms} = \sum_{j=1}^{r} \zeta_{ij} \zeta_{kj} \zeta_{mj} \zeta_{sj} \tag{15.44}$$

Eqns. 15.38 and 15.39 form a system of n + (n(n+1)/2) equations with the same number of unknown values b_k and b_{ik}. These equations are linear so that a solution can easily be found by successive elimination of variables.

The fitting of plane and quadratic surfaces can be generalized to more complex functions in a straightforward manner. In particular, general polynomials and gaussian functions have received attention.

REFERENCES

S. Arpadjan, K. Doerffel, K. Holland-Letz, H. Much and M. Paunach, Z. anal. Chem., 270 (1974) 257.
G.S.C. Beveridge and R.S. Schechter, Optimization : Theory and Practice, McGraw-Hill, New York, 1970.
G.E.P. Box and K.B. Wilson, J. Roy. Statist. Soc., 13 B (1951) 1.
S.H. Brooks, Oper. Res., 7 (1959) 430.
K.M. Cellier and H.C.T. Stace, Appl. Spectrosc., 20 (1966) 26.
W.G. Cochran and G.M. Cox, Experimental Designs, Wiley, New York, 1957.
O.L. Davies, The Design and Analysis of Industrial Experiments, Oliver and Boyd, Edinburgh, 1956.
T.L. Koehler, Chem. Eng., 25 (1960) 142.
E. Leuenberger, P. Guitard and H. Sucker, Pharmazie in unserer Zeit, 5 (3) (1976) 65.
S.L. Morgan and S.N. Deming, Anal. Chem., 46 (1974) 1170.
A.S. Olansky and S.N. Deming, Anal. Chim. Acta, 83 (1976) 241.
I.B. Rubin, T.J. Mitchell and G. Goldstein, Anal. Chem., 43 (1971) 717.

PART III

COMBINATORIAL PROBLEMS

Chapter 16

AN INTRODUCTION TO COMBINATORIAL PROBLEMS IN ANALYTICAL CHEMISTRY

16.1. INTRODUCTION

In the preceding part, we discussed the relatively simple problem of how to optimize a response as a function of a relatively small number of well defined variables. Citing a remark made by Nalimov, we stated that a trend in modern science is to pass from the study of well organized systems to diffuse systems. In this part of the book, we consider the optimization of even more diffuse systems than in Part II. The problems to be solved in this part are of a combinatorial nature. Most of them can be solved by pattern recognition methods or related techniques and operational research. In fact, these methods will be seen to overlap (for example, graph theory, which is considered to be part of operational research, can be used for unsupervised learning techniques, a subdivision of pattern recognition).

In a few instances, other techniques can be applied, such as the reduction of large matrices, in conjunction with least-squares techniques and information theory. These techniques are used for solving problems concerning the selection of preferred sets. For this reason, they are discussed together in Chapter 17. Here too, we must note some overlap (see the section on the relation between hierarchical clustering and information theory in Chapter 18).

Shoenfeld and De Voe (1976) noted that the classification of applications of statistical and numerical methods to analytical chemistry resulted in much frustration. According to them, this is due to the non-uniformity of nomenclature, and also to the lack of fundamental understanding of the basic principles that underlie the diverse applications of these mathematical techniques. This is certainly true in a rapidly developing area such as that described here.

Many chemists working in this area are not so much interested in the application of mathematical techniques in analytical chemistry as in solving a particular problem such as one of the five cited in section 16.2. They find some technique - often in another domain of science - and publish it, using the terminology of this particular area of science. Adding to this the overlap between methods and techniques mentioned above, it is clear that it is very difficult to arrive at a complete classification of methods. We do not attempt to derive a consistent nomenclature here, but some of the interrelationships between the different methods are given.

Most mathematical methods of this part of the book have been introduced into analytical chemistry rather recently. Some of them have not been used at all in this field, but we feel that it is probable that such theories or methods as game theory and PERT will be applied in analytical chemistry and therefore they are introduced in following chapters. After all, it would have seemed improbable 10 years ago that information theory and hierarchical clustering methods would find real uses in analytical chemistry.

In the following section, five typical combinatorial problems, the solutions of which are given in later chapters of Part III, are described in order to show that many such optimization problems occur in analytical chemistry. As mentioned above, Part III is devoted largely to the discussion of pattern recognition techniques and operational research. Therefore, section 16.3 gives an introduction to pattern recognition and as some of the more important of these techniques assume normal distributions, section 16.4 gives a short account of multivariate statistics. Section 16.5 introduces operational research.

16.2. SOME EXAMPLES OF COMBINATORIAL PROBLEMS

(1) A GLC example (I)

There are many stationary liquid phases available for use in gas-liquid chromatography (GLC). McReynolds (1970) published a set of 226 phases. Many workers have pointed out that there is a need for the selection of a set of preferred phases. In fact, such sets have been proposed several times. One means of arriving at such a set was explored by Dupuis and Dijkstra (1975) and later by Eskes et al. (1975). They noted that the purpose of GLC (and of any other analytical method) is to produce information (see also Chapter 8), and they tried to assess the information that could be produced by a set of 2, 3, ..., n phases, so as to determine which set yields the most information ; this would then constitute the preferred set of phases. The selection of optimal sets of analytical attributes or features (GLC stationary phases, wavelengths for spectrophotometry, mass spectrometric positions, etc.) is discussed in Chapter 17.

(2) A GLC example (II)

Another approach to the same problem is to try to classify the GLC phases. It seems evident that a preferred set should be composed of phases with different characteristics. A preliminary step before the selection of phases should then be to classify them. Ideally, if one needs a set of 10 phases, one should divide the 226 phases into 10 groups and select one phase (the best) out of each of the groups. This application stresses the fact that one does not merely look for the 10 individual best phases (this could have been done by using the evaluation methods described in Part I), but for the optimal combination of 10 phases.

For each phase, one obtains the retention times of 10 test substances (probes) and these are used to carry out the classification. The 10 retention times given for each phase constitute a pattern. Pattern recognition methods can therefore be used.

(3) Milk and thyroid examples

One of the more important applications of analytical chemistry is to discriminate between two or more kinds of samples, for example between cows' and goats' milk. One obtains a set of results for different parameters for a number of samples known to belong to one of the categories between which one should make a discrimination. In the milk example, one determines the fatty acid percentages for several fatty acids (typically about 20) for a number of samples that are known to be cows' milk, and the same for a number of samples of goats' milk. The optimization problem is to select a combination of, for example, 5 out of the original 20 parameters in such a way that the best possible discrimination is obtained.

In the thyroid example, one carries out five clinical chemical tests to make a diagnosis of the thyroid state of a patient, for example, in order to make a distinction between euthyroid (normal) and hypothyroid cases. These tests cost money and time and the optimization problem is to establish whether all five tests are necessary, and if not, to make a selection of, say, three tests so that the discrimination obtained is as good as possible. These problems also are solved by pattern recognition methods.

(4) Ion-exchange problem

The ion-exchange separation of mixtures of several metal ions into individual components can usually be achieved in several ways. The analytical chemist has at his disposal a data set consisting of distribution coefficients of the metal ions to be separated and his task is to elaborate an optimal flow scheme. Should one first elute ion a using solvent A and then ion b using solvent B, or should one rather start by eluting c and d together with the purpose of separating them in a second step ? One can construct a network or graph containing all of these possibilities. The problem is then to determine which way one should choose of the many possible alternatives in the network. This is a typical operational research problem.

(5) Clinical laboratory example

There is an overwhelming array of apparatus available for clinical chemistry laboratories. These laboratories have often a rather limited programme. They only perform a few different determinations and one can usually guess with some precision how many of each of these determinations they will have to carry out. Knowing this and knowing what apparatus (manual or automated, 1-, 2-, or n-channel apparatus) is available and the costs of buying and operating each of these instruments, it is possible to determine which set (combination) of apparatus satisfies the requirements of the laboratory most economically. This also is a typical operational research problem.

16.3. PATTERN RECOGNITION AND RELATED TECHNIQUES

Modern analytical methods allow the determination of many substances simultaneously and computers provide facilities for storing or handling the large amounts of data obtained in this way. As many data are obtained at a time, they should be used together, particularly because the results obtained for the parameters are often related to each other.

Clinical chemistry is an area of analytical chemistry where one often determines several chemical parameters for the same sample. However, one uses the results individually (e.g., if a parameter is higher than normal the patient should be suspected of having a certain disease) or successively (e.g., if parameter a is high, but b is normal, certain conclusions can be drawn). As the data are obtained simultaneously and concern one sample, it would be preferable to use them simultaneously instead of individually, and Fig. 16.1, taken from Winkel (1973), illustrates one of the advantages of doing so. The ellipsoidal contour line delineates the region expected to contain 68.3% of the samples, as determined from a two-dimensional gaussian distribution. The two parameters are correlated as the main axis of the ellipsoid is not parallel to the abscissa. In chapter 3, it was seen that the correlation coefficient is

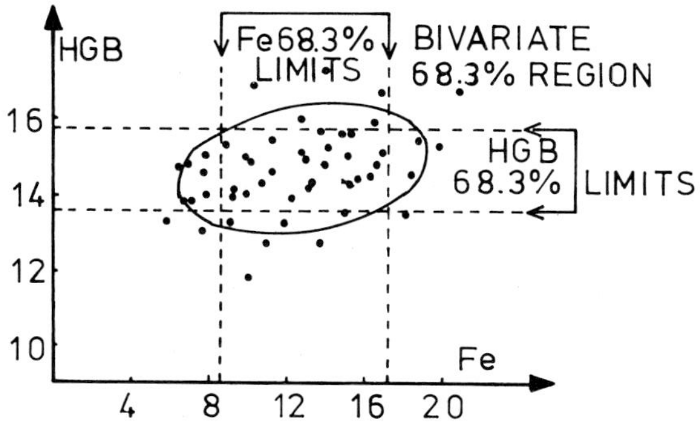

Fig. 16.1. A bivariate distribution. The parameters HGB (B-haemoglobin) and FE (S-iron, transferrin bound) were measured for 52 healthy men (from Winkel, 1973).

one of the parameters that characterizes a bivariate distribution. If one uses the parameters individually (i.e., two separate univariate gaussian distributions), the normal region would be the rectangle. Several points fall outside the rectangle and would be declared abnormal by the univariate thinking person, whereas in fact they are normal, as would be recognized by the bivariate thinking observer.

Another, and in the present context still more important, advantage is that specific patterns can be observed for different kinds of samples. Let us consider, for example, the milk problem mentioned in section 16.2 and suppose that two groups of samples, cows' (C) and goats' (G) milk have to be differentiated and two parameters are used. The parameters of this example are hypothetical and are therefore called parameters 1 and 2 (Fig. 16.2). They might represent, for example, the content of butyric acid and stearic acid, respectively. The two parameters define a two-dimensional space and the C and G groups are found to occupy different locations in this space. They are said to form clusters and, by determining to which of the two clusters an unknown sample belongs, one is able to decide whether it is a C or a G sample.

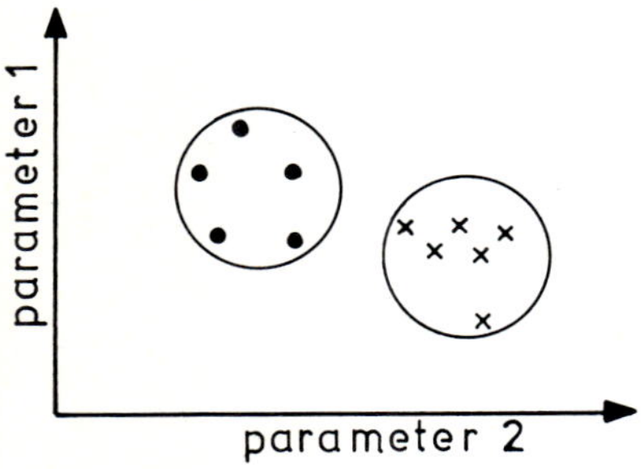

Fig. 16.2. The formation of clusters in a two-dimensional space.

This reasoning can be generalized to more dimensions. Consider again the milk example. A particular milk fat sample is characterized by its fatty acid distribution or pattern. If there are 20 fatty acids, each milk fat sample can be viewed as a point in 20-dimensional space, its coordinates x_i being the percentage fatty acid concentrations. Such a point is conveniently represented by a vector (pattern vector)

$$\vec{x} = (x_1, x_2, \ldots, x_i, \ldots, x_{20})$$

The vector is composed of d measurement results (20 in the milk sample) constituting a set of d scalar values. The d parameters define the pattern space. In the GLC example, the pattern space consists of 10 dimensions and contains 226 points. In the thyroid example, there are five dimensions and each individual patient result is represented by five scalar values, the results of the five clinical tests.

If one were able to observe the pattern space visually, as was possible in the two dimensional case in Fig. 16.2, one would note that the points tend to form groups or clusters. For example, cows' milk samples would cluster together and so would goats' milk samples. In the same way, GLC phases with similar characteristics or patterns will tend to form clusters. The recognition of similar patterns or the isolation of the clusters is therefore of great analytical interest. As d-dimensional points cannot be observed visually, one needs mathematical methods in order to deal with the patterns and clusters in an d-dimensional space.

The general term "pattern recognition" refers to automatic procedures for classifying individual observations into discrete groups on the basis of a multivariate data matrix (Shoenfeld and De Voe, 1976). Pattern recognition has been the object of much study in the last few years. It is one of the most important techniques of the discipline of chemometrics. This term, which seems to have been coined by Kowalski, is defined by this author (1975) as including the application of mathematical and statistical methods to the analysis of chemical measurements. Both the milk and the GLC problem can be solved by pattern recognition methods, but there is an essential difference between them. In the milk problem, the groups between which a classification must be carried out are known (goats and cows). One calls this supervised learning or pattern recognition (in the restricted sense). In the GLC problem, one does not know the groups (one does not even know how many groups to expect). This is called unsupervised learning or pattern cognition.

One of the difficulties in pattern recognition is what has repeatedly been called "the curse of dimensionality". If many variables are present, the classification problem may become too complex, and a reduction in the number of dimensions can help to make it more manageable. Further, if one is able to represent the data originally present in d dimensions in two or three dimensions in such a way that the similarities and dissimilarities between the data points are conserved, at least partially, this can be a valuable aid in arriving at a better understanding of the data.

In general, the human observer is often a better pattern recognizer than the automatic methods described in this part of the book, at least when the data are represented in two or three dimensions. Therefore, it is interesting to be able to make at least a preliminary evaluation of the data present using one of the low-dimensional representation (display) methods. In this context, two multivariate statistical methods must be investigated, namely principal components and factor analysis. Not all workers consider that these techniques are pattern recognition methods. They are, however, certainly related methods. Most books on pattern recognition, such as those by Duda and Hart (1973) and Andrews (1972), contain at least references to them. Many typical pattern recognition data sets have also been investigated by these two techniques (the GLC problem, for instance). The principal components method formed the basis of one of the more interesting pattern recognition techniques (SIMCA, Wold, 1974).

These methods reduce dimensionality by forming linear combinations of the features that determine the original dimensions. Their principal object is therefore to condense the more essential information present in the data and, due to interdependent variables, in such a way that one obtains a few more "fundamental" variables. In the GLC example, there are originally 10 variables (the retention indices of the 10 test compounds) and the information is condensed into two major variables, the first of which, for example, is identified as the more fundamental parameter, polarity. As only two variables remain, a two-dimensional representation is possible. Principal components and factor analysis are discussed in Chapter 19.

In some classifications of pattern recognition, one considers two successive steps, namely feature extraction and classification (see, for example, Young and Calvert, 1974). In the feature extraction, the pattern vector $\vec{x}_d$ is transformed into a feature vector $\vec{x}_r$, so that the dimensions of $\vec{x}_r$ are less than those of $\vec{x}_d$. Display methods are part of feature extraction. Another category of feature extraction is feature selection : one selects from the d variables (dimensions) present r variables that seem to be the most discriminating. The features obtained therefore correspond to some of the given measurements while in the display methods the dimensionality reduction is obtained by combining some of the variables into a new variable. Feature selection constitutes a means of choosing sets of optimally discriminating variables and, if these variables are the results of analytical tests, this consists in fact in the selection of an optimal combination of analytical tests or procedures. This subject is therefore clearly of special importance in the context of this book.

The second step in the pattern recognition procedure is the classification step, which means that one tries to place the samples characterized by their individual patterns in the category to which they belong. This classification is based on distances between points in the r-dimensional space. The smaller the distance between points, the more probable it is that they belong to the same category. Distances are discussed in Chapter 18.

The way in which the classification is carried out depends firstly on whether one is concerned with a supervised or an unsupervised learning problem. In the supervised problem, one knows the categories in which the samples can be classified. With groups consisting of samples with known classification (learning groups), one develops classification rules (decision functions) that permit one to allocate individual samples to the correct category.

The development of the rules is called the learning or training step. When one knows how to combine the variables in order to obtain an optimal classification, one can calculate the contribution of each of the parameters to the discrimination. Clearly, if one wants to select an optimal combination of three analytical tests, it will consist of the three that have been found to contribute most to the discrimination.

One usually makes a distinction between statistical methods in which the data follow a multivariate normal distribution and distribution-free (non-parametric) methods. There is a tendency to reserve the term pattern recognition for the latter category. In Chapter 20, both the parametric methods and the non-parametric techniques are discussed.

Unsupervised learning or clustering methods have been used less in chemistry. Clustering consists in the generation of clusters or classes, when the classes are undefined *a priori*. The only applications in analytical chemistry known to us are aimed at classifying analytical procedures (or their attributes). This is an important step in the selection of optimal procedures and these methods are therefore discussed in detail in Chapter 18.

16.4. MULTIVARIATE STATISTICAL TECHNIQUES

16.4.1. Introduction

Several definitions have been proposed for multivariate statistical techniques. The one suggested here is the most general and seems to us to be the most appropriate. Multivariate statistical techniques are those which are applied when either more than one independent variable or more than one dependent variable are to be considered simultaneously. It can be observed that this definition also includes two-way and higher-way ANOVA, which have traditionally been excluded from these techniques. The reason for this is that usually multivariate statistical techniques are defined as techniques that require the use of matrices.

This survey does not include multivariate frequency distributions and probability functions, which have already been mentioned in Chapter 3.

In the following sections, a survey will be given of the main multivariate statistical techniques, excluding two-way and greater ANOVA and multiple regression, which have been considered extensively in Chapters 4 and 15.

For the reader interested in the mathematical details of the techniques explained here and in the following chapters, the book by Harris (1975) will provide a clear description. Further reading on the subject should include the books by Morrison (1967) and Kendall and Stuart (1968). A book for users of multivariate statistical techniques and in which only the most necessary mathematical details are given was written by Kendall (1975).

16.4.2. Hotelling's T^2

It was seen in section 3.2.4.2.1 that when two populations are considered and a single variable is measured for elements of the two populations, a t-test makes it possible to test whether there is a significant difference between the mean values of the variable for the two populations. Often, however, there are two or more dependent variables for each of the two populations and it can be queried whether a significant difference exists for any of the dependent variables. For this, a linear combination of the dependent variables is computed by associating weights to each of them. With p dependent variables this gives

$$W = w_1 y_1 + w_2 y_2 + \dots + w_p y_p \tag{16.1}$$

For each element i under examination, the value W_i of the new dependent variable W is given by

$$W_i = w_1 y_{1i} + w_2 y_{2i} + \dots + w_p y_{pi} \tag{16.2}$$

This makes it possible to compute a univariate t-value for the difference between the two populations based upon the new dependent variable W. As this t-value will depend upon the chosen weights, a matrix method is used to compute the largest t-value for any set of weights. The square of the t-value for the optimal set of weights is called Hotelling's T^2 and it has been shown that it has an F-distribution. It provides a means of comparing two populations in the presence of two or more dependent variables.

16.4.3. Multivariate analysis of variance (MANOVA)

Just as Hotelling's T^2 generalizes the t-test, one-way multivariate analysis of variance (one-way MANOVA) generalizes one-way ANOVA. One-way MANOVA can be used whenever the influence of the different levels of a factor on more than one dependent variable is being studied. As with Hotelling's T^2, the set of p dependent variables is reduced to a single variable in the same way as in eqn. 16.1 and for each element i an equation identical with eqn. 16.2 is obtained. Again, as for Hotelling's T^2, the set of weights is determined in such a way that the F-value used for testing the equivalence of variable W for the different levels or populations is as large as possible. The distribution of the statistic obtained in this way is complex and the details will not be discussed here. Another approach for testing this hypothesis is based on the determinants of the covariance matrices obtained by considering the different populations. These methods were discussed extensively by Harris (1975). In the same way as for one-way ANOVA, a generalization exists for each ANOVA model.

16.4.4. Measures of correlation

In the presence of one independent variable and one dependent variable, the classical measure of linear relationship is given by the correlation coefficient (see section 3.2.6.3). In the multiple regression problem there is still one dependent variable but several independent variables. The measure of association is provided by the estimation of the regression coefficients, β. It is equal to the correlation between the independent variables and the predicted dependent variable which is obtained using the least-squares estimations of β as coefficients of the independent variables. This correlation is called the multiple correlation coefficient. It can be shown that it provides the largest correlation of the dependent variable with any linear combination of the independent variables.

When both several independent variables and several dependent variables are considered, a further generalization is necessary. Again, linear combinations of the independent variables and of the dependent variables are considered. For this we define

$$W = \Sigma_j w_j x_j \qquad (16.3)$$

and

$$V = \Sigma_j v_j y_j \qquad (16.4)$$

A measure of the relationship between the two sets of variables is obtained by computing the correlation coefficient between V and W and maximizing over all possible weights v_j and w_j. This measure is called the canonical correlation coefficient.

16.4.5. Analysis of covariance

Analysis of covariance (ANCOVA) was briefly mentioned as a particular case of the general linear model in section 4.2.1.1. It arises when there is one dependent variable and when the independent variables are divided into a set of usually continuous variables (the multiple regression situation) and a set of variables which indicate that the elements of the sample are divided into several groups in a one-way or higher-way classification (the ANOVA situation).

Multivariate analysis of covariance (MANCOVA) is an extension of ANCOVA in the presence of several dependent variables.

As the general linear model generalizes multiple regression, one-way and higher-way ANOVA and ANCOVA, a model has been proposed that generalizes canonical correlation, MANOVA and MANCOVA.

16.4.6. Techniques for reducing a set of variables

All of the techniques described so far are concerned with the relationship between a set of independent variables and a set of dependent variables. In this section, two techniques are outlined for reducing the number of variables by replacing the original set with a smaller one ; usually the new variables do not belong to the original set.

In principal components analysis, linear combinations of the original d variables are considered. These have the general form of eqn. 16.4. The weights are determined in such a way that the variance of W should be as large as possible, subject to the condition that the sum of the squares of the weights be unity as increasing the weights indefinitely will also increase the variance of W in the same way.

The variable W obtained in this way is called the first principal component The second principal component is again a linear combination of the form given by eqn. 16.4 found using the following conditions : its variance is to be maximized, the sum of the squares of the weights is unity and it is uncorrelated with the first principal component.

This process is continued until r principal components have been extracted. It can easily be seen that the variance of successive principal components have non-increasing values and that their sum is equal to the sum of the variances of the original variables. It can also be seen that if some or all of the variables are strongly interconnected a small number of principal components will yield allmost all of the variance of the original set of variables. This smaller set of new variables can then be used for any subsequent statistical analysis instead of the original much larger set.

Factor analysis is a method very similar to principal components analysis.

The main difference is that in factor analysis, the variables of the smaller set are not required to be uncorrelated. These variables are called factors. The various factor analysis methods concentrate either on explaining the percentages of the variance of each original variable held in common with other variables given the number of factors or on finding the number of factors given these percentages. Both methodologies and their relation are reviewed by Kruskal (1978). They lead to an attractive geometrical interpretation as planes of closest fit to data points in measurement space (see further Chapter 19).

16.5. OPERATIONAL RESEARCH

Ackoff and Sasieni (1968) defined operational research (OR) as "the application of scientific method by interdisciplinary teams to problems involving the control of organized (man-machine) systems so as to provide solutions which best serve the purposes of the organization as a whole". Clearly, such a method (or rather, collection of methods) is suited for the purpose that is the central theme of this book, namely optimization.

The term "organization" is important in this context. As stated by Goulden (1974) in his article entitled "Management studies and techniques for application in analytical research, development and service", there are three essential components of much human endeavour : the work to be undertaken ; the organization necessary to effect that work and the people by whom the work will be done. Analytical chemists tend to pay more attention to the work to be accomplished and the tools with which to do it than to the two other components. This becomes evident when one considers the clinical laboratory example (section 16.2). Thousands of articles have been published on how to determine a biochemical parameter in an efficient way, but only a few on how to design an optimal configuration for a clinical laboratory !

It is a characteristic of organizations that they are complex systems (see also Part V) and the optimization therefore usually consists in a comparison of many different alternatives. OR techniques are used to find the optimal solution for problems in which many combinations are possible. This is a very common situation in analytical chemistry and for this reason OR techniques should be of general value in this field (see also Massart and Kaufman, 1975).

Many of the problems discussed are problems of organization in the true sense, but others, such as the selection of representative GLC probes (Chapter 23), are not. In this instance, however, there is usually an organizational analogue (in the GLC probe example, the location of supermarkets). In the chapters on OR (Chapters 21-24), the organizational analogue is always used to explain the problem and the solution method.

It may appear surprising to find several chapters devoted to techniques from the management sciences in a book such as this. These techniques are, however, certainly relevant to our purpose. The manager's job is to make the best possible use of the resources at his disposal in order to achieve a certain goal (usually commercial). This formula, however, describes equally well the task of most analytical chemists, even of those who are involved only with research. Hence it is reasonable for the techniques used by modern managers to help them in their decisions to be useful to analytical chemists.

It is very important to note here that we have written "to help them with their decisions" and not "to make their decisions". Although OR methods are mathematical methods, they rarely offer exact and ready-made solutions for real-life problems. OR methods use models, which can rarely be sufficiently precise to cover all factors. Therefore, the solutions obtained should be understood more as a guide for evaluating realistic solutions.

This is usually not understood by analytical chemists, who generally assume that as OR techniques are mathematical techniques, they should lead to exact and unrefutable solutions. When they find that the solution obtained is obviously not the ideal one, for reasons that were not incorporated in the model, they tend to conclude that OR methods are worthless. For example, by carrying out the branch and bound procedure for the selection of GLC probes (Chapter 22), one arrives at the conclusion that propionaldehyde should be one of the selected probes. This substance, however, is not very stable and therefore is not suitable as a GLC probe. When the results of the calculations were presented at a chromatographic symposium (De Clercq et al., 1976) this remark was made by one of the audience and it was very clear that, to him, this rendered the whole model worthless. In fact, it was never the purpose to state that this probe should be used in practice, but rather that, according to the criteria of the model, it was the best available. For practical use, one should then choose a probe that resembles propionaldehyde as closely as possible but with more desirable properties from the point of view of practical application. This difficulty in applying OR results is not restricted to analytical chemistry but is also encountered in more classical applications. For instance, the optimal solution of a job allocation problem in industry may be rejected by management on the grounds of possible difficulties with trade unions.

As indicated above, OR consists of a collection of mathematical techniques. Some of these are linear programming, integer programming, queuing theory, dynamic programming, graph theory, game theory and simulation. The prototype problems that can be solved are the following according to Ackoff and Sasieni (1968) :

1. Allocation
2. Inventory
3. Replacement
4. Queuing
5. Sequencing and coordination
6. Routing
7. Competitive
8. Search

Many, but not all of these prototype problems have been applied in analytical chemistry. In this book, we have gathered the applications into four categories, each of which is discussed in a separate chapter. This classification is highly arbitrary and is used more for convenience than for scientific reasons. In the first chapter (Chapter 21), we discuss some of the oldest methods of OR around the central theme : "how many and which apparatus (or methods) should be used ?". The first problem considered is an allocation problem, which means that a limited amount of facilities must be allocated among different jobs in order to maximize one or other economic function. It is solved by the technique of linear programming. This is followed by a discussion of game theory (which was placed in the category "competitive" by Ackoff and Sasieni) and which is discussed in Chapter 21, because game theory is related to linear programming. In allocation problems one often considers that the jobs can be carried out simultaneously. In practice, this is rarely true. Apparatus may not be immediately available, it may break down or, at certain moments, so much work can be presented that the capacity of the apparatus is temporarily exceeded. Delays in execution result and queues of jobs are formed. The means of minimizing the cost of this under one or other constraint such as a cost constraint is the subject of queuing theory. As the mathematics of queuing theory rapidly become too involved when complex models are studied, one must often resort to simulation techniques.

Chapter 22 discusses allocation problems of a special type, namely those in which the results must be expressed in integers. In the clinical laboratory problem (section 16.2), linear programming could lead to results such as that a combination of 1.61 three-channel apparatus and 0.45 six-channel apparatus is optimal. Clearly this is of no practical use. Techniques of integer programming enable one to arrive at results expressed in integers. Very often, one uses partial enumeration techniques.

Chapter 23 contains some problems for which graphs are used. These are applied first to routing problems, such as how to find the shortest pathway between two locations. It is shown that the ion-exchange problem in section 16.2 can be solved in this way. The same problem is also solved using dynamic programming, a technique which can be applied principally to allocation, inventory and replacement problems. Dynamic programming is not necessarily carried out with the use of graphs, but they often permit the technique to be followed more easily. In the second example in section 23.3 no graphs are used, however. Graphs are always used, however, in the sequencing and control methods known as PERT and CPM. An heuristic method for solving a particular sequencing problem is given in the same chapter.

Chapters 21-23 all discuss models in which one criterion, such as cost, time or variance, must be optimized. However, most solutions to real-life problems are judged according to more than one criterion. Chapter 24 discusses the recent science of multi-criteria analysis, which is designed to give answers to problems of this type.

REFERENCES

R.L. Ackoff and M.W. Sasieni, Fundamentals of Operations Research, Wiley, New York, 1968.

H.C. Andrews, Introduction to Mathematical Techniques in Pattern Recognition, Wiley-Interscience, New York, 1972.

H. De Clercq, M. Despontin, L. Kaufman and D.L. Massart, J. Chromatogr., 122 (1976) 535.

R.O. Duda and P.E. Hart, Pattern Classification and Scene Analysis, Wiley, New York, 1973.

P.F. Dupuis and A. Dijkstra, Anal. Chem., 47 (1975) 379.

A. Eskes, F. Dupuis, A. Dijkstra, H. De Clercq and D.L. Massart, Anal. Chem., 47 (1975) 2168.

R. Goulden, Analyst, 99 (1974) 929.

R.J. Harris, A Primer of Multivariate Statistics, Academic Press, New York, 1975.

M.G. Kendall, Multivariate analysis, Charles Griffin, London, 1975.

M.G. Kendall and A. Stuart, The Advanced Theory of Statistics, Vol. 2, Inference and Relationship, revised ed., Haufner, London, 1968.

B.R. Kowalski, Anal. Chem., 47 (1975) 1152A.

J. Kruskal, Factor Analysis and Principal Components Analysis, the Bilinear Methods, in : Encyclopedia of Statistics, The Free Press, 1978.

D.L. Massart and L. Kaufman, Anal. Chem., 47 (1975) 1244A.

W.O.J. McReynolds, J. Chromatogr. Sci., 8 (1970) 685.

D.F. Morrison, Multivariate Statistical Methods, McGraw Hill, New York, 1967.

P.S. Shoenfeld and J.R. De Voe, Anal. Chem., 48 (1976) 403R.

P. Winkel, Clin. Chem., 19 (1973) 1329.

S. Wold, Technical Rept. No. 357, Dept. of Statistics, Univ. of Wisconsin, Madison, U.S.A., 1974.

T.Y. Young and T.W. Calvert, Classification, Estimation and Pattern Recognition, Elsevier, Amsterdam, 1974.

Chapter 17

PREFERRED SETS - SOME SELECTION PROCEDURES

17.1. QUANTITATIVE MULTICOMPONENT ANALYSIS

In general, analytical problems cannot be solved by measuring one signal. Only in special cases and/or by taking certain precautions will the measurement of one signal be sufficient for solving the analytical problem. The quantitative analysis of complex samples, even if one is interested in one component only, can be attacked either by employing specific or selective procedures or by the use of non-selective or non-specific procedures that have been made selective or specific through a combination with a masking or separation step prior to the measurement. It also is possible to dilute the sample in order to decrease interferences (for instance, the borax technique in X-ray fluorescence spectroscopy) or to apply special calibration techniques (such as the standard additions method). The procedure, usually indicated by the term multicomponent analysis, is used when interferences are present and either all or some of the components in the sample are to be determined.

The purpose of this section is to introduce a mathematical model of the multicomponent analysis and to show its applicability and limitations. In subsequent sections the use of the model for some optimization problems will be discussed. The mathematics of the multicomponent analysis have been treated extensively by Herschberg (1964), Neuer (1971), Junker and Bergmann (1974), Parczewski and Rokosz (1975) and Parczewski (1976 a,b), and Kaiser (1972) used the model to define selectivity, specificity and sensitivity (see also Chapter 7). The model is a generalization of, for instance, the two-component analysis by spectrophotometry where the spectra of the two components overlap. It may be possible that the main applications of the multicomponent model are to be found in spectrophotometry (infrared and ultraviolet-visible), where usually it can be assumed that the absorbance of a mixture of light-absorbing compounds can be considered as the sum of the absorbances of the pure compounds. Moreover, the system normally behaves linearly (validity of the Lambert and Beer laws).

It is clear that for the analysis of an n-component mixture, at least n independent measurements are required, provided that linearity and additivity can be assumed. Then the mathematical model is represented by a set of m linear equations ($m \geqslant n$). If, for the solution of the analytical problem, use is made of spectra, absorbances are measured at m wavelengths. If the absorptivities are known, the concentrations can be determined. The model consists of the following equations (see also Chapter 7)

$$\begin{aligned} y_1 &= S_{11} x_1 + S_{12} x_2 \cdots\cdots\cdots S_{1n} x_n \\ y_2 &= S_{21} x_1 + S_{22} x_2 \cdots\cdots\cdots S_{2n} x_n \\ &\cdots\cdots\cdots\cdots\cdots\cdots\cdots\cdots\cdots\cdots \\ y_m &= S_{m1} x_1 + S_{m2} x_2 \cdots\cdots\cdots S_{mn} x_n \end{aligned} \tag{17.1}$$

In spectrophotometry, y_j represents the absorbance at wavelength j, x_i the concentration of component i and S_{ji} the absorptivity of component i at wavelength j (provided that the optical pathlength is 1 cm). In general, the coefficients S_{ji} are to be regarded as partial sensitivities. For a one-component system, the set of equations can be reduced to one equation and the remaining constant is the calibration constant or the sensitivity of the procedure. Obviously the partial sensitivities (or calibration constants) have to be determined by a calibration, employing either pure substances or mixtures of known composition. For this calibration m x n measurements are required (n samples at m wavelengths in spectrophotometry). Eqn. 17.1 can conveniently be written by using matrices as follows

$$\begin{bmatrix} y_1 \\ y_2 \\ \cdot \\ y_m \end{bmatrix} = \begin{bmatrix} S_{11} & S_{12} & \cdots\cdots\cdots & S_{1n} \\ S_{21} & S_{22} & \cdots\cdots\cdots & S_{2n} \\ \cdots & \cdots & \cdots\cdots\cdots & \cdots \\ S_{m1} & S_{m2} & \cdots\cdots\cdots & S_{mn} \end{bmatrix} \begin{bmatrix} x_1 \\ x_2 \\ \cdot \\ x_n \end{bmatrix} \tag{17.2}$$

or, in abbreviated form

$$\vec{Y}_m = S_{mxn}.\vec{X}_n \quad \text{or} \quad \vec{Y} = S.\vec{X} \tag{17.3}$$

where $\vec{Y}_m$ is a column vector of dimension m and $\vec{X}_n$ a column vector of dimension n. The matrix S_{mxn} of dimension mxn defines the relationship between the measurements and the composition and is called the calibration matrix (Chapter 7). In a geometric sense, S_{mxn} links the m-dimensional space of measurements, R^m, with the n-dimensional space of compositions, R^n. This is illustrated schematically in Fig. 17.1. $\vec{X}_n$ and $\vec{Y}_m$ represent compositions and sets of measurements in these spaces.

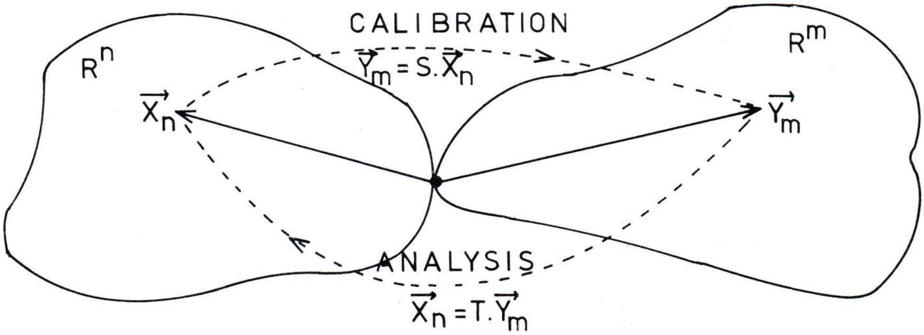

Fig. 17.1. Schematic representation of the processes of calibration and analysis.

In order to be analytically useful, the matrix S_{mxn} should uniquely relate $\vec{X}_n$ and $\vec{Y}_m$. Each set of measurements should correspond to a certain composition (neglecting at present the influence of errors). Conversely, each composition should be uniquely related to a particular set of measurements. We shall not elaborate here on the mathematical details associated with this uniqueness (see Kaiser, 1972). It is sufficient to observe that analysis and calibration in a sense are inverse processes, also from a mathematical point of view. This implies that there is an inverse (or reciprocal) relationship given by

$$\vec{X}_n = T_{nxm}.\vec{Y}_m \quad \text{or} \quad \vec{X} = T.\vec{Y} \tag{17.4}$$

as an abbreviated form of an equation that can be considered as the reciprocal of eqn. 17.2. The matrix T_{nxm} is of dimension nxm. The elements T_{ij} of T are related to the partial sensitivities S_{ji}. The relationships between T and S and between T_{ij} and S_{ji} will be explored in the next section.

To conclude this section, it should be remarked that, in principle, similar models can be envisaged for non-linear and for non-additive systems. However, the description of such systems requires many more calibration constants (compared with the linear system) and consequently the model is more difficult to handle. In some instances the relationship between limited regions of the spaces can be expressed by linear equations and then the model as described in this section is applicable to samples that vary little in composition. A survey of (non-linear) calibration functions in X-ray fluorescence analysis was given by Rasberry and Heinrich (1974).

17.2. LEAST-SQUARES SOLUTION

If the number of (independent) measurements, m, is larger than the number of components, n, the system is said to be overdetermined. In practice, the presence of experimental errors will cause errors in the composition (concentrations). Every set of n equations selected from the m available will yield different values for the composition. The most probable composition can be found by application of the least-squares method to all m measurements. A brief treatment of the least-squares technique as applied to multicomponent analysis will be given here ; a detailed discussion was given by the authors cited in section 17.1.

The set of eqns. 17.1 can be rewritten as follows

$$y_j = S_{j1} x_1 + S_{j2} x_2 + \dots + S_{ji} x_i + \dots + S_{jn} x_n + e_j \qquad (17.5)$$

with j = 1, ..., m and e_j being the unknown error of measurement y_j, provided that all y_j are measured with the same precision (for spectrophotometry this is an approximation as the precision generally depends on the absorbance). As seen in previous chapters (Chapters 3 and 15), the least-squares technique requires $\sum_j e_j^2$ to be minimized. Thus

$$\min_{x_1,\dots x_n} \sum_{j=1}^{m} (y_j - S_{j1} x_1 - S_{j2} x_2 - \dots - S_{ji} x_i - \dots - S_{jn} x_n)^2 \tag{17.6}$$

By differentiating equation 17.6 with respect to $x_1, x_2, \dots, x_i, \dots, x_n$ and setting all differentials to zero, the m equations (eqns. 17.5) are reduced to n equations from which the most probable values of $x_1, x_2, \dots, x_i, \dots, x_n$ can be calculated. These equations are

$$\sum_{j=1}^{m} S_{ji} y_j = x_1 \sum_{j=1}^{m} S_{j1} S_{ji} + x_2 \sum_{j=1}^{m} S_{j2} S_{ji} + \dots$$
$$+ x_i \sum_{j=1}^{m} S_{ji}^2 + \dots + x_n \sum_{j=1}^{m} S_{jn} S_{ji} \qquad (i=1, \dots, n) \tag{17.7}$$

and the values of x_i are given by the quotient of two determinants, i.e.

$$x_i = \frac{\begin{vmatrix} \Sigma S_{j1}^2 & \Sigma S_{j2} S_{j1} & \dots & \Sigma S_{j1} y_j & \dots & \Sigma S_{jn} S_{j1} \\ \Sigma S_{j1} S_{j2} & \Sigma S_{j2}^2 & \dots & \Sigma S_{j2} y_j & \dots & \Sigma S_{jn} S_{j2} \\ \cdots & \cdots & \cdots & \cdots & \cdots & \cdots \\ \Sigma S_{j1} S_{ji} & \Sigma S_{j2} S_{ji} & \dots & \Sigma S_{ji} y_j & \dots & \Sigma S_{jn} S_{ji} \\ \cdots & \cdots & \cdots & \cdots & \cdots & \cdots \\ \Sigma S_{j1} S_{jn} & \Sigma S_{j2} S_{jn} & \dots & \Sigma S_{jn} y_j & \dots & \Sigma S_{jn}^2 \end{vmatrix}}{\begin{vmatrix} \Sigma S_{j1}^2 & \Sigma S_{j2} S_{j1} & \dots & \Sigma S_{ji} S_{j1} & \dots & \Sigma S_{jn} S_{j1} \\ \Sigma S_{j1} S_{j2} & \Sigma S_{j2}^2 & \dots & \Sigma S_{ji} S_{j2} & \dots & \Sigma S_{jn} S_{j2} \\ \cdots & \cdots & \cdots & \cdots & \cdots & \cdots \\ \Sigma S_{j1} S_{ji} & \Sigma S_{j2} S_{ji} & \dots & \Sigma S_{ji}^2 & \dots & \Sigma S_{jn} S_{ji} \\ \cdots & \cdots & \cdots & \cdots & \cdots & \cdots \\ \Sigma S_{j1} S_{jn} & \Sigma S_{j2} S_{jn} & \dots & \Sigma S_{ji} S_{jn} & \dots & \Sigma S_{jn}^2 \end{vmatrix}} \tag{17.8}$$

(all Σ are sums over j).

In order to illustrate the least-squares calculation we shall consider the chlorine-bromine system (see Chapter 7). For an optical pathlength of 1 cm, x_1 and x_2 being the concentrations of chlorine and bromine, respectively, and y_1, y_2, y_3, y_4, y_5 and y_6 the measurements at the wavenumbers 22, 24, 26, 28, 30 and 32 x 10^3 cm^{-1}, the set of (calibration) equations reads

$$\begin{aligned} y_1 &= 4.5\ x_1 + 168\ x_2 \\ y_2 &= 8.4\ x_1 + 211\ x_2 \\ y_3 &= 20\ x_1 + 158\ x_2 \\ y_4 &= 56\ x_1 + 30\ x_2 \\ y_5 &= 100\ x_1 + 4.7\ x_2 \\ y_6 &= 71\ x_1 + 5.3\ x_2 \end{aligned} \tag{17.9}$$

Eqns. 17.7 become

$$\begin{aligned} \Sigma\, S_{j1}\, y_j &= 4.5\ y_1 + 8.4\ y_2 + 20\ y_3 + 56\ y_4 + 100\ y_5 + 71\ y_6 \\ &= 18667.81\ x_1 + 8214.7\ x_2 \\ \Sigma\, S_{j2}\, y_j &= 168\ y_1 + 211\ y_2 + 158\ y_3 + 30\ y_4 + 4.7\ y_5 + 5.3\ y_6 \\ &= 8214.7\ x_1 + 98659.18\ x_2 \end{aligned} \tag{17.10}$$

and the solution will be

$$x_1 = \frac{\begin{vmatrix} \Sigma\, S_{j1}\, y_j & \Sigma\, S_{j2}\, S_{j1} \\ \Sigma\, S_{j2}\, y_j & \Sigma\, S_{j2}^2 \end{vmatrix}}{\begin{vmatrix} \Sigma\, S_{j1}^2 & \Sigma\, S_{j2}\, S_{j1} \\ \Sigma\, S_{j1}\, S_{j2} & \Sigma\, S_{j2}^2 \end{vmatrix}} = \frac{\begin{vmatrix} \Sigma\, S_{j1}\, y_j & 8214.7 \\ \Sigma\, S_{j2}\, y_j & 98659.18 \end{vmatrix}}{1774269531} \tag{17.11}$$

$$x_2 = \frac{\begin{vmatrix} \Sigma\, S_{j1}^2 & \Sigma\, S_{j1}\, y_j \\ \Sigma\, S_{j1}\, S_{j2} & \Sigma\, S_{j2}\, y_j \end{vmatrix}}{\begin{vmatrix} \Sigma\, S_{j1}^2 & \Sigma\, S_{j2}\, S_{j1} \\ \Sigma\, S_{j1}\, S_{j2} & \Sigma\, S_{j2}^2 \end{vmatrix}} = \frac{\begin{vmatrix} 18667.81 & \Sigma\, S_{j1}\, y_j \\ 8214.7 & \Sigma\, S_{j2}\, y_j \end{vmatrix}}{1774269531} \tag{17.12}$$

These solutions can also be written in matrix notation as follows

$$\begin{bmatrix} x_1 \\ x_2 \end{bmatrix} = \begin{bmatrix} T_{11} & T_{12} & T_{13} & T_{14} & T_{15} & T_{16} \\ T_{21} & T_{22} & T_{23} & T_{24} & T_{25} & T_{26} \end{bmatrix} \begin{bmatrix} y_1 \\ y_2 \\ y_3 \\ y_4 \\ y_5 \\ y_6 \end{bmatrix} \qquad (17.13)$$

This equation corresponds to eqn. 17.4, with $\vec{X} = T.\vec{Y}$. The elements T_{ij} of T have been calculated from the partial sensitivities. For the reader who is familiar with matrix algebra, it is easier to condense the conversion of S to T into the expression

$$T = (S'.S)^{-1}.\ S' \qquad (17.14)$$

In other words, T is obtained by pre-multiplying S by its transpose S', inverting this product and subsequently post-multiplying this inverse by S'. Alternatively, T is the left inverse of S and, when m = n (square matrix), T is simply the inverse of S. By taking the same numerical example as used above, it can easily be verified that eqn. 17.4 leads to the T matrix as calculated by the least-squares technique.

The transpose of S is given by

$$S' = \begin{bmatrix} 4.5 & 8.4 & 20 & 56 & 100 & 71 \\ 168 & 211 & 158 & 30 & 4.7 & 5.3 \end{bmatrix} \qquad (17.15)$$

and the product

$$S'.S = \begin{bmatrix} 18667.81 & 8214.7 \\ 8211.7 & 98659.18 \end{bmatrix} \qquad (17.16)$$

Inverting this product leads to the matrix

$$(S'.S)^{-1} = \frac{1}{1\ 774\ 269\ 531} \begin{bmatrix} 98659.18 & -8214.7 \\ -8214.7 & 18667.81 \end{bmatrix} \qquad (17.17)$$

and finally we obtain

$$(S'.S)^{-1} . S' = \begin{bmatrix} -0.00053 & -0.00051 & 0.00038 & 0.00298 & 0.00554 & 0.00392 \\ 0.00175 & 0.00218 & 0.00157 & 0.00006 & -0.00041 & -0.00027 \end{bmatrix} \tag{17.18}$$

It can be expected that, in general, the precision of the procedure increases with an increasing number of measurements. To some extent the effect of using an overdetermined system is the same as the effect of repeated measurements on the precision. For the multicomponent analysis (by using spectrophotometry), the analyst can choose between the use of an overdetermined system or reduce the number of measurements to the minimum ($m = n$) required for the determination. For a treatment of the relationships between the number of measurements, the errors of the measurements and the precision of the procedure, the reader is referred to Herschberg (1964), Parczewski and Rokosz (1975) and Parczewski (1976 a). Obviously it is advantageous, when considering the precision of the procedure, to select wavelengths at which the absorptivities are large even when an overdetermined system is used. Optimization with respect to the precision has also been described by Sustek (1974) and Parczewski (1976 b). In a number of instances, the application of the least-squares procedure may lead to a negative value for one or more of the concentrations and modifications of the least-squares procedure have therefore been proposed (see, for instance, Leggett, 1977).

17.3. CHOOSING AN OPTIMAL SET OF WAVENUMBERS

In section 17.1 it was observed that for the determination of n components not more than n measurements are required, provided that the multicomponent system can be represented by a set of linear equations for which $y = 0$ if $x = 0$. The question arises of which set of wavenumbers (in spectrophotometry) is to be preferred. The search for such an optimal set requires the use of an optimization criterion. Provided that the experimental errors (in an absolute sense) in the determination of absorbances are the same at every wavenumber, the sensitivity is a suitable criterion. It is clear that for a one-component analysis this results in choosing the wavenumber where the absorption peak has its maximum. This results in a minimal (relative) error and, in general, in a minimal limit of detection.

As has been shown, multicomponent analysis is characterized by a set of partial sensitivities and maximizing one of these does not necessarily result in the maximization of the others. However, according to Kaiser (1972), it is possible to define the sensitivity of a multicomponent procedure as the absolute value of the determinant of the calibration matrix. This is possible only for a number of measurements equal to the number of the components ($m = n$). Then

$$|S| = \begin{vmatrix} S_{11} & S_{12} & \cdots & S_{1n} \\ S_{21} & S_{22} & \cdots & S_{2n} \\ \cdots & \cdots & \cdots & \cdots \\ S_{n1} & S_{n2} & \cdots & S_{nn} \end{vmatrix} \qquad (17.19)$$

A maximum sensitivity corresponds to a determinant with large diagonal elements and low off-diagonal elements, or in more general terms, with one that can be converted into such a determinant by an interchange of rows (which leaves the absolute value of the determinant unchanged). In fact, a high sensitivity runs partly parallel with a high selectivity (see Chapter 7), which means that a highly sensitive procedure is a procedure in which each measurement is largely dependent on the concentration of only one of the components. The sensitivity is therefore a parameter that can be used for comparing different sets of wavenumbers. A maximum sensitivity corresponds with a maximum precision.

In order to illustrate the principle of using the sensitivity as an optimization parameter, we again choose the chlorine-bromine system. From the six wavenumbers it is possible to choose several pairs, to be exact $\binom{m}{n} = \binom{6}{2} = 15$ pairs. For each of these pairs it is possible to calculate the sensitivity, i.e., the absolute value of the determinant of the absorptivities. The values of these determinants are given in Table 17.I.

Table 17.I

Sensitivities for the system chlorine-bromine in chloroform with combinations of two wavenumbers

	Wavenumber ($\times 10^3$ cm^{-1})					
	22	24	26	28	30	32
22	0	460	1650	9300	16800	12000
24		0	2890	11600	21050	15000
26			0	8250	15705	11000
28				0	2740	1830
30					0	189
32						0

Of the several possibilities, the combination of 24 x 10^3 and 30 x 10^3 cm^{-1} appears to be the best. This is not surprising when one considers the corresponding table of absorptivities or the spectra (see Chapter 7, Table 7.I and Fig. 7.1). Bromine has its absorption maximum at 24 x 10^3 cm^{-1} and chlorine at 30 x 10^3 cm^{-1}. The principle of a maximal sensitivity corresponding to (on average) minimal errors also holds for larger numbers of components where spectra overlap in a complicated way and where it is impossible to choose a set simply by looking at the spectra.

Although the optimization procedure is relatively simple, in practice its application requires an appreciable number of calculations and thus computer time. With m wavenumbers from which a set of n is to be chosen, $\binom{m}{n} = m!/(m-n)!\,n!$ sensitivities have to be compared. For the relatively simple situation of m = 30 and n = 6, the number of determinants to be calculated is 593775. Therefore, it can be stated that the straightforward procedure as described here cannot be applied, even when using a modern computer. Junker and Bergmann (1974) have developed an optimization procedure that requires fewer calculations. In order to introduce this method a generalization of the sensitivity concept of Kaiser has to be presented. Junker and Bergmann (1974) defined the sensitivity by

$$|S| = \sqrt{|(S'.S)|} \qquad (17.20)$$

The sensitivity, defined as the root of the determinant of the product of the calibration matrix and its transpose, can also be used when m $\neq$ n. |S| as defined by eqn. 17.20 is the determinant of a square matrix with n^2 elements and equals $|S|$ defined by eqn. 17.19 if m = n. The sensitivity is at a maximum when each measurement is largely determined by one component ; a high sensitivity runs parallel to a high selectivity. Thus, the sensitivity defined by eqn. 17.20 can be used for a comparison and selection of overdetermined systems.

The optimization procedure of Junker and Bergmann (1974) consists in first calculating the sensitivities of the m possible combinations of m-1 positions. Of these m combinations, that with the highest sensitivity is retained and, as a consequence, the position which influences the sensitivity least is dropped. The procedure is repeated by next considering the m-1 combinations of m-2 combinations of m-2 positions each. Again, the set with the highest sensitivity is retained. The procedure is repeated until the required number of wavelengths remains (of course, m is always greater than or equal to n). It is clear that the number of determinants to be calculated is greatly reduced in comparison with the "complete" selection procedure. For m = 30 and n = 6, this number is 444. However, the determinants to be calculated are, on average, larger in comparison with those required for the "complete" procedure. Junker and Bergmann (1974) quoted a reduction in computer time of about 1000-fold.

As indicated above, the optimization procedure of Junker and Bergmann (1974) can be terminated at any number of wavenumbers that one wishes to retain (in order to increase the precision it may be required to choose m > n). For m = n as well as for m > n, it leads to a near-optimal set with regard to sensitivity and precision. It is difficult, or probably impossible, to prove whether or not the real optimum has been found. To put it differently : the wavenumbers that are selected by application of the procedure of Junker and Bergmann (1974) need not necessarily correspond to the largest value of all possible determinants. Another reason for not finding the true optimum is the following. An infrared or ultraviolet-visible spectrum consists of several hundreds of independent measurements and it is essential to pre-select some tens of wavenumbers in order to avoid spurious calculations. Junker and Bergmann (1974) therefore suggested a "manual" pre-selection prior to the "automatic" selection as described above.

Fig. 17.2 is a reproduction of the selection procedure applied to o-, m- and p-xylene. The figure should be read together with Table 17.II. Junker and Bergmann (1974) have "identified" the selected positions by ascribing these positions to the components in the mixture (o = o-xylene, etc.). In this particular example, this may be a valid procedure ; for more components this "identification" is impossible and not relevant.

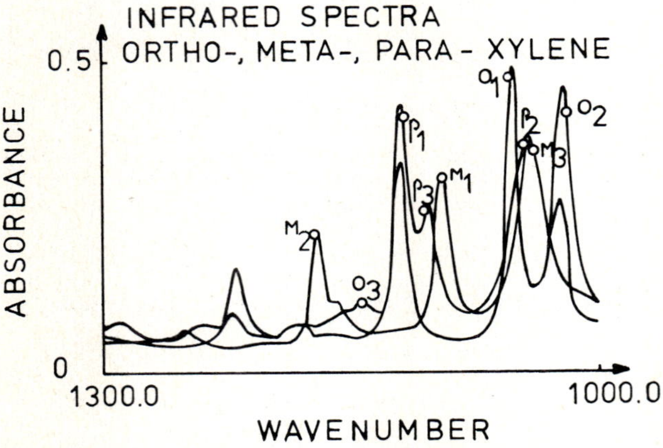

Fig. 17.2. Choice of an optimal set of wavenumbers (Junker and Bergmann, 1974).

Table 17.II

Result of optimization procedure of Junker and Bergmann (1974). (See also Fig. 17.1).

Number of wavenumbers	Dropped wavenumbers		Sensitivity of remaining set
	Wavenumber (cm^{-1})	Identification	
13	1071	P_5	0.1216
12	1227	O_4	0.1200
11	1145	O_3	0.1183
10	1159	M_4	0.1150
9	1221	P_4	0.1091
8	1105	P_3	0.0948
7	1037	M_3	0.0795
6	1020	O_2	0.0656
5	1172	M_2	0.0529
4	1044	P_2	0.0350
optimal combination of three wavenumbers	1095 1054 1119	M_1 O_1 P_1	

17.4. THE INFORMATION CONTENT OF COMBINED PROCEDURES

In Chapter 8 it was observed that the information content of a combination of procedures is usually smaller than the sum of the information contents of the individual procedures. This is due to correlations between the physical quantities (signals) from which the identities of the unknown compounds (or concentrations in quantitative analysis) are derived. As a result of these correlations, the measurement of two (or more) physical quantities yields partly the same information (also called mutual information).

For instance, both melting points and boiling points on average increase with increasing molecular weight. When a high melting point has been observed, the boiling point is also expected to be high. If the correlation between melting point and boiling point were perfect, it would make no sense to determine both quantities for identification purposes. However, the correlation is not perfect because melting and boiling points are not determined solely by the size of the molecule but are also governed by factors such as the polarity of the molecule. From this crude physical description, it is clear that the measurement of the boiling point will yield an additional amount of information even if the melting point is known. However, this additional amount of information is smaller than that obtained in the case of unknown melting point.

If the amount of information obtained from a combination of procedures is to be calculated, use has to be made of an information theoretical model that takes into account the correlation between the signals that are used for gathering the information. In principle, two different ways of calculating the information contents of combined procedures can be distinguished. The first way has already been indicated in Chapter 8. Eqn. 8.10 takes into account all possible combinations of (two) signals. In fact, each combination is considered as one (composite) signal. The probability for each combination is introduced into Shannon's equation and thus the influence of the correlation upon the information content is implicitly taken into account.

Another way of calculating information contents has been applied to combinations of stationary phases for gas chromatography by Dupuis and Dijkstra (1975) and Eskes et al. (1975). In a slightly different way, the same procedure has been applied by Van Marlen and Dijkstra (1976) to mass spectrometry (combination of mass peaks). Dupuis and Dijkstra showed that the distribution of retention indices for a large number of substances approximately follows a normal distribution. This has been verified by application of the χ^2-test. Introducing such a normal distribution into Shannon's equation leads to an information content equal to

$$I = \frac{1}{2} \log_2 \frac{s_t^2 + s_e^2}{s_e^2} \tag{17.21}$$

(see eqn. 8.15), where s_t^2 is the estimated variance of the "true" retention indices and s_e^2 the estimated variance of the errors. It was assumed that the n-dimensional distribution of the retention indices for n stationary phases follows an n-dimensional normal distribution that can be represented by

$$p(y_1, y_2, \ldots, y_n) = \frac{\exp\{\frac{1}{2} (\vec{Y}-\vec{\bar{Y}})' \, C^{-1} \, (\vec{Y}-\vec{\bar{Y}})\}}{(2\pi)^{n/2} \, |C|^{\frac{1}{2}}} \tag{17.22}$$

where $\vec{Y}$ and $\vec{\bar{Y}}$ are the column vectors of the variables $y_1, \ldots, y_n$ and the averages $\bar{y}_1, \ldots, \bar{y}_n$ and C and |C| are the covariance matrix and its determinant. The covariance matrix is given by

$$C = \begin{bmatrix} s_{11} & c_{12} & \cdots & c_{1n} \\ c_{21} & s_{22} & \cdots & c_{2n} \\ \cdots & \cdots & \cdots & \cdots \\ c_{n1} & c_{n2} & \cdots & s_{nn} \end{bmatrix} \tag{17.23}$$

with $s_{11} = s_i^2$ and $c_{ij} = c_{ji}$ being the estimated variances and covariances.

The assumption of a multinormal distribution is difficult to verify because application of a χ^2-test requires an unreasonably large number of retention indices. Roughly, one can state that if for a test for one dimension n substances are required, the amount of substances required for two dimensions will be n^2, for three n^3, etc. Thus the assumption of a n-dimensional normal distribution has to be considered as an approximation. The n-dimensional equivalent of Shannon's eqn. 8.14 leads tc an information content

$$I(1,2,\ldots,n) = \qquad (17.24)$$
$$- \int_{y_1}\int_{y_2} \cdots \int_{y_n} p_m(y_1,y_2,\ldots,y_n)\ \log_2 p_m(y_1,y_2,\ldots,y_n)\ dy_1,dy_2,\ldots,dy_n$$
$$+ \int_{y_1}\int_{y_2} \cdots \int_{y_n} p_e(y_1,y_2,\ldots,y_n)\ \log_2 p_e(y_1,y_2,\ldots,y_n)\ dy_1,dy_2,\ldots,dy_n$$

where p_m and p_e are the n-dimensional distribution functions of the measurements and errors, respectively. Integration of eqn. 17.24 leads to

$$I(1,2,\ldots,n) = \frac{1}{2} \log_2 \frac{|C|_m}{|C|_e} \qquad (17.25)$$

$|C|_m$ and $|C|_e$ are the determinants of the covariance matrices of measurements (true values plus errors) and errors.

The model used for calculating information contents expressed mathematically by eqn. 17.25 thus explicitly takes into account the correlations between the retention indices for the several stationary phases.

17.5. SELECTION OF AN OPTIMAL SET OF STATIONARY PHASES

The problem of selecting an optimal set of stationary phases also requires the choice of an optimization criterion. In section 8.5 the information content as a criterion was discussed. In the context of combined procedures, the information content is a criterion including the spread of the retention indices and the correlations between the retention indices. The choice of n stationary phases from a total number m can in a way be compared with the selection of wavelengths for multicomponent analysis. Here again, a large number of determinants have to be calculated in order to find the optimal set of phases.

Eskes et al. (1975), in a comparative study of selection by information theory and taxonomy (see Chapter 18), used the 16 stationary phases listed in Table 17.III.

Table 17.III

List of stationary phases

Column No.	Stationary phase	Column No.	Stationary phase
1	Squalane	9	Tricresyl phosphate
2	Apiezon L	10	Diglycerol
3	SE-30	11	Zonyl E7
4	Diisodecyl phthalate	12	QF-1
5	Polyphenyl ether (6 rings)	13	Hyprose SP80
6	Bis(ethoxyethyl) phthalate	14	Triton X-305
7	Carbowax 20M	15	XF 1150
8	Diethyl glycol succinate	16	Quadrol

By using eqn. 17.25 and assuming that the errors for all stationary phases are the same and not correlated (covariances of the errors are zero), Eskes et al. (1975) calculated information contents for combinations of two and three columns, requiring the calculation of $\binom{16}{2} = 120$ and $\binom{16}{3} = 560$ determinants, respectively. The results of a comparison of these combinations are given in Table 17.IV.

Table 17.IV

Best single stationary phases and best combinations in general and for classes of alcohols, aldehydes/ketones, esters (Eskes et al., 1975)

Class of compounds	Best stationary phase	Information	Best combination of two	Information	Best combination of three	Information
General	13	6.8 bit	10,12	13.5 bit	1,8,10	19.2 bit
Alcohols	8	6.4	2,8	12.0	8,10,12	15.8
Aldehydes/ketones	8	6.4	10,12	12.9	1,8,10	17.9
Esters	12	6.4	1,8	12.1	2,8,11	16.0

Without giving a detailed discussion, two remarks should be made before we treat the selection procedure to be used for the combination of a greater number of phases. Firstly, it should be observed that for different classes of compounds, different combinations of stationary phases emerge. This is not surprising - it merely confirms (in an objective way) what is felt intuitively or what is to be expected by considering the molecular interactions. Secondly, the best phase does not always belong to the best set of two or best set of three.

When the number of stationary phases increases, the number of calculations becomes prohibitive for comparing all possible combinations. Dupuis and Dijkstra (1975) introduced the following selection procedure that avoids the need to calculate all determinants. The first phase selected is the one that yields the largest amount of information. The second phase is added by using the criterion $I_2(1-r_{21})$. A maximal value of this criterion corresponds to a large information content I_2 for the second phase and a low correlation with the first. In general, the kth stationary phase is selected by using the criterion

$$\text{Max } I_k \left(1 - \frac{\sum_{i=1}^{k} |r_{ki}|}{k-1}\right) \qquad (17.26)$$

where I_k is the information content of phase k and $\sum_{i=1}^{k} |r_{ki}| / (k-1)$ is the "average" correlation of the kth phase with those already selected. The sequence of stationary phases determined in this way is not optimal in the sense that the first k phases of this sequence do not necessarily yield the best set of k phases (where "best" is identical with the highest information content). It merely should be optimal in the sense that on plotting the information content against the number of phases selected, the largest increase is obtained when one more phase is added. However, the sequence found by application of criterion 17.26 only approximates the optimal sequence. A better approximation can be obtained by performing some additional calculations.

Dupuis and Dijkstra considered the determinant of the covariance matrix corresponding to the sequence determined by application of criterion 17.26. This determinant can be written as

$$\begin{vmatrix} s_{11} & c_{12} & c_{13} & \cdots \\ c_{21} & s_{22} & c_{23} & \cdots \\ c_{31} & c_{32} & s_{33} & \cdots \\ \cdots & & & \end{vmatrix} \tag{17.27}$$

It can be converted into a determinant of which all elements in the lower triangle are zero, for instance by application of the Gauss elimination method. Following this method, we subtract from the elements of the second row the corresponding elements of the first row multiplied by c_{21}/s_{11}, from the third row the elements of the first multiplied by c_{31}/s_{11}, from the fourth, etc. Then the following determinant is obtained

$$\begin{vmatrix} s_{11} & c_{12} & c_{13} & \cdots \\ 0 & s'_{22} & c'_{23} & \cdots \\ 0 & c'_{32} & s'_{33} & \cdots \\ 0 & \cdots & & \end{vmatrix} = \begin{vmatrix} s_{11} & c_{12} & c_{13} & \cdots \\ 0 & s_{22} - \dfrac{c_{21}c_{12}}{s_{11}} & c_{23} - \dfrac{c_{21}c_{13}}{s_{11}} & \cdots \\ 0 & c_{32} - \dfrac{c_{31}\cdot c_{12}}{s_{11}} & s_{33} - \dfrac{c_{31}\cdot c_{13}}{s_{11}} & \cdots \\ 0 & \cdots & \cdots & \cdots \end{vmatrix} \tag{17.28}$$

The elements of the first row are the same as those of the original determinant. The elements of the first column are zero, apart from the first element of that column. The value of the new determinant is the same as that of the original one.

Next, the procedure is repeated by subtracting multiples of the elements of the second row from those of the third and subsequent rows, in order to obtain a determinant with the same value, all elements of the second column (apart from the first and the second element) being zero. In this step, the elements of the second row remain unchanged. The elimination procedure is repeated until all elements in the lower triangle are zero. The value of this finally obtained triangulated determinant is equal to the product of the diagonal elements. Hence, the information content of the combination of stationary phases is proportional to the logarithm of the product of the diagonal elements of the triangulated determinant. From the elimination procedure it is also clear that the information content of the first n stationary phases is related to the product of the first n diagonal elements. The value of each diagonal element is influenced only by the preceding rows of the determinant and not by the following rows.

An optimal sequence would correspond to a triangulated determinant with the diagonal elements arranged according to decreasing magnitude. An earlier selected phase should contribute more to the information content than a phase which is selected later. If the sequence appears to be non-optimal as judged from the value of the diagonal elements, an interchange of phases is required. This interchange corresponds to an interchange of rows and columns of the determinant. It is obvious that the Gauss elimination procedure has to be repeated with the determinant corresponding to the adjusted sequence of stationary phases. Usually one adjustment will lead to the optimal sequence ; if necessary, the whole procedure can be repeated.

Apart from the Gauss elimination procedure, other methods are available for the triangulation. In some procedures use is made of the symmetry of the determinant, i.e., $c_{ij} = c_{ji}$ (Wilkinson and Reinsch, 1971).

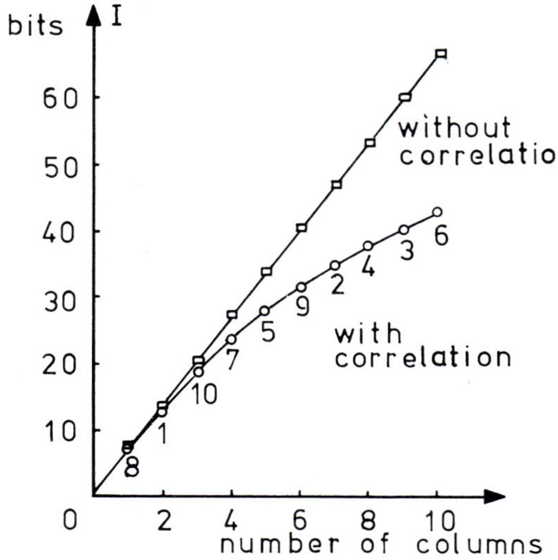

Fig. 17.3. Amount of information as a function of the number of columns (Dupuis and Dijkstra, 1975). Reprinted with permission. Copyright American Chemical Society.

As an example of the procedure described, the sequencing of ten stationary phases (numbers 1-10 in Table 17.IV) is shown in Fig. 17.3, taken from Dupuis and Dijkstra (1975). From this figure, it appears that the influence of the correlation on the information content is appreciable. Each stationary phase added to the combination contributes less to the information content than the former.

17.6. DISCUSSION

In this chapter on the selection of preferred sets, two main themes have emerged. The first theme is the construction of a model that defines the optimal or preferred set of signals or measurements to be used for the analysis. Secondly, attention has been paid to the determination of that optimum. To establish this optimum, a large number of calculations are usually required and it is of great importance to define algorithms in order to keep this number within reasonable limits. In this section, we shall restrict the discussion to some of the models that have been used. With respect to the calculations, we observe that there are probably other means of achieving the same goal. As yet it is impossible to state that the algorithms presented are the best. We have discussed only those methods which have been used so far to solve problems in analytical chemistry that require the calculation of a large number of determinants.

Obviously, the selection procedures introduced in the preceding sections can be applied only if the selection criterion can be shaped into a determinant. The value of the determinant represents the sensitivity of the multicomponent procedure or the information content of a combination of procedures. We shall not question the use of these criteria here. The information content as a selection criterion has already been discussed in Chapter 8. For the selection of wavenumbers, criteria other than the sensitivity can also be used (see Junker and Bergmann, 1976 a,b).

The limitations to the use of the sensitivity as a criterion are clearly set by the nature of the system which has to be linear. Optimization of non-linear systems is much more complicated and it is doubtful whether such systems can be treated and optimized by using procedures similar to those described in this chapter. In some instances, non-linear systems can be converted into linear systems by using suitable transformations. It should be noted once again that the procedure of Junker and Bergmann (1974) leads essentially to a set of measurements that is as selective as possible for the components to be determined. The rational approach therefore runs parallel to the approach that would be, but not always can be, followed intuitively.

The model used by Dupuis and Dijkstra for calculating the information contents of combinations of stationary phases assumes that the retention indices measured on n stationary phases follow an n-dimensional normal distribution. It is difficult to verify this assumption because of the large number of indices required for testing. As a result, the information contents calculated may be in error. Although it is not possible to discuss in detail the consequences of the assumption, it is possible to state that the information contents are to be considered as maximal values. In the n-dimensional model, only interactions (correlations) between two stationary phases have been taken into account. As a result of these correlations, the information content of combined stationary phases will be lower than the sum of the information contents of the individual phases. If higher order interactions are taken into account the information content might decrease even further. However, it is safe to state that the effect of such higher order interactions causing a deviation from the n-dimensional normal distribution will be small in comparison with the effects of the correlations between the retention indices of two stationary phases.

In analytical chemistry, many more problems might be solved if it were possible to calculate the information content of combinations of signals. Not only retention indices can be and are used for identification purposes ; physical properties such as melting points, boiling points and peaks of spectra (mass, infrared, magnetic resonance) can also be used. The problem can be formulated as the determination of the combination of signals (retention indices, melting points or peaks of spectra) that will yield the largest amount of information and thus be the most useful for identification purposes. When small numbers of compounds are to be distinguished by their physical properties or spectra, the approach to be followed should be similar to that for selecting the best TLC system (Chapter 8). When a large number of compounds is involved, the approach necessarily needs to be statistical, i.e., it runs parallel to the selection of stationary phases as treated in this chapter. Then assumptions with respect to the distribution of signals have to be made in order to be able to calculate information contents and to select properties that are useful for large retrieval systems.

Van Marlen and Dijkstra (1976) considered the special case of binary coded peak intensities in mass spectrometry. Information contents of individual peaks can be calculated by using eqn. 7 in Chapter 8. The approximate value of the information content of the combination of peaks can be obtained by using eqn. 17.25. However, the assumption of a multinormal distribution for a set of binary peaks is not justified and Van Marlen and Dijkstra introduced a correction factor in order to account for the non-normal distribution and consequently to arrive at a better approximation of the information content. The results of this study, consisting in the selection of an optimal set of peaks and the corresponding values of the information content, without taking into account the experimental errors, are represented in Fig. 17.4 and Table 17.V. The general picture resembles that of the set of stationary phases for gas chromatography. It is interesting to observe that from the information theoretical point of view it makes no sense to store all binary coded mass peaks for retrieval purposes. It appears that 120 peaks yield the same amount of information as the total set of 300 peaks from which those peaks were selected.

Table 17.V

Sequence of selected masses with amounts of information

Selected masses										Information for n masses (bits)	n
77	69	27	50	45	40	57	75	81	44	8.4	10
115	85	15	91	93	73	58	31	127	32	14.4	20
98	38	61	55	87	105	18	65	53	59	19.3	30
89	43	86	71	119	139	28	66	101	72	23.6	40
26	83	30	107	64	79	113	131	60	103	27.3	50
100	70	74	39	78	14	152	52	47	56	30.6	60
95	67	37	84	80	63	76	169	51	90	33.3	70
46	143	109	106	88	36	99	121	102	54	35.7	80
165	114	117	92	94	126	189	42	128	29	37.3	90
82	97	49	104	138	111	112	133	16	19	39.0	100
188	33	149	222	141	135	68	25	62	155	40.1	110
125	41	167	120	123	178	212	262	145	163	40.9	120

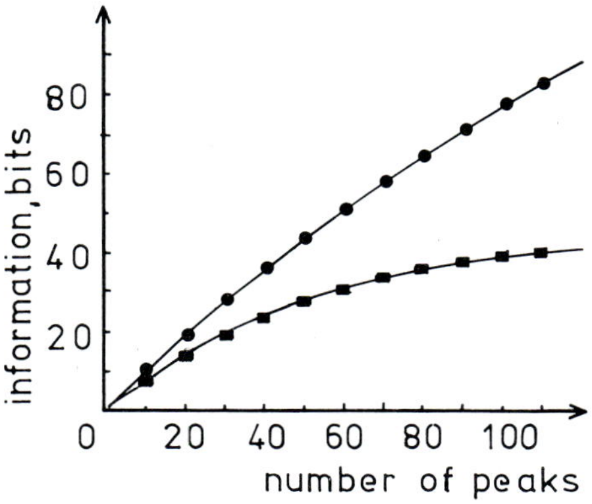

Fig. 17.4. Information vs. number of peaks (van Marlen and Dijkstra, 1976). (•) Without correlation, (▪) with correlation. Reprinted with permission. Copyright American Chemical Society.

17.7 MATRIX ALGEBRA

17.7.1. Introduction

The simultaneous use of several values of a variable or of several variables leads to complicated notations and long equations. In statistics, one is frequently confronted with problems in which such complicated notations make the sometimes simple results difficult to interpret. Matrix algebra makes it possible to make statistical notations simpler and therefore easier to interpret.

17.7.2. Some definitions

17.7.2.1. Matrices and vectors

A matrix A or A_{mxn} is a rectangular table of mxn elements

a_{ij} (i = 1,2,...,m ; j = 1,2,...,n) :

$$A = A_{mxn} = \begin{bmatrix} a_{11} & a_{12} & \cdot & \cdot & \cdot & a_{1n} \\ a_{21} & a_{22} & \cdot & \cdot & \cdot & a_{2n} \\ \vdots & \vdots & \vdots & \vdots & \vdots & \vdots \\ a_{m1} & a_{m2} & \cdot & \cdot & \cdot & a_{mn} \end{bmatrix} \qquad (17.29)$$

The numbers of rows m and columns n define the dimensions of the matrix. A matrix containing 1 row (m = 1) or 1 column (n = 1) is called a vector. If m = 1 it is called a row-vector and if n = 1 a column-vector. A row-vector can be represented as

$$\vec{B} = \vec{B}_n = \begin{bmatrix} b_1, & b_2 & \ldots & b_n \end{bmatrix} \qquad (17.30)$$

and a column-vector as

$$\vec{C} = \vec{C}_m = \begin{bmatrix} c_1 \\ c_2 \\ \vdots \\ c_m \end{bmatrix} \qquad (17.31)$$

If the number of rows is equal to the number of columns (m = n), the matrix is called a square matrix. The principal diagonal of a square matrix A is the set of elements $\{a_{11}\ a_{22}\ a_{33}\ \cdot\ \cdot\ \cdot\ a_{mm}\}$.

If in a square matrix for each i and j a_{ij} is equal to a_{ji} the matrix is called symmetric. In the representation as a table the principal diagonal of such a matrix is a line of symmetry.

A square matrix in which all the elements on one side of the principal diagonal are zero is called triangular.

$$a_{ij} = 0 \quad \text{for} \quad i > j$$

or

$$a_{ij} = 0 \quad \text{for} \quad i < j \tag{17.32}$$

If both conditions are verified, the matrix is called diagonal

$$a_{ij} = 0 \quad \text{for} \quad i \neq j \tag{17.33}$$

A diagonal matrix is, of course, symmetric.

A square matrix is called an identity matrix if all elements outside the principal diagonal are zero and all elements of the principal diagonal are equal to unity

$$\begin{aligned} a_{ij} &= 0 \quad \text{for} \quad i \neq j \\ a_{ii} &= 1 \end{aligned} \tag{17.34}$$

An identity matrix of m rows and columns is represented by I_m

$$I_m = \begin{bmatrix} 1 & 0 & . & . & . & 0 \\ 0 & 1 & . & . & . & 0 \\ \vdots & \vdots & \vdots & \vdots & \vdots & \vdots \\ 0 & 0 & . & . & . & 1 \end{bmatrix} \tag{17.35}$$

A matrix of which all elements are equal to zero is represented by 0_{mxn}.

17.7.2.2. Transpose of a matrix and a vector

The transpose of a matrix A_{mxn} is a matrix obtained through a permutation of its rows and columns. It is represented by A'_{nxm} and its elements are given by

$$a'_{ij} = a_{ji} \tag{17.36}$$

The transpose of a square matrix of dimension m is a square matrix of dimension m. The transpose of a symmetric matrix is the matrix itself. The transpose of a row-vector is a column-vector, and vice versa.

Examples

Consider a matrix A

$$A = A_{3x4} = \begin{bmatrix} 2 & 4 & 0 & -1 \\ 3 & 4 & 7 & 2 \\ 4 & -5 & 0 & 0 \end{bmatrix}$$

The transpose is given by

$$A' = \begin{bmatrix} 2 & 3 & 4 \\ 4 & 4 & -5 \\ 0 & 7 & 0 \\ -1 & 2 & 0 \end{bmatrix}$$

Consider a row-vector $\vec{C}$

$$\vec{C} = \vec{C}_5 = \begin{bmatrix} 2 & 0 & 1 & -2 & 3 \end{bmatrix}$$

Its transpose is given by

$$\vec{C}' = \begin{bmatrix} 2 \\ 0 \\ 1 \\ -2 \\ 3 \end{bmatrix}$$

17.7.2.3. Determinant of a square matrix

With each square matrix, A corresponds a real value called its determinant and written as $|A|$. To compute the determinant of a square matrix A of dimension m, the determinants of a number of submatrices of dimension (m-1) contained in A are used. To make this possible it must first be clear how to compute the determinants of small square matrices.

For m = 1 the determinant is equal to the unique element of the matrix

$$\text{If } A = [a_{11}] \quad \text{then} \quad |A| = a_{11} \tag{17.37}$$

For m = 2 the determinant is given by

$$|A| = \begin{vmatrix} a_{11} & a_{12} \\ a_{21} & a_{22} \end{vmatrix} = a_{11}\, a_{22} - a_{12}\, a_{21} \tag{17.38}$$

For m = 3 the formula becomes more complicated

$$|A| = \begin{vmatrix} a_{11} & a_{12} & a_{13} \\ a_{21} & a_{22} & a_{23} \\ a_{31} & a_{32} & a_{33} \end{vmatrix} = \begin{array}{l} a_{11}a_{22}a_{33} + a_{12}a_{23}a_{31} + a_{13}a_{21}a_{32} \\ - a_{31}a_{22}a_{13} - a_{32}a_{23}a_{11} - a_{33}a_{21}a_{12} \end{array} \tag{17.39}$$

This formula can also be written as

$$a_{11}\,(a_{22}a_{33} - a_{23}a_{32}) - a_{12}\,(a_{21}a_{33} - a_{31}a_{23}) + a_{13}\,(a_{21}a_{32} - a_{31}a_{22})$$

To obtain this formula, one considers the first row of A : take the first element of the row a_{11} and consider the determinant of the matrix obtained by removing from A the row and the column containing a_{11}. It is given by

$$\begin{vmatrix} a_{22} & a_{23} \\ a_{32} & a_{33} \end{vmatrix} = a_{22}a_{33} - a_{32}a_{23}$$

This determinant is called the minor of a_{11}. By multiplying the minor of an element a_{ij} with a factor $(-1)^{i+j}$ one obtains the cofactor of the element, which is written as A_{ij}. The determinant is given by

$$a_{11}A_{11} + a_{12}A_{12} + a_{13}A_{13} = \sum_j a_{1j}A_{1j} = a_{11}\,(a_{22}a_{33} - a_{32}a_{23})$$

$$+ a_{12}(-1)^{1+2}\,(a_{21}a_{33} - a_{23}a_{31}) + a_{13}(-1)^{1+3}\,(a_{21}a_{32} - a_{31}a_{22})$$

$$= a_{11}a_{22}a_{33} - a_{11}a_{32}a_{23} - a_{12}a_{21}a_{33} + a_{12}a_{23}a_{31} + a_{13}a_{21}a_{32} - a_{13}a_{31}a_{22}$$

It can be shown that the same result is obtained by taking any row or column of the matrix, multiplying each element of this row or column by its cofactor and adding the results

$$|A| = \sum_{i=1}^{m} a_{ij}\,A_{ij} = \sum_{j=1}^{m} a_{ij}\,A_{ij} \qquad (17.40)$$

The cofactors of the elements of a square matrix can again be considered as elements of a square matrix, the dimension of which is the same as the dimension of the original matrix. The transpose of this matrix is called the adjoint matrix of the original matrix ; it is written as Adj A

$$\text{Adj } A = \left[A_{ij}\right]' \qquad (17.41)$$

17.7.2.4. Geometric interpretation of matrices

Consider a matrix A_{3x2} given by

$$\begin{bmatrix} 2 & 4 \\ -1 & 0 \\ 3 & 2 \end{bmatrix}$$

By considering the rows of A as separate row-vectors the matrix is equivalent to three row-vectors of two elements each : (2 4), (-1 0) and (3 2). These vectors can also be represented in a plane as

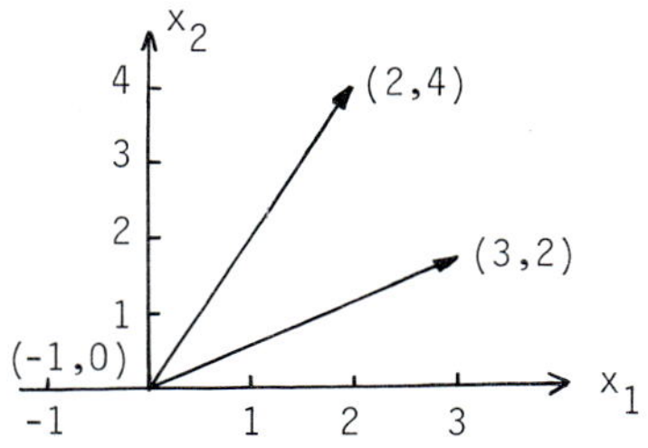

In the same way, the columns of A can be considered as two separate column -vectors of three elements each

$$\begin{bmatrix} 2 \\ -1 \\ 3 \end{bmatrix} \quad \text{and} \quad \begin{bmatrix} 4 \\ 0 \\ 2 \end{bmatrix}$$

These vectors can also be represented as vectors of a three-dimensional space. When considering a square matrix A_{mxm} it is possible to regard it as a set of m row-vectors in m-dimensional space or as a set of m column-vectors also in m-dimensional space (R^m).

17.7.2.5. Rank of a matrix

Consider a matrix A_{mxn}. The rank of A is defined as the dimension of the largest square matrix contained in A and with a non-zero determinant. Therefore, the rank of a matrix is at most equal to the smallest dimension. The rank of a matrix A is written as $r\left[A\right]$

$$r\left[A_{mxn}\right] \leq \min(m,n) \tag{17.42}$$

A matrix for which

$$r\left[A_{mxn}\right] < \min(m,n) \tag{17.43}$$

is called singular. If $r\left[A_{mxn}\right] = \min(m,n)$ it is called regular. A square matrix is therefore regular if its determinant is different from zero. Geometrically, the rank of a matrix can be defined as the maximal number of linear independent rows or columns that it contains. It can also be seen as the dimension of the smallest subspace containing either all of its rows or all of its columns.

17.7.3. Operations on matrices

17.7.3.1. Equality of matrices

Two matrices with the same dimensions are called equal if each element of one matrix is equal to the corresponding element of the second matrix.

$$A_{mxn} = B_{mxn} \qquad (17.44)$$

if and only if $a_{ij} = b_{ij}$ for all i and j.

17.7.3.2. Sum of matrices

The sum of two matrices with the same dimensions is a new matrix obtained by adding the corresponding terms of the two matrices

$$C_{mxn} = A_{mxn} + B_{mxn} \qquad (17.45)$$

if and only if $c_{ij} = a_{ij} + b_{ij}$ for all i and j.

17.7.3.3. Product of a matrix by a constant

The product of a matrix by a constant is a new matrix obtained by multiplying each element of the matrix by the constant

$$B_{mxn} = k\ A_{mxn} \qquad (17.46)$$

if and only if $b_{ij} = a_{ij} \times k$ for all i and j.

17.7.3.4. Product of two matrices

The product of two matrices can only be defined if the number of columns of the first matrix is equal to the number of rows of the second matrix. An element c_{ij} of the resulting matrix C is obtained by considering the ith row of the first matrix A_{mxp} and the jth column of the second matrix B_{pxn}. The ith row of A_{mxp} contains p elements and so does the jth column of B_{pxn}. The elements of the ith row are multiplied by the elements of the jth column and the results are added. This can also be expressed in the following way

$$C_{mxn} = A_{mxp} \times B_{pxn} \qquad (17.47)$$

if and only if $c_{ij} = \sum_{k=1}^{p} a_{ik} \cdot b_{kj}$ for all i and j.

It can be observed that the product of two square matrices with identical dimensions gives a new matrix with the same dimensions and that in general the result depends upon the order of the two matrices

$$A_{mxm} \times B_{mxm} \neq B_{mxm} \times A_{mxm} \qquad (17.48)$$

The product of a row-vector by a column-vector is a matrix containing a single element

$$A_{1xp} \times B_{px1} = C_{1x1} = c = \sum_{i=1}^{p} a_i b_i \qquad (17.49)$$

When the product is zero, the two vectors are said to be orthogonal. The length of a row-vector is defined as the square root of the product of this vector by its transpose

$$\sqrt{A_{1xp} \times A'_{px1}} \qquad (17.50)$$

17.7.3.5. Inverse of a square matrix

The inverse of a square matrix is a new square matrix of the same dimension such that the product of the two matrices in any order is equal to the identity matrix : the inverse of a matrix A is called A^{-1}

A^{-1}_{mxm} is the inverse of A_{mxm} if and only if

$$A^{-1}_{mxm} \times A_{mxm} = A_{mxm} \times A^{-1}_{mxm} = I_m \qquad (17.51)$$

It can be shown that only regular matrices have an inverse and that the inverse is given by

$$A^{-1}_{mxm} = \frac{\text{adj } A_{mxm}}{|A_{mxm}|} \qquad (17.52)$$

17.7.4. Eigenvalues and eigenvectors

17.7.4.1. Eigenvalues

Consider a square matrix A_{mxm}. Suppose that λ is an unknown value and consider the matrix $A-\lambda.I$. This matrix is obtained by subtracting λ from all diagonal elements of A. An eigenvalue of the matrix A is a value of λ for which the determinant of the resulting matrix is zero

$$|A-\lambda.I| = 0 \qquad (17.53)$$

The computation of this determinant yields an equation depending upon λ and of the mth degree. In general, this equation can be written in the following way

$$(-\lambda)^m + c_{m-1}(-\lambda)^{m-1} + c_{m-2}(-\lambda)^{m-2} + \ldots + c_1(-\lambda) + c_0 = 0 \qquad (17.54)$$

The coefficients c_{m-1}, c_{m-2}, ..., c_1, c_0 depend upon the elements of the matrix A. This equation has m solutions which can be real or imaginary ; they will be called $\lambda_1, \lambda_2, \ldots, \lambda_m$. When the matrix is symmetric the eigenvalues are real. In this instance one has the interesting property that the sum of the eigenvalues is equal to the sum of the diagonal elements of A. This value is also called the trace of A

$$\sum_{i=1}^{m} \lambda_i = \sum_{i=1}^{m} a_{ii} = \text{tr } A$$

The product of the eigenvalues is then equal to the determinant of A

$$\prod_{i=1}^{m} \lambda_i = |A|$$

Variance-covariance matrices have the property that their eigenvalues are all positive or zero.

Example

Consider the following matrix

$$B = \begin{bmatrix} 1 & 2 & 2 \\ 0 & 0 & 2 \\ 0 & 0 & 1 \end{bmatrix}$$

$$B - \lambda . I = \begin{bmatrix} 1-\lambda & 2 & 2 \\ 0 & -\lambda & 2 \\ 0 & 0 & 1-\lambda \end{bmatrix}$$

$$|B - \lambda . I| = (1 - \lambda)\ (-\lambda)\ (1 - \lambda) = 0$$

This equation has three real solutions : $\lambda_1 = 0$, $\lambda_2 = 1$ and $\lambda_3 = 1$.

17.7.4.2. Eigenvectors

To each eigenvalue λ_i a column-vector $\vec{C}(i)$ can be associated that is orthogonal on the matrix $A - \lambda_i . I$

$$(A - \lambda_i \ I) \times \vec{C}(i) = 0 \tag{17.55}$$

This equality can also be written as

$$A \times \vec{C}(i) = \lambda_i \ \vec{C}(i) \tag{17.56}$$

When multiplying $\vec{C}(i)$ by a constant another eigenvector is obtained. For this reason, the vector $\vec{C}(i)$ is reduced to unit length by dividing it by its length. This yields a new vector $\vec{V}(i)$

$$\vec{V}(i) = \frac{\vec{C}(i)}{\sqrt{\vec{C}(i)' \times \vec{C}(i)}} \tag{17.57}$$

Consider the example of the previous section again. $\lambda = 1$ is an eigenvalue of matrix B given by

$$B = \begin{bmatrix} 1 & 2 & 2 \\ 0 & 0 & 2 \\ 0 & 0 & 1 \end{bmatrix}$$

Eigenvector $\vec{C}$ is found using eqn. 17.56

$$\begin{bmatrix} 1 & 2 & 2 \\ 0 & 0 & 2 \\ 0 & 0 & 1 \end{bmatrix} \begin{bmatrix} C_1 \\ C_2 \\ C_3 \end{bmatrix} = 1 \quad . \quad \begin{bmatrix} C_1 \\ C_2 \\ C_3 \end{bmatrix}$$

This gives the following equations

$$\begin{aligned} C_1 + 2\ C_2 + 2\ C_3 &= C_1 \\ 2\ C_3 &= C_2 \\ C_3 &= C_3 \end{aligned}$$

A solution is given by

$$\vec{C} = \begin{bmatrix} 1 \\ 0 \\ 0 \end{bmatrix}$$

As this vector has unit length it is equal to the corresponding vector $\vec{V}$.

17.7.4.3. Linear transformations

Consider a square matrix A_{mxm} and a column-vector $\vec{X}$ of dimension m. Multiplying matrix A by vector $\vec{X}$ yields a new column-vector $\vec{Y}$ of dimension m. Therefore, a matrix A_{mxm} defines a transformation on the set of column-vectors of dimension m. The transformation is linear because the elements of the resulting vector $\vec{Y}$ are obtained from $\vec{X}$ only by multiplication with constants and addition. From the definition of eigenvectors it can be seen that a non-zero vector $\vec{X}$ is an eigenvector of matrix A if its transformation through A yields a vector on the same straight line as $\vec{X}$. Therefore, the resulting vector is equal to $\vec{X}$ multiplied by a constant

$$A \vec{X} = \lambda \vec{X}$$

The constant λ is the eigenvalue. It can be seen that to each eigenvector there corresponds exactly one eigenvalue but the same constant can be an eigenvalue of many eigenvectors.

Consider the example of the previous section. Multiplying matrix B by its eigenvector $\vec{C}$ gives

$$\begin{bmatrix} 1 & 2 & 2 \\ 0 & 0 & 2 \\ 0 & 0 & 1 \end{bmatrix} \begin{bmatrix} 1 \\ 0 \\ 0 \end{bmatrix} = \begin{bmatrix} 1 \\ 0 \\ 0 \end{bmatrix}$$

This vector is on the same straight line as $\vec{C}$. The entire straight line of the eigenvector is left unchanged by the linear transformation. On the other hand, a vector not on this line will not be transformed in the same way. For example, considering the vector $\vec{D}$

$$\vec{D} = \begin{bmatrix} 1 \\ 1 \\ 1 \end{bmatrix}$$

It is transformed into

$$\begin{bmatrix} 1 & 2 & 2 \\ 0 & 0 & 2 \\ 0 & 0 & 1 \end{bmatrix} \begin{bmatrix} 1 \\ 1 \\ 1 \end{bmatrix} = \begin{bmatrix} 5 \\ 2 \\ 1 \end{bmatrix}$$

which is not on the same line as $\vec{D}$. The straight line containing $\vec{D}$ is transformed into a different straight line.

REFERENCES

F. Dupuis and A. Dijkstra, Anal. Chem., 47 (1975) 379.
A. Eskes, F. Dupuis, A. Dijkstra, H. De Clercq and D.L. Massart, Anal. Chem., 47 (1975) 2168.
I.S. Herschberg, Z. anal. Chem., 205 (1964) 180.
A. Junker and G. Bergmann, Z. anal. Chem., 272 (1974) 267.
A. Junker and G. Bergmann, Z. anal. Chem., 278 (1976 a) 191.
A. Junker and G. Bergmann, Z. anal. Chem., 278 (1976 b) 273.
H. Kaiser, Z. anal. Chem., 260 (1972) 252.
D.J. Leggett, Anal. Chem., 49 (1977) 276.
H. Neuer, Z. anal. Chem., 253 (1971) 337.
A. Parczewski, Chem. Anal. (Warsaw), 21 (1976 a) 321.
A. Parczewski, Chem. Anal. (Warsaw), 21 (1976 b) 593.
A. Parczewski and A. Rokosz, Chem. Anal. (Warsaw), 20 (1975) 267.
S.D. Rasberry and K.F.J. Heinrich, Anal. Chem., 46 (1974) 81.
J. Sustek, Anal. Chem., 46 (1974) 1676.
G. van Marlen and A. Dijkstra, Anal. Chem., 48 (1976) 595.
J.H. Wilkinson and C. Reinsch, in F.L. Bauer et al. (Editors), Handbook for Automatic Computation, Vol. II, Linear Algebra, Springer, Berlin, 1971.

Chapter 18

PREFERRED SETS - THE CLASSIFICATION APPROACH

18.1. THE CLASSIFICATION PROBLEM

In this chapter we discuss GLC example (II) of Chapter 16, together with some related problems. A large number of liquid GLC phases (more than 700) have been proposed in the literature, of which at least 200 have been used by at least a few workers. It is clear that some reduction of this large number of phases is necessary because it should eliminate redundant phases and leave a restricted number with really different characteristics, i.e., a restricted preferred set. This can be achieved by grouping (classifying) liquid phases with analogous retention behaviour for the same test substances (probes). Let us suppose that such a classification is attempted by using only two probes (for instance, benzene and ethanol). The (imaginary) retention indices for a number of liquid phases (called A, B, ..., J) of these two substances are shown in Fig. 18.1.

Clearly, phases E and D have very similar retention properties and, if some of the phases have to be eliminated, either E or D should be one of these. A classification of these phases permits one to distinguish first two groups (or classes or clusters), namely ABCDE and FGHIJ. If only two phases are to be retained from the original ten, it seems logical to take one of the first group and one of the second. On closer observation, one notes that the first group can be divided into two sub-groups, namely ABC and ED, and that in the second one can discern further two sub-groups, namely FGHI and J. If four phases are to be selected, one of each of the sub-groups should be included. One could therefore re-state the problem of selecting a restricted set of liquid phases in the following way : classify the existing liquid phases in such a way that groups of liquid phases with analogous characteristics are formed and select from each group one (or more) liquid phases. This approach to the selection of a restricted set of phases for GLC (and TLC) was developed in a series of

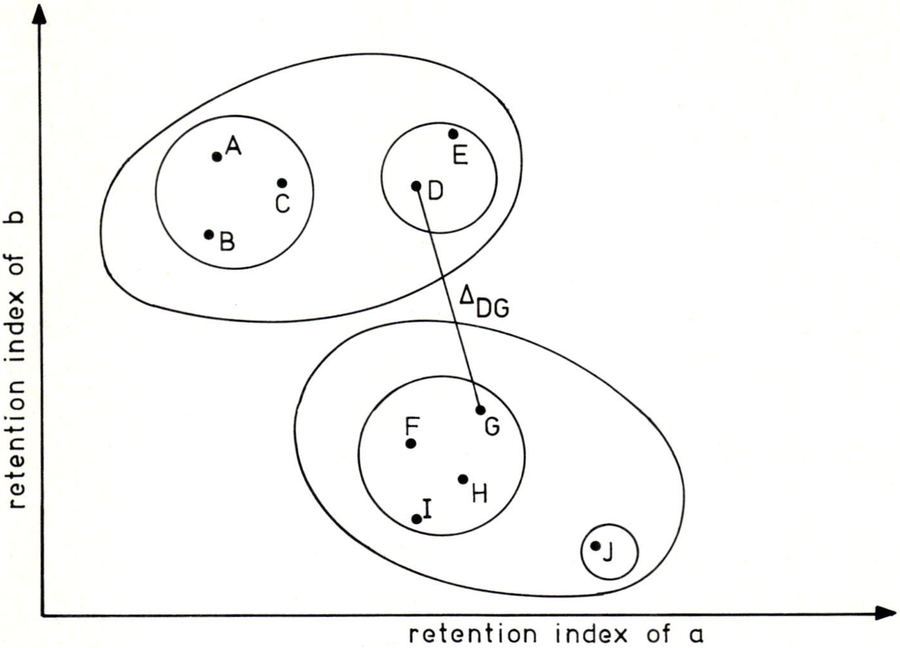

Fig. 18.1. Plot of the retention indices of substances a and b for ten liquid phases (imaginary values are used).

papers by Massart and co-workers (Massart and De Clercq, 1974 ; Massart et al., 1974 ; De Clercq et al., 1975 ; De Clercq and Massart, 1975).

18.2. CLASSIFICATION TECHNIQUES

The solution that is proposed here is the elaboration of a hierarchical classification of liquid phases. It is hierarchical because large groups are divided into smaller ones (for instance, in Fig. 18.1 group ABCDE into groups ABC and ED). These can be split up again until eventually each group consists of only one liquid phase. The resulting classification can also be depicted as in Fig. 18.2. This kind of representation is called a dendrogram.

To arrive at this classification, one has to detect clusters of points in a pattern space (two-dimensional in this instance). These techniques are therefore called hierarchical clustering techniques and are part of the group of unsupervised learning techniques.

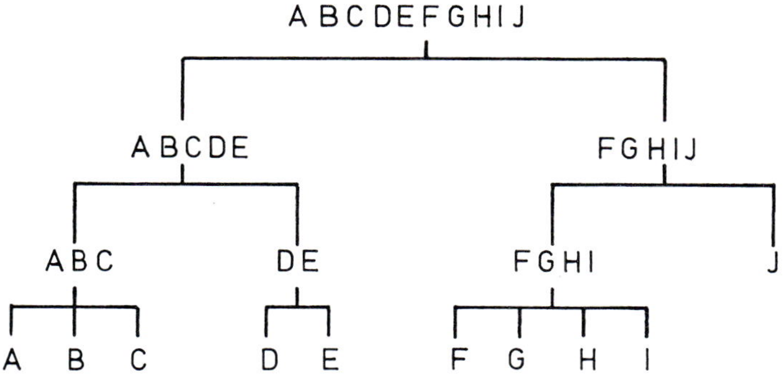

Fig. 18.2. Dendrogram for the classification of the liquid phases represented in Fig. 18.1.

A classification such as that shown in Fig. 18.2 resembles closely a biological classification in which all living species are first divided into regna and then into phyla, classes, subclasses, etc., down to the individual species. In the last 15-20 years, taxonomists have used numerical techniques (in fact, clustering methods) to arrive at this result. The collection of these techniques has been called numerical taxonomy. A standard book on this subject has been written by Sneath and Sokal (1973). The terms numerical taxonomy and hierarchical clustering methods are interchangeable. Massart and De Clercq (1974) in papers on the classification of TLC and GLC systems preferred the term numerical taxonomy.

In this chapter we shall follow mainly the terminology used by Sneath and Sokal. For example, we call the objects to be classified operational taxonomic units (OTUs). In the examples of interest to us they represent the individual GLC liquid phases or TLC systems and in biology the individual living species. These OTUs are classified according to the values taken by a number of parameters, called the characteristics. In the GLC problem, these are the retention indices of the probes. The comparison and classification of OTUs is carried out in five steps, as follows :

(a) The construction of a data matrix. This matrix can consist simply of the original data or of data transformed, for example, by scaling (see Chapter 20). In the GLC problem (Massart et al., 1974), a 226 x 10 matrix was used, consisting of the retention indices of 10 probes for 226 liquid phases as given by McReynolds (1970).

(b) The measurement of resemblance. In Fig. 18.1, the clustering was carried out on the basis of a geometrical distance. The larger the distance between two liquid phases, the less they resemble each other. Several measures of resemblance are possible and are discussed in sections 18.3 and 18.5.2.

(c) The clustering procedure. A wide variety of possibilities exists. An enumeration of some of these is given in section 18.4.1 and two methods are discussed in more detail in sections 18.4.2 and 18.4.3.

(d) The display of the classification. The display used with the two clustering methods is discussed in sections 18.4.2 and 18.4.3, in which these methods are introduced.

(e) The selection of the preferred set. This is discussed in section 18.4.4.

18.3. MEASURES OF RESEMBLANCE

To compare pairs of OTUs, a measure of resemblance must be defined that serves to quantify the resemblance between the values of the characteristics for two OTUs in the data matrix. The data x_{ik} are recorded in an i x k matrix with i = 1, ..., d and j = 1, ..., n, d being the number of characteristics and n the number of OTUs. The resemblance or similarity must be computed between each pair of columns in this matrix. Many coefficients of resemblance, often created for specific classification problems, have been proposed in several, sometimes very diverse, domains of science. The two types of coefficients that have been employed in most analytical applications are the distance and the correlation coefficient.

The first coefficients measure a distance between two OTUs in a pattern space. The smaller the distance between two OTUs, the more they resemble each other. Two identical OTUs coincide so that the distance between them is zero. The geometrical Euclidean distance, also called the taxonomical distance, between the two OTUs D and G in Fig. 18.1 is given by

$$\Delta_{DG} = \sqrt{(x_{1D} - x_{1G})^2 + (x_{2D} - x_{2G})^2} \quad (18.1)$$

This can be generalized for d characteristics, i.e., d-dimensional space

$$\Delta_{kl} = \left[\sum_{i=1}^{d} (x_{il} - x_{ik})^2 \right]^{1/2} \quad (18.2)$$

The Euclidian distance is a special case of the more general Minkowski distance (between elements) or Mahalanobis distance (between groups).

The similarity can also be expressed as a correlation coefficient. In this instance, one determines this coefficient between all pairs of columns of the data matrix. When the coefficient is 1, the OTUs behave in the same way and the closer the coefficient approaches 0, the less two OTUs resemble each other. The correlation coefficient and the distance do not measure the same property. This can be understood from the data in Table 18.I, where three (imaginary) TLC systems are compared. The characteristics are the R_f values of the substances to be separated.

Table 18.I

R_f values of imaginary TLC systems

Substance	System		
	A	B	C
1	0.30	0.60	0.15
2	0.20	0.40	0.25
3	0.10	0.20	0.35
4	0.25	0.50	0.20
5	0.35	0.70	0.10

The Euclidian distance between systems A and B is large, so that when this coefficient of similarity is used, A and B are very different. The values for A and B are, however, completely correlated, so that when using this measure

A and B are identical. If one wants to classify the systems according to eluting power, the Euclidean distance should be used. If, on the other hand, one wants to obtain different elution orders one should use the correlation coefficient. It should be noted that in this instance the sign of the correlation coefficient is not important. The correlation coefficient for systems C and A is -1. The order of elution is exactly the inverse for C compared with A. For practical separation purposes, one may consider that C does not yield other separation possibilities than A so that A and C are considered to be identical. One should then use the absolute value of the correlation coefficient.

18.4. CLUSTERING PROCEDURES

18.4.1. Types of clustering procedures

With the help of one of the coefficients of similarity mentioned above, the similarity matrix is constructed. This is a symmetrical $n \times n$ resemblance or similarity matrix. Let us suppose, for example, that the systems A, B, C, D and E in Table 18.II have to be classified. Using eqn. 18.2 one obtains the similarity matrix in Table 18.III.

Table 18.II

Example of data matrix

System	Retention index			
	probe a	probe b	probe c	probe d
A	100	80	70	60
B	80	60	50	40
C	80	70	40	50
D	40	20	20	10
E	50	10	20	10

Because the matrix is symmetrical, only half of this matrix need be used. The clustering procedure consists in isolating the clusters from such a matrix. There is a wide variety of these procedures available and it is impossible to discuss all of them here. Readers are referred to Chapter 5 of the book by

Sneath and Sokal (1973) for more details.

Table 18.III

Similarity matrix for the distances obtained from Table 18.II

	A	B	C	D	E
A	0				
B	40.0	0			
C	38.7	17.3	0		
D	110.4	70.7	78.1	0	
E	111.4	72.1	80.6	14.1	0

We shall discuss here only the so-called SAHN techniques (sequential, agglomerative, hierarchical, non-overlapping). These techniques are called :

(a) agglomerative because they start from individual OTUs which are taken together in small sets, after which those sets merge with other small sets or individual OTUs ;

(b) sequential because the grouping of the OTUs is obtained using a sequential grouping algorithm (and not a simultaneous grouping of all the OTUs in classes) ;

(c) hierarchical because, when the classification is represented as in Fig. 18.2, there are always less classes at each ascending level ;

(d) non-overlapping because the classes are mutually exclusive at each level. This means that the OTUs that are members of a certain class cannot simultaneously be members of another group.

The branch and bound technique explained in Chapter 22 is, on the contrary, a divisive, simultaneous, non-hierarchical, non-overlapping method because it divides the set of all OTUs into subsets in one operation. There is only one level in the classification and members of one group do not belong to any other group. Several algorithms have been proposed for SAHN clustering methods. The two algorithms discussed here are the average and single linkage grouping. For the latter, an operational research model is used.

18.4.2. The average linkage algorithm

In the similarity matrix, one seeks the smallest Δ_{kl} value (or whatever other similarity coefficient is used). Let us suppose that it is Δ_{qp}, which means that of all of the OTUs to be classified, q and p are the most similar. They are considered to form a new combined OTU p^*. The resemblance matrix is thereby reduced to (n - 1) x (n - 1). The similarities between the new OTU and all the others are obtained by averaging the similarities of q and p with these other OTUs. For example

$$\Delta_{kp^*} = (\Delta_{kq} + \Delta_{kp}) / 2 \tag{18.3}$$

This process is repeated until all OTUs are linked together in one hierarchical classification system, which is represented by a dendrogram. This procedure can now be explained using the data in Table 18.III. The smallest Δ is 14.1 (between D and E). D and E are combined first.

Table 18.IV

Successive reduced matrices for the data in Table 18.III

(a)	A	B	C	D*
A	0			
B	40.0	0		
C	38.7	17.3	0	
D*	110.9	71.4	79.3	0

D* is the OTU resulting from the combination of D and E

(b)	A	B*	D*
A	0		
B*	39.3	0	
D*	110.9	75.3	0

B* is the OTU resulting from the combination of B and C

(c)	A*	D*
A*	0	
D*	93.1	0

A* is the OTU resulting from the combinations of A and B*

(d) The last step consists in the junction of A* and D*
The resulting dendrogram is given in Fig. 18.3

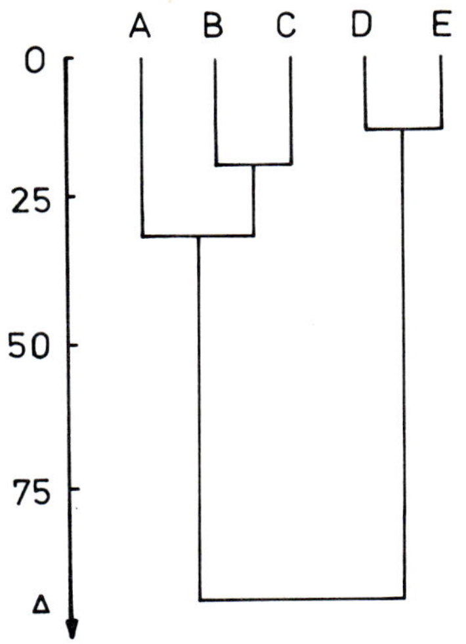

Fig. 18.3. Dendrogram for the data in Table 18.III.

The average linkage method explained here is probably the most commonly used. It does not, however, constitute the only possible average linkage method. One can use other than arithmetic averages. Moreover, the averaging procedure described is a weighted method. Consider, for example, the union of an OTU q with another OTU p^* which arose itself from the union of two OTUs p and o. When making the averages, p^* and q have the same weights but as p^* consists of p and o, the latter have only half the weight of q. The method used here is therefore called a weighted pair group method using arithmetic averages (WPGMA), in contrast with UPGMA methods (unweighted pair, etc.) where the distances are calculated with equal weights for each of the original OTUs. If the OTU resulting from the union of p^* and q is called q^*, then

$$\Delta_{kq^*} = (\Delta_{kp} + \Delta_{ko} + \Delta_{kq}) / 3 \qquad (18.4)$$

18.4.3. Operational research techniques

In the average linkage procedure the similarity coefficient between an OTU and a class consisting of two or more OTUs is computed as an average (eqn. 18.3).

In the single linkage procedure, the similarity is expressed as the similarity between the OTU and the nearest (most similar) OTU of the class. In Table 18.IV (a) the distance between A and D^* would not be 110.9 but 110.4, the shortest of the distances between A and D and A and E, D and E being the OTUs of class D^*. The single linkage procedure can be carried out according to the same kind of algorithm as the average linkage procedure.

It can also be carried out conveniently using a graph-theoretical algorithm, namely the calculation of the minimal spanning tree in a network. One of the simplest algorithms was derived by Kruskal (1956). The terminology used in graph theory is explained in section 23.1. Suppose that seven towns must be connected to each other through highways (or a production unit to six clients through a pipeline). This must be done in such a way that the total length of the highway is minimal. The distances between the cities are known. In Fig. 18.4 two possible configurations are given. Clearly (a) is a better solution than (b). Both (a) and (b) are graphs that are part of the complete graph containing all possible links and both are connected graphs (all of the nodes are linked directly or indirectly to each other). These graphs are called trees and the tree for which the sum of the values of the links is minimal is called the minimal spanning tree. This minimal spanning tree is also the optimal solution for the highway problem.

Let us now consider how to find the minimal spanning tree. There are several algorithms that can be used to achieve this. Kruskal's (1956) algorithm is the simplest when the number of nodes is not too large. Kruskal's algorithm can be stated as follows : "add to the tree the edge with the smallest value which does not form a cycle with the edges already part of the tree". In Table 18.V the distances between the nodes of Fig. 18.4 are given. According to this algorithm, one selects first the smallest value in the Table (link BC, value 23). The next smallest value is 28 (link AB). The next smallest values are 29 and 30 (links EF and EG). The next smallest value in the table is 32 (link AC). This would, however, close the cycle ABC and is therefore eliminated. Instead,

Table 18.V

Distance between points in Fig. 18.4 (Massart and Kaufman, 1975. Reprinted with permission from the American Chemical Society)

	A	B	C	D	E	F	G
A	0						
B	28	0					
C	32	23	0				
D	35	40	60	0			
E	100	80	103	75	0		
F	119	104	128	90	29	0	
G	127	105	126	105	30	35	0

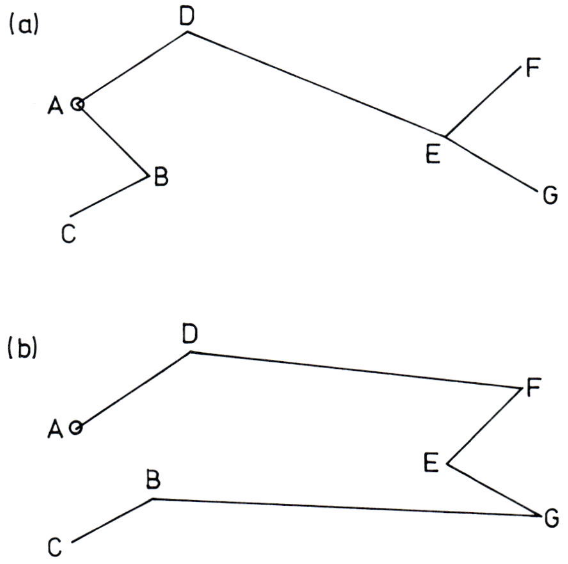

Fig. 18.4. Examples of trees in a graph. (a) is the minimal spanning tree (Massart and Kaufman, 1975. Reprinted with permission from the American Chemical Society).

the next link that satisfies the conditions of Kruskal's algorithm is AD and the last one is DE. The minimal spanning tree obtained in this way is that given in Fig. 18 (a). By careful inspection of this figure, one notes that two clusters can be obtained in a formal way by breaking the longest edge (DE). Various other possibilities have been proposed by Zahn (1971).

If A, B, ..., G are chromatographic systems, then this allows one to classify these systems into two classes. When a more detailed classification is needed, one breaks the second longest edge, etc., until the desired number of classes is obtained. De Clercq and Massart (1975) showed how this can be applied to thin-layer chromatography.

18.4.4. The selection of the preferred set from the classification

When a dendrogram has been obtained, one isolates the clusters or classes by breaking the links of lowest similarity. In the example in Fig. 18.3, one breaks first the link at Δ = 85.2, then that at Δ = 28.0, etc. This procedure resembles closely the breaking of the longest edge in a minimal spanning tree. Once the clusters have been isolated, one can choose from each of them the best OTU, the definition of which obviously depends on the criteria used. For the selection of liquid phases in GLC, one could think of price, stability, availability and other practicability criteria. If it is the purpose to develop an optimal set for qualitative analysis, one could also select from each of the clusters the liquid phase with the best separation characteristics, i.e., the one which yields the largest amount of information.

18.5. INFORMATION AND CLASSIFICATION

18.5.1. A comparison of the information theoretical and numerical taxonomic approaches

Eskes et al. (1975) applied both information theory and numerical taxonomy to the selection of preferred sets of 2-5 liquid phases from a set of 16. The amount of information obtained for these liquid phases was calculated using the retention indices of a data set of 248 substances from eqn. 17.25. The numerical taxonomy was carried out by taking the liquid phases (columns) as OTUs and the retention indices as the characteristics. The correlation coefficient was used as the measure of resemblance. In Table 18.VI, the results are compared.

Table 18.VI

Combination of columns yielding a maximal amount of information and the classification of columns obtained by numerical taxonomy (for names of columns, see Eskes et al., 1975)

No. of columns	Best combination of columns	Total amount of information (bits)	Classification of columns
1	13	6.8	-
2	10, 12	13.5	1-6,9,11,12/7,8,10,13-16
3	1, 8, 10	19.2	1-6,9,11,12/7,8,13-16/10
4	2, 8, 10, 11	24.4	1-6,9/7,8,13-16/11,12/10
5	2, 8, 10, 11, 13	29.1	1-6,9/7,13-16/11,12/10/8

There is complete agreement between the two approaches. Consider, for example, the preferred set of five columns (Nos. 2, 8, 10, 11 and 13) obtained by using information theory. They should be present in different groups in the numerical taxonomic classification. This is indeed the case. The agreement between the results is not surprising because the information per column varies only slightly, so that the correlation is also the determining factor in the information theoretical approach. Nevertheless, this agreement demonstrates that both methods are different but equivalent approaches to the problem of selecting optimal sets of tests for qualitative analysis. Both apply implicitly or explicitly the following two rules : each test of the selected set should be as good as possible, and each test of the selected set should be as different as possible. We shall see that this is also the case for the third approach, discussed in Chapter 20, namely the use of linear discriminant analysis.

18.5.2. Classification using an information theoretical criterion

In the preceding section, we concluded that the numerical taxonomic and information theoretical approaches are different expressions of the application of the same basic selection rules. In this section, we shall show that the two approaches can be combined into one, in the sense that the amount of information can be used as a measure of resemblance in a numerical taxonomic application.

Let us return, therefore, to the example with which we introduced information theory, namely the evaluation of the information content of individual TLC chromatographic systems. We now want to develop an optimal combination of TLC chromatographic systems. First, we re-write Shannon's eqn. 8.3 for the amount of information of chromatographic system k.

As shown in Chapter 8, the information content of a chromatographic system k can be obtained by dividing the R_F scale into m groups, so that in each group there is a number $n_i(k)$ of the total number n_o of substances, the separation of which is being considered ($0 \leqslant n_i(k) \leqslant n_o$, $\sum_{k=1}^{m} n_i(k) = n_o$), and by using eqn. 8.3

$$I(k) = -\sum_{i=1}^{m} \frac{n_i(k)}{n_o} \cdot \log_2 \left(\frac{n_i(k)}{n_o}\right) \qquad (18.5)$$

If the information in t different systems were completely uncorrelated then the total amount of information or joint information would be

$$I(1,2,\ldots,t) = \sum_{k=1}^{t} I(k) \qquad (18.6)$$

It is well established that chromatographic systems are often very similar : one can say that they are highly correlated, yield highly correlated information or, to use classification or numerical taxonomical language, they have a high similarity coefficient. The higher the correlated or redundant information, the more similar the systems are and it is logical to use the amount of redundant information as a similarity coefficient in the same way as we have used Euclidian distances or linear correlation coefficients as similarity coefficients (section 18.3). In biological numerical taxonomy, this has been done, for example, by Orloci (1969) and in analytical chemistry by Ritter et al. (1976) in a pattern recognition application (concerning IR spectra).

The effect of correlation is that less information is obtained when combining t chromatographic systems than can be calculated from eqn. 18.6 for the joint information

$$I\,(1,2,\ldots,t) = \sum_{k=1}^{t} I(k) - I(1;2;\ldots;t) \qquad (18.7)$$

where I(1;2;...;t) is called "mutual information". The mutual information depends on the total amount of information in the sense that when the latter is higher, that the values for the mutual information also have a tendency to be higher. Therefore, a relative measure such as Rajski's coherence coefficient can be of more value. For two chromatographic systems i and j, this coefficient is given by

$$R(k,l) = \left[1 - d^2\,(k,l)\right]^{1/2} \qquad (18.8)$$

where

$$d(k,l) = 1 - \frac{I(k;l)}{I(k,l)} \qquad (18.9)$$

Table 18.VII

R_F values in two TLC systems for the identification of food dyes (from Massart and De Clercq, 1974 and Hoodless et al., 1973)

Dye	System 1	System 2
Amaranth	0.6	0.3
Bordeaux B	0.2	0.6
Carmoisine	0.3	0.7
Eosine	0.2	1.0
Erythrosine	0.1	1.0
Fast red E	0.4	0.7
Ponceau 4R	0.7	0.5
Ponceau 6R	0.8	0.2
Ponceau MX	0.2	0.7
Ponceau SX	0.4	0.7
Red 2G	0.6	0.6
Red 6B	0.4	0.3
Red 10B	0.2	0.5
Red FB	0.0	0.3
Rhodamine B	0.5	1.0
Scarlet GN	0.9	0.7
Auramine	0.3	1.0
Acid Yellow	0.7	0.6
Chrysoidine	0.1	0.8
Chrisoine S	0.5	0.8
Naphthol Yellow S	0.6	0.7
Orange G	0.8	0.7
Orange GGN	0.6	0.7
Orange I	0.4	0.8
Sunset Yellow	0.6	0.7
Tartrazine	0.8	0.4

To show how this is done in practice, Table 18.VII gives the R_F values of two TLC systems. The information content of each of the systems is calculated by using eqn. 18.5. To obtain the joint information, one first makes the contingency table (Table 18.VIII).

Table 18.VIII

Contingency table relating the R_F values of systems 1 and 2 in Table 18.VII

		Class values in system 1											
	i →	1	2	3	4	5	6	7	8	9	10	11	Sum
Class values in system 2 j ↓		0	0.1	0.2	0.3	0.4	0.5	0.6	0.7	0.8	0.9	1.0	Sum
1	0	0	0	0	0	0	0	0	0	0	0	0	0
2	0.1	0	0	0	0	0	0	0	0	0	0	0	0
3	0.2	0	0	0	0	0	0	0	0	1	0	0	1
4	0.3	1	0	0	0	1	0	1	0	0	0	0	3
5	0.4	0	0	0	0	0	0	0	0	1	0	0	1
6	0.5	0	0	1	0	0	0	0	1	0	0	0	2
7	0.6	0	0	1	0	0	0	1	1	0	0	0	3
8	0.7	0	0	1	1	2	0	3	0	1	1	0	9
9	0.8	0	1	0	0	1	1	0	0	0	0	0	3
10	0.9	0	0	0	0	0	0	0	0	0	0	0	0
11	1.0	0	1	1	1	0	1	0	0	0	0	0	4
	Sum	1	2	4	2	4	2	5	2	3	1	0	26

One now considers as classes the cells obtained in this table. This represents in fact the two-dimensional chromatogram with m^2 cells that would have been obtained with systems k and l (k = 1, l = 2, in this particular instance). In each cell or class there are n_{ij} (k,l) substances and Shannon's equation now becomes

$$I(k,l) = - \sum_{i=1}^{m} \sum_{j=1}^{m} \frac{n_{ij}(k,l)}{n_o} \cdot \log_2 \left(\frac{n_{ij}(k,l)}{n_o} \right) \qquad (18.10)$$

From eqns. 18.6 and 18.7 we have

$$I(k;l) = I(k) + I(l) - I(k,l) \qquad (18.11)$$

so that the mutual information content I(k;l) is known and can be entered in eqns. 18.8 and 18.9, yielding R(k,l).

In the present example

$I(k) = 2.67$ bits
$I(l) = 3.15$ bits
$I(k,l) = 4.44$ bits
$I(k,l) = 1.38$ bits
$d(k,l) = 0.69$
$R(k,l) = 0.72$

By doing this for all pairs of systems and considering $R(i,j)$ as a similarity coefficient, a similarity matrix is obtained. This can be reduced in the usual way, for example using an average linkage WPGMA method.

18.6. APPLICATIONS

Several applications of numerical taxonomy for the determination of optimal sets in thin-layer chromatography have been published. The first (Massart and De Clercq, 1974) described the selection of optimal combinations for the identification of 26 yellow, orange and red synthetic food dyes. A set of three TLC systems was chosen from 10 candidate systems so that an unambiguous identification could be obtained of all the dyes (with the exception of two that cannot be separated in any of the 10 systems).

Other applications include the identification of 22 sulphonamides with 51 candidate systems. The best set of three systems allowed the unambiguous identification of all of the sulphonamides (De Clercq et al., 1977). The best sequence of four from seven candidate systems for the identification of 139 basic drugs was also obtained. This sequence allowed the unambiguous identification of 87% of the substances compared with 72% for the published sequence (Massart and De Clercq, 1978). The identification of a smaller set of basic drugs was discussed by De Clercq and Massart (1975). All these applications are carried out according to the same general scheme. A dendrogram or a minimal spanning tree using the distance as the similarity measure is obtained, the lowest branches of the dendrogram (or the longest edges of the tree) are broken so that groups of systems are isolated and in each of the groups the best system is selected using the information content (Chapter 8).

The GLC problem (II) (see Chapter 16) was also solved by numerical taxonomy. The data matrix here consists of 226 liquid phases (the OTUs) for which the retention indices of 10 substances (the characteristics) are given. In a first publication (Massart et al., 1974) the classification was based on distances and the average linkage-weighted pair clustering technique and in a second paper (De Clercq et al., 1975) the average linkage-unweighted pair method was used. The latter is preferred, but there is little difference between them.

No "best set" is proposed, because it is argued that criteria such as stability of the phase, cost and availability should be taken into account and that this selection should therefore be left to specialists. However, it is shown that the classification is valid, as all of the phases which are known to be special in their interactions with solutes are found in classes of a single phase, and as in those instances where it is possible to make a classification by purely chemical argumentation the classification obtained by numerical taxonomy appears to be logical. For example, in the class of the apolar phases all of the saturated hydrocarbons are found in one sub-class. The same is true for the Apiezons, two groups of silicones, one with the methylsilicones and one with the methylphenyl or ethylphenylvinyl derivatives, a group of three esters and a group of three fluorinated hydrocarbons, each of these forming one sub-class of the apolar phases. A classification of a set of 121 more recent phases was also given (Massart and De Clercq, 1978).

All of these applications and the theory are described in more detail in the review by Massart and De Clercq (1978).

18.7. CORRELATION AND DISTANCE

In order to examine the structure of a set of variables or the elements of a population, it is necessary to have means of measuring the similarity relationship between variables or elements. When studying a set of variables, the usual way of describing the relationship between two variables is by means of a coefficient of association or of correlation. Such a coefficient measures the similarity between the values taken by the two variables for a given data set or population.

When comparing elements of a data set or population, one has to combine scores of several variables in order to obtain a similarity measure. Such a combination is then used to measure a "distance" between two elements. In the literature, several definitions of correlation and distance have been proposed, each having specific properties. In the following sections some of these definitions are discussed.

18.7.1. Measures of distance between elements

When examining the relation between two elements of a population, it is common to define a distance. In the following section, the scores of the elements l_1 and l_2 for the different variables i (i = 1,2,..., m) will be called x_{i1} and x_{i2}.

18.7.1.1. The Minkowski distance

The Minkowski distance between two elements l_1 and l_2 is defined as

$$D_p(l_1,l_2) = \left(\sum_{i=1}^{m} |x_{i1} - x_{i2}|^p\right)^{1/p} \quad (18.12)$$

where p ⩾ 1. By choosing various values of p, many different distances are obtained. The Euclidean distance is obtained for p = 2

$$D_2(l_1,l_2) = \sqrt{\sum_{i=1}^{m} (x_{i1} - x_{i2})^2} \quad (18.13)$$

The Manhattan distance is given by p = 1

$$D_1(l_1,l_2) = \sum_{i=1}^{m} |x_{i1} - x_{i2}| \quad (18.14)$$

18.7.1.2. The Lance and Williams distance

The Lance and Williams distance between two elements l_1 and l_2 is defined as

$$D_{LW}(l_1,l_2) = \frac{\sum_{i=1}^{m} |x_{i1} - x_{i2}|}{\sum_{i=1}^{m} (x_{i1} + x_{i2})} \qquad (18.15)$$

The numerator is the Manhattan distance D_1 and the denominator the total magnitude of the two elements. In this way the distance does not depend on the magnitude of x_{i1} and x_{i2}.

18.7.1.3. The Calkoun distance

The Calkoun distance is based on the ordering of elements for each variable. It has the particular property that the distance between two elements also depends on the other elements. It requires the following definitions : N_1 is the number of elements that fall between the two points on at least one variable ; N_2 is the number of elements not in N_1 but which tie in value on at least one variable with one of the two elements ; and N_3 is the number of elements not in N_1 or N_2 but tie in value on at least one variable with both elements.

The Calkoun distance is then defined as

$$D_C(l_1,l_2) = 6\ N_1 + 3\ N_2 + 2\ N_3 \qquad (18.16)$$

The normalized Calkoun distance is given by

$$D_{CN}(l_1,l_2) = \frac{6\ N_1 + 3\ N_2 + 2\ N_3}{6\ (N-2)} \qquad (18.17)$$

where N is the total number of data points.

18.7.2. Correlation and distance based on correlation

A correlation coefficient can be computed between two elements of a population (see also section 3.2.5). It is given by

$$r(l_1,l_2) = \frac{\sum_{i=1}^{m} (x_{i1}-\bar{x}_{.1})(x_{i2}-\bar{x}_{.2})}{\sqrt{\sum_{i=1}^{m} (x_{i1}-\bar{x}_{.1})^2 \sum_{i=1}^{m} (x_{i2}-\bar{x}_{.2})^2}} \qquad (18.18)$$

where

$$\bar{x}_{.j} = \frac{1}{m} \sum_{i=1}^{m} x_{ij} \qquad (j = 1,2)$$

It can be used as a measure of resemblance (see section 18.3).

By defining

$$A_j = \sqrt{\sum_{i=1}^{m} (x_{ij}-x_{.j})^2} \qquad (j = 1,2) \qquad (18.19)$$

eqn. 18.18 becomes

$$r(l_1,l_2) = \sum_{i=1}^{m} \left(\frac{x_{i1}-\bar{x}_{.1}}{A_1}\right)\left(\frac{x_{i2}-\bar{x}_{.2}}{A_2}\right) \qquad (18.20)$$

and by introducing reduced variables given by

$$x'_{ij} = \frac{x_{ij}-\bar{x}_{.j}}{A_j}$$

the equation becomes

$$r(l_1,l_2) = \sum_{i=1}^{m} x'_{i1} \cdot x'_{i2} \qquad (18.21)$$

A method of measuring the distance between the two elements is to use the Euclidean distance between the transformed variables x'_{i1} and x'_{i2}. It is given by

$$D(l_1,l_2) = \sqrt{\sum_{i=1}^{m} (x'_{i1} - x'_{i2})^2} \qquad (18.22)$$

It can be shown that

$$D(l_1,l_2) = 2\ (1 - r(l_1,l_2)\ ) \tag{18.23}$$

In addition to the measures of correlation and distance described in the last section, many others have been proposed in the literature. A clear and complete discussion of these measures was given by Anderberg (1973).

18.7.3. Distance between groups

When the set of variables or elements of a population is divided into groups or subsets, a generalized distance between groups can be defined. For these generalized distances the following notations will be used :

x_{ijk} is the measurement of variable i for element j of group k :

$i = 1,2,\ldots,m$
$j = 1,2,\ldots,n_k$
$k = 1,2,\ldots,K$

K is the number of groups ;

n_k is the number of elements in group k $(k = 1,2,\ldots,K)$;

m is the number of variables ;

$n = \sum_{k=1}^{K} n_k$ is the total number of elements ;

$\overline{x_{i..}} = \frac{1}{n} \sum_{k=1}^{K} \sum_{j=1}^{n_k} x_{ijk}$ is the mean value for variable i ;

$\overline{x_{i.k}} = \frac{1}{n_k} \sum_{j=1}^{n_k} x_{ijk}$ is the mean value for variable i in group k ;

$\vec{x}_k = \begin{bmatrix} \overline{x_{1.k}} \\ \overline{x_{2.k}} \\ \vdots \\ \overline{x_{m.k}} \end{bmatrix}$ is the vector of mean values in group k.

A first measure of the distance between two groups k_1 and k_2 is given by the distance between the two vectors $\vec{x}_{k_1}$ and $\vec{x}_{k_2}$, equal to

$$\sqrt{\sum_{i=1}^{m} (\bar{x}_{i.k_1} - \bar{x}_{i.k_2})^2} \tag{18.24}$$

The square of this distance can also be written as

$$(\vec{x}_{k_1} - \vec{x}_{k_2})' \cdot (\vec{x}_{k_1} - \vec{x}_{k_2})$$

or as

$$(\vec{x}_{k_1} - \vec{x}_{k_2})' \cdot I_m . (\vec{x}_{k_1} - \vec{x}_{k_2}) \tag{18.25}$$

where I_m is the identity matrix of dimensions m x m.

With the discovery of discriminant analysis (see Chapter 20), it was proved that a more efficient distance is found by introducing a different matrix instead of I_m in eqn. 18.25. This new matrix in fact gives a set of weights for each product of two elements from vector $(\vec{x}_{k_1} - \vec{x}_{k_2})$.

It was shown that the optimal set of weights is given by the matrix T^{-1}, the inverse of the total covariance matrix T of the data given by

$$t_{i_1 i_2} = \frac{1}{n} \sum_{k=1}^{K} \sum_{j=1}^{n_k} (x_{i_1 jk} - \bar{x}_{i_1 ..}) (x_{i_2 jk} - \bar{x}_{i_2 ..}) \tag{18.26}$$

The distance defined in this way is called the Mahalanobis distance and is given by

$$D^2(k_1,k_2) = (\vec{x}_{k_1} - \vec{x}_{k_2})' \cdot T^{-1} . (\vec{x}_{k_1} - \vec{x}_{k_2}) \tag{18.27}$$

REFERENCES

M.R. Anderberg, Cluster Analysis for Applications, Academic Press, New York, 1973.
H. De Clercq, D.L. Massart and L. Dryon, J. Pharm. Sci., 66 (1977) 1269.
H. De Clercq and D.L. Massart, J. Chromatogr., 115 (1975) 1.
H. De Clercq, D. Van Oudheusden and D.L. Massart, Analusis, 3 (1975) 527.
A. Eskes, F. Dupuis, A. Dijkstra, H. De Clercq and D.L. Massart, Anal. Chem., 47 (1975) 2168.

R.A. Hoodless, K.G. Pitman, T.E. Stewart, J. Thomson and J.E. Arnold, J. Chromatogr., 54 (1973) 393.
J.B. Kruskal, Proc. Amer. Math. Soc., 7 (1956) 48.
D.L. Massart and H. De Clercq, Anal. Chem., 46 (1974) 1988.
D.L. Massart and H. De Clercq, in J.C. Giddings, E. Grushka, J. Cazes and P.R. Brown (Editors), Advances in Chromatography, Vol. 16, Marcel Dekker, New York, 1978.
D.L. Massart and L. Kaufman, Anal. Chem., 47 (1975) 1244A.
D.L. Massart, M. Lauwereys and P. Lenders, J. Chromatogr. Sci., 12 (1974) 617.
W.O. McReynolds, J. Chromatogr. Sci., 8 (1970) 685.
L. Orloci, in A.J. Cole (Editor), Numerical Taxonomy, Academic Press, London, 1969, p. 148.
G.L. Ritter, S.R. Lowry, H.B. Woodruff and T.L. Isenhour, Anal. Chem., 48 (1976) 1027.
P.H.A. Sneath and R.R. Sokal, Numerical Taxonomy, Freeman, San Francisco, 1973.
C.T. Zahn, IEEE Trans. Comput., C-20 (1) (1971) 68.

Chapter 19

FACTOR AND PRINCIPAL COMPONENTS ANALYSIS

Svante Wold, Research Group for Chemometrics, Umeå University, Sweden

19.1. INTRODUCTION

The chemist is often confronted with the problem of finding order and structure in a seemingly hopelessly large table of data. This problem was introduced in section 16.3 using a GLC example. To illustrate the present treatment, we shall use part of the McReynolds (1970) data matrix (see also section 16.2). We have used the data for 20 of the 226 liquid phases (LP) (see Table 19.I).

Several questions can be posed with respect to this data table. One might wish to find out how many "factors" influence the retention indices of chemical compounds - factors such as "polarity", "hydrophobicity" and "charge separation". One might wish to find similarities and dissimilarities between LPs to facilitate the choice of column for a particular separation problem. Finally, one might be interested to find similarities and dissimilarities among the behaviour of the 10 test solutes on the LPs.

When the data can be arranged to form a matrix Y (see below for notation), relatively simple, but still very powerful, tools exist for extracting information from the data. Factor analysis (FA) and principal components analysis (PCA) are the best known and most widely used of these tools and they also are important as they form much of the basis of multivariate data analysis (see Kruskal, 1978).

The nomenclature of these methods is thoroughly confusing. Traditionally, FA is used in social sciences to find the correlation patterns among a group of vectors, while PCA is aimed at the description of the variation among the same group of vectors. The two methods are, however, based on the same model (see below) and differ only in the assumptions concerning the residuals. In the way FA and PCA are used in chemical applications, the two methods are equivalent.

They are also equivalent to data analytical tools called eigen-vector analysis, eigen-vector decomposition, singular value decomposition, Karhunen-Loewe expansion and others. As this presentation is limited to chemical applications, where all of these names relate to exactly the same method, we shall henceforth treat them all under the name of FPCA and explicitly handle FA and PCA separately only when they differ in results.

Table 19.I

Part of McReynolds' matrix of retention indices. The LPs 121-140 are here renumbered as 1-20

		i =	1	2	3	4	5	6	7	8	9	10
LP no (k)	Column no	Liquid Phase	Δ I Benzene	Δ I Butanol	Δ I 2-Pentanone	Δ I Nitropropane	Δ I Pyridine	Δ I 2-Methyl-2-pentanol	Δ I 1-Iodobutane	Δ I 2-Octyne	Δ I 1,4-Dioxane	Δ I cis-Hydrindane
1	2132	Squalene	152	341	238	329	344	248	140	101	265	64
2	2046	UCON 50-HB-280X	177	362	227	351	302	252	151	130	256	65
3	2131	Polytergent J-300	168	366	227	350	308	266	149	119	255	61
4	2047	Tricresyl Phosphate	176	321	250	374	299	242	169	131	254	76
5	2085	SAIB	172	330	251	378	295	264	147	128	276	54
6	2164	Paraplex G-25	189	328	239	368	312	257	169	124	271	79
7	2302	Ethomeen 18/25	176	382	230	353	323	275	158	118	265	72
8	2180	Polytergent J-400	180	375	234	366	317	270	159	127	265	68
9	2025	Oronite NIW	185	370	242	370	327	267	165	130	275	75
10	2086	QF-1	144	233	355	463	305	203	136	53	280	59
11	2093	PPG Sebacate	196	345	251	381	328	271	176	129	285	83
12	2252	UCON 50-HB-660	193	380	241	376	321	265	166	141	274	75
13	2126	OV-210	146	238	358	468	310	206	139	56	283	60
14	2251	UCON 50-HB-3520	198	381	241	379	323	264	169	144	278	80
15	2021	Ethofat 60/25	191	382	244	380	333	277	168	131	279	73
16	2062	Ethomeen S125	186	395	242	370	339	285	169	127	279	79
17	2261	Igepal CO-630	192	381	253	382	344	277	172	136	288	78
18	2092	LSX-3-0295	152	241	366	479	319	208	144	55	291	64
19	2008	Pluronic P85	201	390	247	388	335	271	172	145	285	82
20	2005	Pluronic P65	203	394	251	393	340	276	174	146	289	83

Reproduced from the Journal of Chromatographic Science, by permission of Preston Publications, Inc.

19.2. A SHORT PRESENTATION OF FPCA

For a data matrix Y with elements y_{ik}, obtained by measuring the values of variables with index i (retention index of test compound i in Table 19.I) on "objects" with index k (LP k in Table 19.I), the purpose of FPCA is essentially to subdivide the variation in the data matrix Y into one part which varies only with the variables i (test compounds), one part which varies only with the objects k (LPs), and a "random" part, the residuals, which describe the non-systematic variation. The parameters varying with the variables are denoted by b_{ip} and referred to as loadings. The parameters varying with the objects are denoted by u_{pk} and called factors or components. The residuals are denoted by e_{ik}. Often, the mean value of each variable, denoted by $\bar{y}_i$, is taken as the point of reference of the variation in the model.

Graphically, we can illustrate this decomposition as the following block structure :

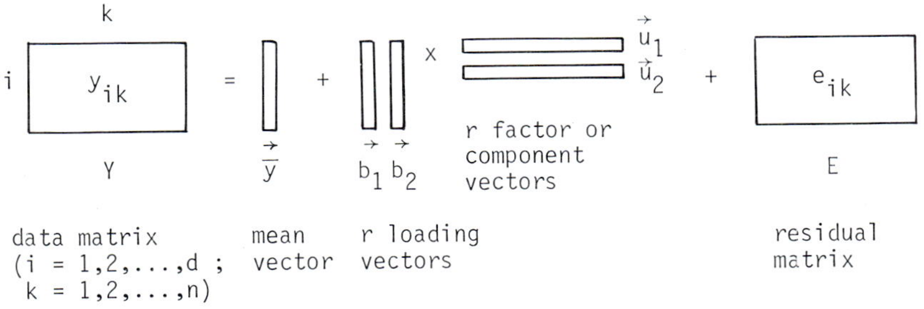

(19.1)

Expressing the model in terms of each individual observation (variable i measured on object k), we have eqn. 19.2, which in matrix form is ecn. 19.3

$$y_{ik} = \bar{y}_i + \sum_{p=1}^{r} b_{ip} u_{pk} + e_{ik} \qquad (19.2)$$

$$Y = \vec{\bar{y}} + B\ U + E \qquad (19.3)$$

In PCA and in the first phase of FA, the consecutive product terms in the model are determined to explain as much of the variation in Y as possible. This makes the parameters B and U unique but for a multiplicative constant. To anchor the parameter scales, one therefore needs a normalization condition for either b_{ip} or u_{pk}. In FA the usual practice is to use

$$\sum_{i=1}^{d} b_{ip}^2 = L_p \tag{19.4}$$

where L_p is the pth eigen-value of the correlation matrix of Y. In PCA, another normalization condition is sometimes used

$$\sum_{i=1}^{d} b_{ip}^2 = 1 \tag{19.5}$$

The normalization 19.4 is due to the fact that the eigen-value L_p is related to the proportion of the variance in Y that is explained by the pth product term in the factor model (eqns. 19.1-19.3). Also, traditionally the parameter vector $\vec{b}$ is estimated as the pth eigen-vector of the correlation matrix of Y.

When FA and PCA are applied to the same data matrix with regularized (normalized) variables (see section 19.3.1), they give numerically equivalent results, but with the normalizations shown above, we see that the b-values of FA will be $\sqrt{L_p}$ times the b-values of PCA. Analogously, the u-values of FA will be $\sqrt{L_p}$ times smaller than the u-values of PCA, but the product $b_{ip}u_{pk}$ will be the same for both methods. In both FA and PCA, the u-values will decrease in size with their decreasing explanatory power of the variance in Y.

19.3. BASIS OF FPCA

19.3.1. Relation to multiple regression (MR)

In MR, one assumes that a measured variable y is explained by a linear combination of r independent variables x_p

$$y_k = \sum_{p=1}^{r} c_p x_{pk} + e_k \tag{19.6}$$

The values of x_{pk} are assumed to be "accurately" known compared with the "error" in y_k, e_k.

FPCA can be seen as another version of the same model (eqn. 19.6), but where the values of the variables x_p are not directly observed. Instead, one assumes that these variables occur "intrinsically" in each of a set of measured variables y_i, so that each vector $\vec{y}_i$ is a realization of the linear combinations of these variables x_p

$$y_{ik} = \sum_{p=1}^{r} a_{ip}\, x_{pk} + e_{ik} \tag{19.7}$$

The data analysis now involves a simultaneous estimation of both the "regression coefficients" a_{ip} (b_{ip} in the FPC model) and the intrinsic variables x_{pk} (u_{pk} in the FPC model). This estimation is possible because of the existence of the multitude of variables y_i ($i = 1, 2, \ldots, d$). It follows that the factors/components u_{pk} can be interpreted as the "fundamental" variables of the data set. Being only indirectly observed, they are often called latent variables.

19.3.2. FPCA as a general model for a group of similar objects

A deeper and more general interpretation of FPCA than the relation to MR is obtained if we assume that the data y_{ik} are observed on a number of objects of limited diversity, i.e., the objects are in some way "similar". Assuming further that the data y can be seen as generated by a smooth, several times differentiable function f, a Taylor expansion of this function leads to the FPCA model (Wold, 1976). As in ordinary Taylor expansions, the more terms that are needed the larger is the variation in the generating function f, i.e., the larger is the diversity among the objects.

The assumption that measured data can be seen as generated by such a function f is natural in chemistry and other branches of natural science. Thus, in the current fundamental theory of chemistry, i.e., quantum theory and statistical mechanics, any observable quantity is an eigen-value to an operator equation (Eyring et al., 1944). This gives observed quantities a function-like behaviour. The assumption of limited diversity is needed to make the number of terms in the Taylor expansion of this function small.

Although this interpretation is not in direct contradiction to the one discussed in the previous section, it emphasizes the need for caution in the interpretation of the loadings and factors/components $\vec{b}$ and $\vec{u}$. Thus, phenomenologically, it is difficult to distinguish between the case when these actually can be seen as linear "latent variables" and the case when they express a linearization of a complex non-linear function observed over a limited domain.

19.3.3. Geometrical interpretation

A simple means of obtaining a geometrical interpretation of FPCA is to construct a d-dimensional space with orthogonal coordinate axes, one for each variable i (d variables altogether). In such a space, the data vector measured on one object is represented as a point.

FPCA can be seen as a method where an r-dimensional hyperplane is least-squares fitted to the points of the objects. r-dimensional hyperplanes are difficult to imagine, but we can easily think about these when r = 1 or r = 2. The coefficients b_{ip} determine the direction of this plane which passes through the point $\vec{\bar{y}}$ defined by the variable mean values. The coefficients u_{pk} describe where on this plane the projection of point k is situated. The standard deviation (SD) of the residual vector $\vec{e}_k$ (eqn. 19.3), finally, measures the orthogonal distance between the plane and point k (see eqn. 19.14 for a definition of this SD, $s(e)_k$).

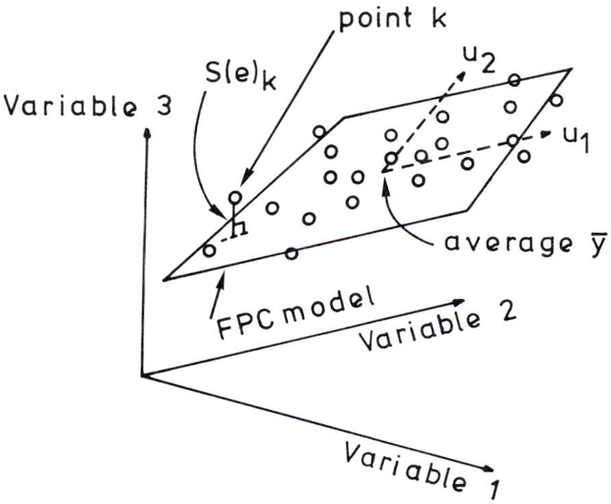

Fig. 19.1. Data points in a d-space with d = 3, with a factor model (eqns. 19.1-19.3) with r = 2 (the plane of closest fit to the data points).

19.3.4. Limitations of the method

Like all methods of data analysis, FPCA is sensitive to inhomogeneities in the data set. FPCA is based on the assumption that all objects included in the study are fairly similar. For example, if one wishes to analyse the variability of the IR spectra of a number of carbonyl compounds, some of which are conjugated and some not, one might first group the spectra into the sub-sets "conjugated" and "non-conjugated". This will also lead to a much simpler FPC model for each sub-group with fewer product terms than obtained if all data are analysed in the same model. Therefore, when the data are obviously grouped, the data set should be divided into sub-sets and each sub-set of more similar objects analysed separately.

In order to obtain stable estimates of the parameters, the number of parameters must not approach the number of available data points. This condition is of great importance in all data analytical methods and was discussed in connection with multiple regression. In FPCA, this condition corresponds to an upper limit on the number of product terms, r, of about a quarter or a third of the smaller dimension of the data matrix.

Finally, all least-squares based methods work best when the residuals are fairly centrally distributed and have fairly equal variance both row-wise and column-wise. Apparent deviations from these conditions are usually removed by data transformations, chemists often taking the logarithm of observations before they enter the data into the analysis.

19.4. ESTIMATION OF THE PARAMETERS IN THE FPCA MODEL

For a given data matrix Y with the variables appropriately scaled (see below), the following parameters are to be estimated to give the model the "closest fit" to the data.

(1) The values $\bar{y}_i$, i.e., the means of each variable

$$\bar{y}_i = \sum_{k=1}^{n} y_{ik} / n \qquad (19.8)$$

(2) The number of significant factors or components, r.

(3) The values of the loadings b_{ip} (p = 1, 2, ..., r) and the factors or components u_{pk} (p = 1, 2, ..., r).

19.4.1. Scaling of data

FA analyses the correlations of the data matrix Y, which is equivalent to an analysis of regularised data, i.e., the average of each variable $\bar{y}_i$ (eqn. 19.8) is first subtracted and then each element of the data is divided by the data standard deviation $s(y)_i$. Denoting the normalized data by y_{ik}'

$$y_{ik}' = (y_{ik} - \bar{y}_i) / s_i(y) \qquad (19.9)$$

with

$$s(y)_i^2 = \sum_{k=1}^{n} (y_{ik} - \bar{y}_i)^2 / (n-1) \qquad (19.10)$$

The results of PCA are dependent on the data scaling. If not all of the variables are measured in the same units as in the GC-LP example, the recommended practice is to make PCA equivalent to FA by using regularized data in the PCA.

In the GC-LP example, one might wish to minimize the residuals when these are expressed in the same units as the original raw data. In such a case, the PCA is performed on unscaled data. Below, the analysis is made in both ways as it is of interest to compare the results.

19.4.2. The number of factors, r

In FPCA, the first important parameter to be estimated is the number of significant product terms (factors) in the model. This estimation corresponds to finding out how much of the variation in Y is systematic and how much is "random noise". The former is described by the parameters $\bar{y}$, b_p and u_p ($p = 1, 2, \ldots, r$) and the latter by the residuals e_{ik}. Hence, one must use a "stopping rule" to determine when the number of product terms are sufficient for the purpose of the data analysis or, alternatively, when the next term makes an insignificant contribution to the explanation of the variation in Y.

Most of the variation in Y is usually described by the first few product terms, and the following terms each describe very little of this variation. In the GC-LP example, factor 5 and beyond together explain less than 7.5% of the SD of Y (see Fig. 19.2). It is clear that beyond a certain r, the factors have neither statistical nor chemical significance.

One must be clear about the difference between statistical and chemical significance. A parameter is statistically significant when the probability that the parameter has a value different from zero is fairly large. That parameter can still be chemically utterly insignificant, however. The important thing is that a parameter must be statistically significant in order to be chemically significant. Thus, statistical significance is a necessary but not sufficient condition for chemical significance.

The number of statistically significant factors in FPCA increases when the

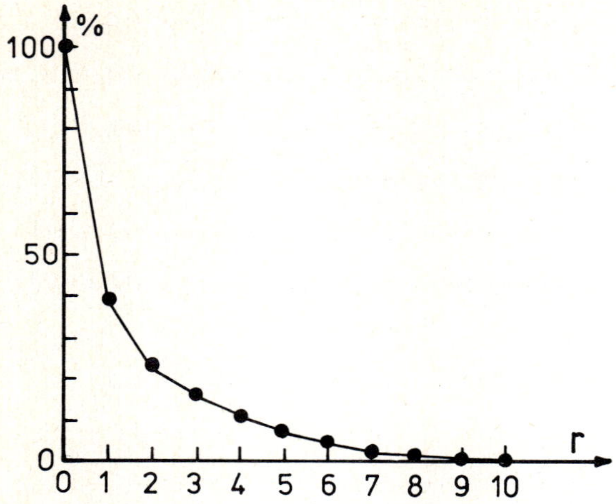

Fig. 19.2. Percent of the variation in Y, measured as SD, remaining after r product terms in the GC-LP example (unscaled data).

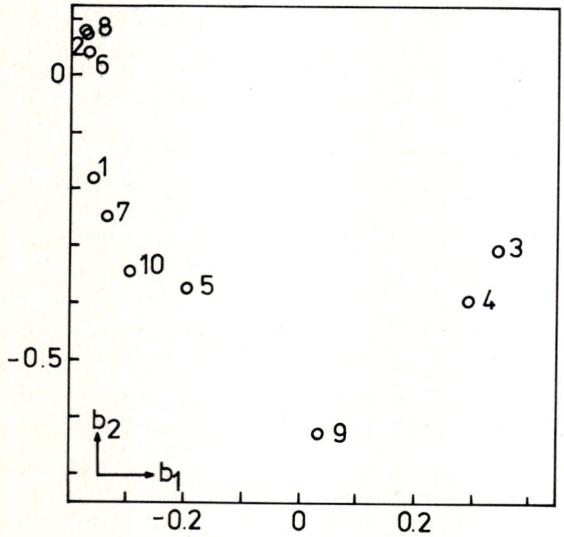

Fig. 19.3. Plot of b_{i2} against b_{i1} (values from lower part of Table 19.III) showing the similarity between the variables 1-10.

dimensions of the matrix increase. Hence, a statistical "stopping rule" will provide the maximal number of significant product terms. The common sense of the chemist decides whether the number of chemically significant terms is smaller.

Four main rules are used in chemical applications of FPCA.

(1) Use as many factors as necessary such that a large part of Y, usually 95% of the variance, is described by the model.

(2) Use as many factors as necessary such that the residuals (e_{ik}) have a variance corresponding to the "known" precision of the data.

(3) Use only factors with eigen-values (L_p) of the correlation matrix which are larger than unity. This ensures that factors included in the model contain contributions from at least two variables in Y.

(4) Use only factors that increase the predictive power of the factor model. To assess this predictive power, part of the data matrix is deleted and the parameters B and U are estimated from the remaining reduced data matrix. For each value of r one calculates predicted values for the deleted points by means of the model and parameter values. These predicted values are compared with the actual values of the deleted points. The value of r is chosen which gives the smallest average difference between predicted and actual values.

Considering these four rules, it can be said that the first two have less statistical significance. Although the idea of reproducing the data within the error of measurement at first might seem attractive, it is based on the implicit assumption that the FPCA model is "exact" apart from errors of measurement in the data. This is, at best, a very doubtful assumption - empirical models such as the FPCA model are approximations to the data and therefore nearly always contain model errors of unknown magnitude. Rule 3 is incorporated in most FPCA standard programmes, for instance SPSS (Nie et al., 1975) and usually works reasonably well. Rule 4 is often more conservative than rule 3, giving smaller values of r. In this way, it seems to correspond better to chemical common sense.

In the GC-LP example, rules 3 and 4 indicate the presence of four statistically significant factors (Table 19.II). From Fig. 19.2, one observes that the first two are responsible for about 70% of the variation in Y. These two major factors would probably be used to find the chemical (dis)similarities of solutes or LPs. If we wish to predict an external property by the FPC model, we would use four factors to obtain as small prediction errors as possible.

19.4.3. Estimation of loadings and factors (components)

Once the number of significant terms, r, has been determined in relation to the actual application, the remaining estimation problem is merely a technical one, which is easily handled by available statistical programme packages, such as SPSS (Nie et al., 1975). Hence, this section is of less fundamental interest than the previous one.

Table 19.II

Performance of "stopping rules" 1-4 on the 10 x 20 McReynolds data matrix. Note that as FA operates on the correlation matrix, the eigen-value, L_p, in column 4 refers to data scaled by regularization. The PRESS/SS(e) refers to the ratio between the sum of squared prediction errors and the sum of the squared residuals with r being one unit lower. When this ratio is larger than unity, the latest included term is statistically insignificant according to rule 4. Values corresponding to the "best" value of r for each rule are underlined. The error of measurement in the data is approximately 3 units and s(y) of the raw data is 29.4

	Regularized data				Unscaled raw data		
p	$s(e)^2/s(y)^2$ rule 1	s(e) rule 2	L_p rule 3	PRESS/SS(e) rule 4	$s(e)^2/s(y)^2$ rule 1	s(e) rule 2	PRESS/SS(e) rule 4
1	.35	19	10.9	0.45	.15	12.4	0.20
2	.13	12	6.56	0.73	.05	8.1	0.48
3	.05	9	3.74	1.07	.02	6.0	0.86
4	.02	5.7	2.61	-	.01	4.5	0.64
5	.01	4.9	1.18	-	.005	3.5	28.0
6	.003	3.3	1.14	-	.001	2.2	-
7	.001	2.1	0.69	-	.0003	1.3	-
8	.0001	0.9	0.40	-	.0001	0.8	-
9	.00003	0.6	0.13	-	.00002	0.6	-
10	0	0	0.07	-	0	0	-

For a complete data matrix (without missing observations), the standard way of proceeding is first to form the covariance matrix C

$$C = (Y - \vec{y})'(Y - \vec{y}) \qquad (19.11)$$

and, in FA, then normalize this to give the correlation matrix R (s_i, s_j and c_{ij} are elements of C)

$$r_{ij} = \frac{c_{ij}}{\sqrt{s_i \, s_j}} \qquad (19.12)$$

In PCA, the loadings b_{ip} are estimated as the pth eigen-vector of the covariance matrix C, and in FA, the loadings are estimated as the pth eigen-vector of the correlation matrix R. As the covariance and correlation matrices are identical for regularized data, FA and PCA are equivalent in such a case.

From the eigen-vector properties of the loadings $\vec{B}_p$, it follows that these vectors are orthogonal to each other, for $p \neq q$

$$\vec{B}_p \, ' \, \vec{B}_q = 0 \qquad (19.13)$$

The same orthogonality property holds for the vectors $\vec{U}_p$. These can be estimated in a variety of ways by using the fact that model 19.3 is linear once the values of $\bar{y}_i$ and b_{ip} are known.

Other methods can also be used to estimate the values of the parameters b_{ip} and u_{pk} but, regardless of the method used, the values all become the same. Hence, the chemist need not concern himself with the numerical problems ; the important point is that the parameter estimation corresponds to finding the plane of closest fit to the points in d-space (section 19.3.3).

In Table 19.III, the parameters resulting for the GC-LP data are given both for the scaled and unscaled data. The normalization of the $\vec{B}_p$-vectors is that corresponding to PCA (see section 19.3.1). The influence of the regularization of the data on the $\vec{B}_1$ vectors is revealing. In the unscaled data, variable 2 has the largest variation and variable 10 the smallest (row 1 in Table 19.III). The PC model tries to explain as much as possible of the variation of y in each product term. Hence, the coefficient $b_{2,1}$ is large and $b_{10,1}$ small for the unscaled data.

Table 19.III

Resulting parameters $\overline{y}_i$ and b_{ip} for the PCA of the GC-LP data matrix. Upper part, unscaled data ; lower part, regularized data obtained by subtracting $\overline{y}_i$ in row 2 and dividing by $s(y)_i$ in row 1 (eqn. 19.9), i.e., factor scaling.

	i =	1	2	3	4	5	6	7	8	9	10
	$s(y)_i$	18.3	52.1	44.0	39.6	15.1	24.5	13.1	29.4	11.4	8.95
	m_i	179	347	259	385	321	257	160	119	275	71.5
	b_{i1}	0.1669	0.5914	-0.5010	-0.4164	0.0671	0.2713	0.1035	0.3277	-0.0425	0.0575
UN-SCALED	b_{i2}	-0.3333	-0.3362	-0.3206	-0.5733	-0.2780	-0.1742	-0.2498	-0.1755	-0.3315	-0.1836
	b_{i3}	-0.3387	0.3175	0.1680	-0.1929	0.6221	0.0919	-0.2483	-0.4930	0.1118	-0.0792
	b_{i4}	-0.1915	0.5112	0.1022	0.3536	-0.4924	0.0958	-0.3682	-0.0348	-0.1617	-0.3899
	b_{i1}	0.3584	0.3754	-0.3476	-0.2980	0.1952	0.3723	0.3366	0.3757	-0.0371	0.2976
	b_{i2}	-0.1826	0.0760	-0.3038	-0.3923	-0.3724	0.0456	-0.2466	0.0815	-0.6261	-0.3383
SCALED	b_{i3}	-0.3030	0.1894	-0.0712	-0.2493	0.7398	0.1589	-0.3636	-0.1886	0.1151	-0.2270
	b_{i4}	0.1925	0.1983	-0.0112	0.1626	-0.2905	0.3558	-0.1922	0.2263	0.4959	-0.5933
	s_i	0.15	0.26	0.12	0.23	0.093	0.27	0.19	0.23	0.18	0.11

In the regularized data, where all variables have the same variance, b_{i1} is instead a pure measure of the correlation structure in the data and $b_{2,1}$ is of the same size as $b_{10,1}$. These two variables are equally correlated to the majority of the variables in the data matrix. This shows that FPCA of regularized data usually gives results that are simpler to interpret. PCA of regularized data, on the other hand, explains as much as possible of the variation in the observed data. Thus, if one wishes to use the analysis to predict the values of measurements of future objects, the PCA of the unscaled data should be used. This is an example of the general rule of great importance that one must know what the results will be used for, in order to choose the appropriate way of analysing a given data set.

19.5. INFORMATION FROM THE DATA ANALYSIS

Four types of information are extracted from the data set by an FPCA.

(1) The number of significant product terms in the model. This is often central in itself, as in the GC-LP example where it is indeed informative that two major "factors" explain more than 95% of the variance in the data and that no more than two additional, minor but statistically significant, "factors" are found in the data.

In the analysis of spectroscopic data where spectra have been recorded for a number of solutions with the same solutes added in different proportions, this number bears on the number of chemically significant species that need to be postulated to exist in the solutions (Hugus and El-Awady, 1971).

(2) Values of the variable averages $\bar{y}_i$ and the loadings b_{ip}. Often the averages are important for the interpretation - in the GC-LP example they give, for instance, the information that the retention index of benzene (i = 1) is close to that of 1-iodobutane (i = 7), on average.

The loadings b_{ip} describe the correlation structure in the data set. Thus, from b_{i1} of the scaled GC-LP data, we see that the variables 1 and 2 are positively correlated (their b_{i1} values are of the same sign) whereas variables 1 and 3 are negatively correlated (b_{i1} have different signs). Further, we see that variable 9 does not participate in this correlation structure, in as much as $b_{9,1}$ is small.

Fig. 19.3 shows a plot of the first two $\vec{B}$ vectors of the scaled data. One can see a grouping of the solutes, some being close to each other. This indicates a similarity between, on the one hand, variables 2, 6 and 8 (perhaps related to hydrogen bonding ability), and, on the other hand, variables 3 and 4. Variables 1, 5, 7 and 10 seem to form a rather loose group, while variable 9 (dioxane) seems to behave differently from all of the others in this data set. This grouping has earlier been studied by other means (see Chapter 18).

(3) Values of the factors (components) u_{pk}. These parameters relate to the individual objects - LPs in the GC-LP example - and their "positions" in the data set. In the geometrical interpretation (Fig. 19.1), the u_{pk} describe where on the plane of closest fit the kth object is situated. A plot of u_{1k} against u_{2k} (Fig. 19.4) reveals a grouping of the LPs into one major group and one minor group containing the LPs 10, 13 and 18. This can be interpreted in terms of the latter three LPs having different separation properties. Thus FPCA can be useful in the classification and combination of LPs.

(4) The residuals e_{ik}. These numbers describe the non-systematic part of the data - the part unexplained by the model. The residual standard deviation (RSD) for each object, $s(e)_k$, gives information of how much "non-systematic" variation the data vector of the object contains

$$s(e)_k^2 = \sum_{i=1}^{d} e_{ik}^2 \; / \; (d-r) \qquad (19.14)$$

These values of $s(e)_k$ can be used to find "outliers" among the objects, i.e., objects that do not fit the same data structure as the majority of the objects. This is done by an approximate F-test with (d-r) and (d-r)(n-r-1) degrees of freedom, where $s(e)_k$ is compared with the total RSD, s(e)

$$F = s(e)_k^2 \, / \, s(e)^2 \tag{19.15}$$

$$s(e)^2 = \sum_{i=1}^{d} \sum_{k=1}^{n} e_{ik}^2 \, / \, (d-r)\,(n-r-1) \tag{19.16}$$

In the GC-LP data, none of the 20 LPs were outliers according to this F-test with $\alpha = 0.05$.

Analogously, the RSD $s(e)_i$ relates to the amount of non-systematic variation in the ith variable. These values for the regularized data are shown at the bottom of Table 19.III.

$$s(e)_i^2 = \sum_{k=1}^{n} e_{ik}^2 \, / \, (n-r-1) \tag{19.17}$$

In the GC-LP application, all variables have $s(e)_i$ values smaller than 0.3, showing that the FPC model describes most of their variation. In the context of pattern recognition, $s(e)_i$ provides an important criterion for selection of relevant variables (see Chapter 20).

19.6. ROTATIONS AND TRANSFORMATIONS OF THE PARAMETER VECTORS

For $r > 1$, the parameters in the FPC model, eqn. 19.3, are non-unique for transformations. Consider, for simplicity, a case with $r = 2$. The parameters first obtained, $\bar{y}_i$, b_{i1}, b_{i2}, u_{1k} and u_{2k}, are calculated so as to maximize the variance explained by the model in each step. If this condition is relaxed, the solution

$$b_{i1}' = b_{i1} + b_{i2} \tag{19.18}$$
$$b_{i2}' = b_{i1} - b_{i2}$$

and other combinations (rotations) are equally feasible. In FA, a large number of different such transformations are used for various purposes. In chemical problems, there is mainly a need for two types of transformations, the first of which, in our view, is best approached graphically.

19.6.1. Transformations to simplify the parameter vectors

Consider the interpretation of the variables u_{1k} and u_{2k} in the GC-LP example (Fig. 19.4). The introduction of two new coordinate axes indicated in the figure (w_1 and w_2) will make the latter small for most of the LPs and, in addition, make w_1 correspond closer to the "polarity" of the LPs.

We see that this corresponds to a variable transformation to a coordinate system with non-orthogonal axes and a shifted origin. In FA, the so-called rotations are usually made without a shift of origin, mainly for computational convenience. Today, with computers available, this fixing of the origin, which is a serious restriction, is no longer necessary.

Variable transformations of this kind are, of necessity, subjective. A simpler interpretation to one user might be a more complicated interpretation to another.

19.6.2. Transformations to find correlations with external variables

In both the MR and the "similarity" interpretations of FPCA, the variables y_i are explained as linear combinations of intrinsic, latent, variables u_{pk}. If we now bring in a new variable j with the measurements y_{jk}, we might wish to investigate whether this new variable is also related to the same set of latent variables. As the latter are already estimated in the initial FPCA, this involves a simple linear regression

$$y_{jk} = \bar{y}_j + \sum_{p=1}^{r} a_p u_{pk} + e_{jk} \qquad (19.19)$$

If the resulting residuals e_{jk} are small, preferably having an RSD of the same magnitude as the s_is of the initially included variables, we conclude that indeed y_{jk} is explained by the same model and the same set of latent variables. The whole battery of multiple regression methodology can be used to test for the significance of this regression, to estimate confidence intervals of the "regression coefficients" a_p, and so on. Predictions of y_{jk} for cases where this has not been measured, but values of u_{pk} have been estimated, can also be made (see section 19.7).

An analogous analysis to see whether a new object with the data vector y_i^* fits the FPCA model is evident. This involves the regression

$$(y_i^* - \bar{y}_i) = \sum_{p=1}^{r} v_p \, b_{ip} + e_i^* \qquad (19.20)$$

When a new variable j or a new object is introduced, one does not need to have all values in their corresponding data vectors defined by observed values to estimate its loadings or factors (components). Thus, in the regressions 19.19 and 19.20, the vectors y_{jk} and y_i^* need only have sufficient defined elements to give a "stable" estimation of the regression coefficients a_p or v_p. Two types of information are obtained in such a regression. Firstly, the size of the residual standard deviation (RSD) indicates if it is likely that the fitted data are described by the same model as that describing the "old" data. Secondly, if this is the case, one obtains predicted values for the missing elements by means of eqn. 19.19 or 19.20 with the residuals e_{jk} or e_i^* set to zero. The standard formulae of multiple regression can be used to get approximate confidence intervals for these predicted values. This technique was called "floating rotation" by Weiner (1977).

19.7. ADDITIONAL DATA ANALYTICAL APPLICATIONS OF FPCA

Apart from the applications of FPCA discussed above, the methodology is often used to reduce the dimensionality of the matrix of independent variables in multiple regression (MR). The reason is that MR is sensitive to (i) strong correlations among the independent variables x and (ii) a number of independent variables approaching the number of observations n. A solution to both of these problems is to reduce the matrix X to a smaller matrix U containing the most significant factors (components) of X in a FPCA. Thus, the use of u_{pk}s as independent variables is advantageous because one includes only as many as the number of observations n warrants and, secondly, the u-vectors are non-correlated. We see that this is in fact closely analogous to the application discussed in section 19.6.2.

A second useful application of FPCA is in display methods, i.e., to find the "eigen-vector" projection of a data set Y for graphical plots (section 16.3). This corresponds to finding the first two or three u_p vectors of the data set and then using these as coordinate axes in a plot in the same way as in Fig. 19.4. It follows from the least-squares properties of FPCA that this projection of the data set preserves as much of the original variance as possible (Kowalski and Bender, 1973 - see also Chapter 20).

19.8. OTHER ANALYTICAL CHEMICAL APPLICATIONS OF FPCA

The most common multivariate data sets are those in which the variables relate to concentrations of constituents in the sample - in trace element analysis the chemical elements and in chromatographic methods volatile or other components. Data sets resulting when these methods are used to analyse a collection of samples are often collected in order to obtain information about the similarities (dissimilarities) between samples and to obtain information about the type or source of samples. Typical examples are the chemical characterization of inorganic materials such as steels by their trace element constitution and then trying to find whether a given steel is of type one, say corrosive, or two, say, non-corrosive, on the basis of this characterization. If one has a quantitative measure of the corrosiveness, one might instead wish to relate the trace element concentrations to this measure, i.e., an FPCA with relation to an external variable discussed in section 19.6.2 (Wold et al., 1978). In the same way, organic materials are often characterized by means of gas chromatography with the hope of finding "patterns" in the peak heights or peak areas that relate to desired type of information (Elliot et al., 1971).

FPCA of data obtained from samples of a known type or source together with

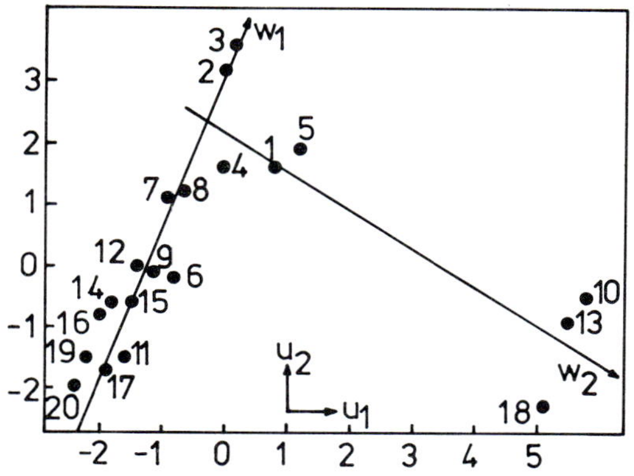

Fig. 19.4. Plot of u_{2k} against u_{1k} for the regularised GC-LP data, indicating the similarities between LP's (k = 1 to 20).

data from samples of an unknown type along the lines indicated above can often provide this desired information to the analytical chemist, if he uses his chemical knowledge to find good and relevant variables to enter into the data analysis (Wold and Sjöström, 1977).

Spectroscopic data (IR, UV, MS, NMR, ESCA, etc.) measured on a number of compounds or mixtures can be subjected to FPCA in order to find regularities in the data. The variables then can be those obtained by digitizing the spectra at regular frequency or wavelength intervals (Rozett and Petersen, 1975). Usually the FPCA works better, however, with the much fewer variables that chemists are accustomed to derive from these spectra such as the positions and absorbances of characteristic peaks (Wold and Sjöström, 1978).

The analysis of digitized waveforms of electrochemical methods - polarography, voltametry, etc. - can be made by FPCA both to characterize the type of waveform for samples and to assess concentrations of species in the analysed solutions. Also, collections of kinetic curves of various kinds lend themselves to FPCA. The resulting information - the number of "factor" curves that are needed to describe the collection, and the shapes of these "factor" curves - is often more useful and easier to interpret than the kinetic parameters obtained by traditional curve-fitting (Howery, 1972).

Nothing prevents the data matrix analysed by FPCA from containing variables of different kinds. One might, for instance, wish to characterize a number of oils both by 20 trace element concentrations and the areas of 100 peaks in gas chromatograms. FPCA of such a "mixed" matrix presents no special problems. In some instances, however, one might wish to keep the sets of variables apart and treat them distinctly. Extensions of FPCA for such an analysis have recently been developed (Wold, 1977) and have also started to be applied in various branches of chemistry (Wold et al., 1978).

An interesting and perennial problem is that of comparing analytical chemical methods or laboratories on the basis of their performance on a number of real samples, sometimes called "round-robin" data. We note that FPCA was originally developed to treat exactly this problem, not in chemistry, but in psychology and education, where samples correspond to students and the analytical methods to ability tests. When applied to a comparison of six different methods for the determination of glucose in blood, FPCA gave information about the precision of each method and also detected systematic errors in two of the methods (Carey et al., 1975 - see also section 3.1.1).

19.9. CONCLUSIONS

Data sets consisting of multiple measurements made on several samples or objects are becoming increasingly common in analytical chemistry. Many of the questions that chemists normally put to such data sets are answered in a straightforward way by FPCA of the data set. Like all methods of data analysis, FPCA is based on a number of assumptions. Recognizing these assumptions and realizing the limitations introduced by them, the chemist can use FPCA to great advantage as a flexible tool for extracting useful information from chemical data. Two important conditions for success are (i) that the chemist knows what kind of information he or she wants from the data and (ii) that the data have been collected in relevant and well performed experiments. FPCA works like an amplifier. When the data are good and the method is used sensibly, one obtains more from the data than without the method. However, if the data are bad and/or the method is used without common sense, one obtains worse results with FPCA than without.

REFERENCES

R.N. Carey, S. Wold and J.O. Westgard, Anal. Chem., 47 (1975) 1824.
S.C. Elliott, N.A. Hartmann and S.J. Hawkes, Anal. Chem., 43 (1971) 1938.
H. Eyring, J. Walter and G.E. Kimball, Quantum Chemistry, Wiley, New York, 1944.
H.H. Harman, Modern Factor Analysis, 2nd edition, The University of Chicago Press, Chicago 1967.
D.G. Howery, Bull. Chem. Soc. Japan, 45 (1972) 2643.
D.G. Howery, Intern. Laborat., March/april (1976) 11.
Z.Z. Hugus Jr. and A.A. El-Awady, J. Phys. Chem., 75 (1971) 2954.
B.R. Kowalski and C.F. Bender, J. Amer. Chem. Soc., 95 (1973) 1973.
J. Kruskal, Factor Analysis and Principal Components Analysis, the Bilinear Methods in Encyclopedia of Statistics, The Free Press, New York, 1978.
W.O. McReynolds, J. Chromatogr. Sci., 8 (1970) 685.
N.H. Nie, C.H. Hall, J.G. Jenkins, K. Steinbrenner and D.H. Bent, SPSS Manual, McGraw-Hill, New York, 1975.
R.W. Rozett and E.M. Petersen, Anal. Chem., 47 (1975) 1301.
P.H. Weiner, E.R. Malinowski and A.R. Levinstone, J. Phys. Chem., 74 (1970) 4537.
P.H. Weiner, Chem. Tech. May (1977) 321.
H. Wold, in Essays in Honour of Oscar Morgenstern, R. Henn and O. Moeschlin eds, Springer, Berlin, 1977.
S. Wold, Pattern Recognition, 8 (1976) 127.
S. Wold and M. Sjöström, SIMCA : A Method for Analyzing Chemical Data in Terms of Similarity and Analogy. Chapter 12 in Chemometrics : Theory and Applications, B.R. Kowalski, ed., American Chem. Soc. Symposium Series no 52, Washington D.C., 1977.
S. Wold and M. Sjöström in Correlation Analysis in Chemistry : Recent Advances, J. Shorter and N.B. Chapman eds., Plenum, London, 1978.
S. Wold et al., Proceedings of the Int. Conf. on Computers and Optimization in Analytical Chemistry, Amsterdam, April 1978, to be published, 1978.

Chapter 20

SUPERVISED LEARNING METHODS *

20.1. INTRODUCTION

Let us consider the milk (or the thyroid) example from Chapter 16. In its simplest form, this can be represented by Fig. 20.1. One can imagine that samples from class K (for example, goats' milk) should be discriminated from samples from class L (for example, cows' milk) according to two variables, x_1 (for example, butyric acid concentration) and x_2 (for example, stearic acid concentration). These variables can be measurements or, for instance, concentrations. In Fig. 20.1, this discrimination can be achieved by drawing lines such as a and b. It should be observed that the solutions are not necessarily unique.

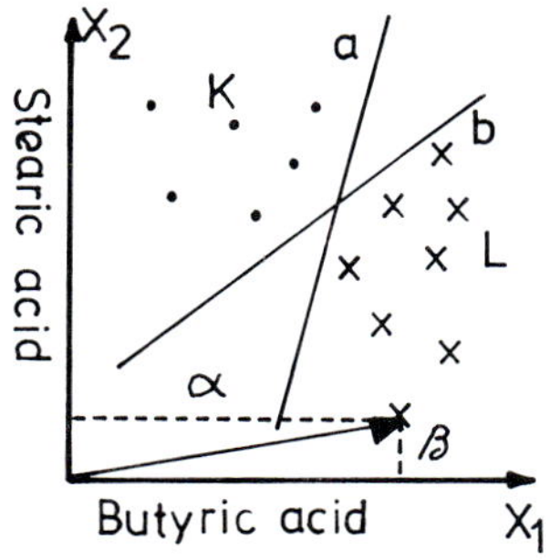

Fig. 20.1. Separation of two classes, K and L, in two-dimensional space.

This can be generalized to situations with more variables. A d-dimensional space is then obtained in which the samples are represented by points. The easiest means of doing this is to characterize them with a vector, called the pattern vector. In the two-dimensional case in Fig. 20.1, the samples for which the vector is shown can now be denoted by $\vec{x} = (\alpha, \beta)$. In general, a sample will be represented in hyperspace by $\vec{x} = (x_1, x_2, \ldots, x_d)$.

---

* This chapter has been written in collaboration with S. Wold, University of Umeå, Sweden.

In this chapter, which considers supervized learning systems, there are two kinds of samples, namely those which constitute the training or learning set and those which have to be classified (the test set). The training set consists of samples for which both the pattern vector and the identity are known. In the training or learning step, one develops a decision function such as line a or b in Fig. 20.1, or a mathematical description of the data structure, such as in SIMCA.

Of the samples that must be classified (the test set), one knows only the pattern vector. The mathematical description from the learning step is used to classify these samples into one of the known classes. Often one distinguishes between two classes (a binary decision) and in many methods this function is adjusted in the training step in such a way that for members of class K the function will be larger than zero and for members of class L smaller than or equal to zero.

One makes a distinction between parametric and non-parametric techniques. In the parametric techniques, statistical parameters of the distribution of the samples are used in the derivation of the decision function (often, but not necessarily, a multivariate normal distribution is assumed). The non-parametric methods are not explicitely based on distribution statistics. Both parametric and non-parametric methods have been used in analytical chemistry. Each has is own advantages and therefore both approaches will be discussed.

The most important advantage of applying non-parametric methods in analytical chemistry, is that it is often impossible to derive a representative distribution because the training sets are too small. When a distribution is obtained it is often not Gaussian. The most important advantage for the parametric and particularly the Gaussian methods is that statistical tests can be carried out to decide whether certain features should be included or not. Another advantage, of less importance in the present context, is that it is possible to quantify the classification risk, i.e., to state a probability of correct classification. However, some non-parametric methods contain possibilities both for feature selection and estimation of the classification risk (see SIMCA, below). Hence, the distinction between parametric and non-parametric methods is not clear.

As there are many variants of pattern recognition, we have decided to discuss only those methods which have become more or less established techniques in analytical chemistry. More details can be found in books or reviews by Duda and Hart (1973), Young and Calvert (1974), Andrews (1972) (general), Jurs and Isenhour (1975), Isenhour et al. (1974) and Kowalski (1975) (non-parametric techniques) and Kendall (1975) (parametric techniques).

We must consider here the terminology involved. Books on multivariate statistics call the parametric technique discussed in the next section linear discriminant analysis. This term is used also in a more general sense in the whole field of pattern recognition. To make the distinction, we call the parametric technique in question "statistical linear discriminant analysis".

An important aspect of pattern recognition is the validation of the classification rules obtained. One distinguishes between recognition and prediction ability. The recognition (or classification) ability is characterized by the percentage of the members of the training set that are correctly classified. The prediction ability is determined by the percentage of the members of the test set, correctly classified by using the decision functions or classification rules developed during the training step. When one determines only the recognition ability, there is a risk that one will deceive oneself into taking an over-optimistic view of the classification success of the problem at hand. In particular, classification methods such as linear discriminant analysis and the learning machine, which try to maximize the differences between classes, and methods of feature selection conditioned on class separation, tend to give an optimistic classification rate if this is calculated from the classification success of the training (learning) set only. It is therefore recommended that the prediction ability is verified also.

20.2. STATISTICAL LINEAR DISCRIMINANT ANALYSIS

20.2.1. Classification

Let us consider again the case of the separation of two classes. This was depicted in Fig. 20.1, but for convenience it is shown again in Fig. 20.2a. The supposedly normal distribution of the two classes to be separated, K and L, is shown in Fig. 20.2b for one of the variables, x_2. The mathematical problem is then to find an optimal decision rule for the classification of these two groups.

Let us consider first the simplest possible case. Two classes, K and L, have to be distinguished using a single variable, x_2. It is clear that the discrimination will be better when the distance between $\bar{x}_{2K}$ and $\bar{x}_{2L}$ (i.e., the mean values of x_2 for classes K and L) is large and the width of the distributions is small. In other words, one must maximize the ratio of the difference between means to the variance of the distribution.

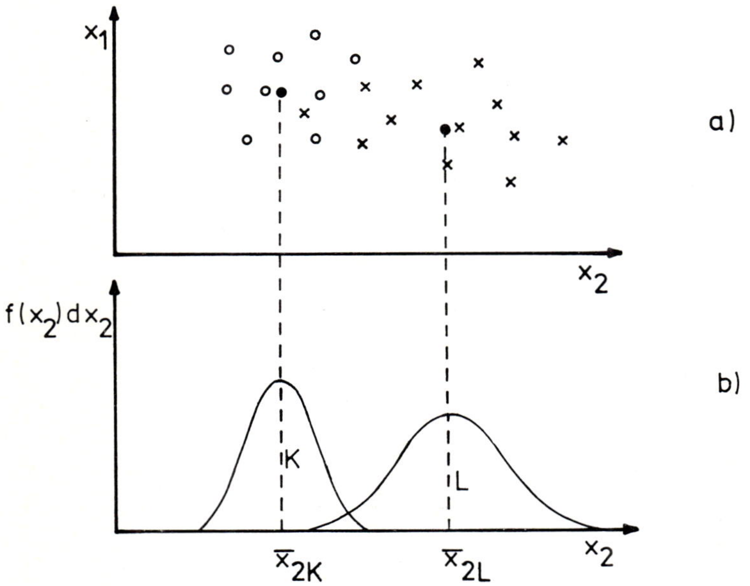

Fig. 20.2. Separation of two classes, K and L, in a two-dimensional space (a) and projection on x_2 (b).

When one considers the situation with the two variables, x_1 and x_2, it is again evident that the discriminating power of the combined variables will be good when the centroids of both sets of samples are sufficiently distant from each other and when the clusters are tight or dense. In mathematical terms this means that the distance between the centroids is large compared with the within-class variance. On the other hand, it is also clear that if the results obtained with the two methods, i.e., the two variables, are highly correlated then the benefit of adding the second variable to the first will be small. When the two variables are absolutely correlated (r = 1), the second is of no help and can be eliminated. One therefore finds that the three mathematical parameters that determine the discriminating effect of two chemical variables are : the distance between centroids (if there are more groups, then the between-class variance), the within-class variance and the correlation between the chemical variables.

In the method of linear discriminant analysis, one seeks a linear function D of the variables x_i :

$$D = \sum_{i=1}^{d} w_i x_i \tag{20.1}$$

which maximizes the ratio between both variances, taking correlation into account ; w_i are weights given to the variables and d is the number of variables. For the case of two classes K and L, this means that one must maximize

$$\frac{\left\{ \sum_{i=1}^{d} w_i (\bar{x}_{iK} - \bar{x}_{iL}) \right\}^2}{\sum_{i=1}^{d} \sum_{j=1}^{d} w_i w_j C_{ij}} \tag{20.2}$$

where $\bar{x}_{iK}$ is the mean of variable x_i for class K and C_{ij} is an element of the pooled or average variance-covariance matrix. The use of a pooled variance-covariance matrix implies that the variance-covariance matrices for both populations are equal. This is often not the case, for example when the populations to be separated are a normal and a clinically abnormal one, as occurs in many applications in clinical analytical chemistry. As was remarked by Winkel and Juhl (1971), the variance of a parameter for an abnormal population is usually larger than for normal patients.

Differentiation with respect to w_i (Kendall, 1975) leads to

$$\bar{x}_{iK} - \bar{x}_{iL} = \frac{\sum_{i=1}^{d} w_i (\bar{x}_{iK} - \bar{x}_{iL}) \sum_{j=1}^{d} C_{ij} w_j}{2 \sum_{i=1}^{d} \sum_{j=1}^{d} w_i w_j C_{ij}} \qquad (20.3)$$

so that w_j can be obtained from

$$\sum_{i=1}^{d} C_{ij}^{-1} (\bar{x}_{iK} - \bar{x}_{iL}) \qquad (20.4)$$

Once the w values have been obtained, one calculates the values of the discriminant function D (also called discriminant scores) for the centroids of classes K and L as

$$D_K = \sum_{i=1}^{d} w_i \bar{x}_{iK} \qquad (20.5)$$

$$D_L = \sum_{i=1}^{d} w_i \bar{x}_{iL} \qquad (20.6)$$

For an individual object u with variable values x_{iu}, the same function

$$D_u = \sum_{i=1}^{d} w_i x_{iu} \qquad (20.7)$$

is obtained and u is classified with K if D_u is situated nearest to D_K compared with D_L. Generalization is possible : when there are d variables, up to d - 1 discriminant functions can be obtained. It is however rarely of interest to obtain more than two.

The "discrimination power" of each variable (as a percentage) within this framework is given by

$$100 \cdot \frac{| w_i (\bar{x}_{iK} - \bar{x}_{iL}) |}{\sum_{i=1}^{d} | w_i (\bar{x}_{iK} - \bar{x}_{iL}) |} \qquad (20.8)$$

Such an equation allows one to assess the importance of each variable in the analytical program. It should be added that such a procedure is not recommended by some statisticians. Other ways of selecting variables are discussed in section 20.4.

20.2.2. Applications

A typical application from the laboratory of one of the authors (Smeyers-Verbeke et al., 1977) involves the classification of milk samples according to their origin : cows', sheeps' or goats' milk (see also the milk problem, Chapter 16). The learning groups consisted of 20 samples of milk fat of each of the three categories and the pattern vectors contained the percentage distribution data of 15 fatty acids. It was found that in this instance the asumptions of the linear discriminant technique (normal distributions, equal variance-covariance matrices) are reasonably correct. An exact classification (i.e., a complete separation) of the three groups was obtained. This was still the case when the learning groups consisted of groups such as pure cows' milk, pure goats' milk and a mixture of 10% cows' milk and 90% goats' milk. All pure milk samples could be separated from samples containing 10% of the milk of another species with 85 - 100% success.

When one or two discriminant functions are used, the results can be shown graphically. In this instance, one can consider the linear discriminant analysis as a feature extraction method (see Chapter 16). An example for the milk problem is given in Fig. 20.3.

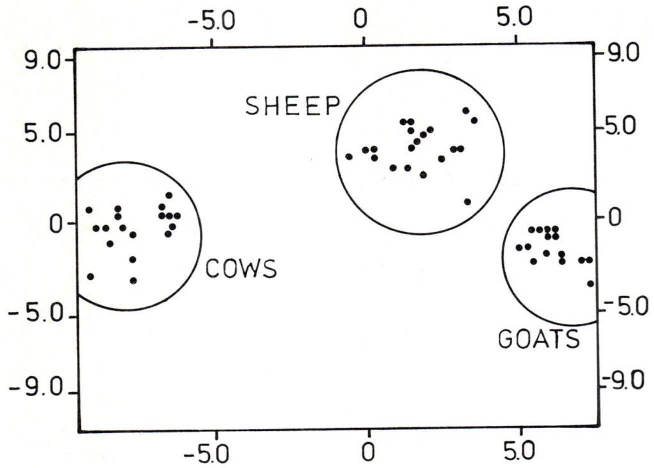

Fig. 20.3. Graphical representation of the classification of milk samples. The coordinates are the values of the discriminant scores of the samples.

The percentage contribution of each of the 15 variables can be calculated for each of the classification problems by using eqn. 20.8. For a typical problem, the separation of pure cows milk to which 10% of sheeps'milk was added, the contribution of the most important parameter (C 10:0, capric acid) is 28.4% and the five most important parameters together contribute 88.4%. These five variables allow a 97.5% correct classification (the recognition ability is 97.5%).

In this way, the original set of 15 parameters to be measured can be reduced to 5, and the GLC analysis for this problem can be shortened, as it is found that only medium- and long-chain fatty acids are of importance.

It should be added, however, that in this work only the classification ability was investigated and that this may give an optimistic view of the prediction ability (see also section 20.1). In the thyroid example introduced in Chapter 16 (Coomans et al., 1978), five different biochemical tests are used, namely RT3U (triiodothyronine resin uptake), T4 (total serum thyroxine), T3RiA (total serum triiodothyronine by radioimmunoassay), TSH (total serum thyroid-stimulating hormone) and ΔTSH (increase of TSH after injection of thyrotrofine-releasing hormone). The results are used to separate samples belonging to hypothyroid subjects from normal samples or to distinguish between hyperthyroid and normal samples. In this instance, the stepwise procedure using Rao's V criterion was used. The results were computed using the SPSS program (Nie et al., 1975). For the hypo-/normal classification, the advantageous result was obtained that the two first selected parameters alone permit a 99.8% classification. In other words, using only two tests, one does not obtain a poorer result than with five tests ; the other three tests appear merely to add noise (see also section 20.3.3).

In analytical chemistry, linear discriminant analysis has been used most frequently for medical applications. Reviews of these applications were given by Radhakrishna (1964) and Romeder (1973). We cannot discuss here all of these applications and we shall therefore confine the discussion to some more recent typical examples. Werner et al. (1972) compared a multi-component method (electrophoresis of protein fractions) with a battery of tests (specific assays for individual proteins). They concluded that the electrophoretic method provides the same classification effectiveness as the test battery. The financial cost per result is, however, markedly lower for the electrophoretic procedure, so that an optimization of the cost parameter is possible without loss of information.

Another example can be found in the work of Winkel et al. (1975). They considered a number of clinical-chemical tests such as albumin, prothrombin-proconvertin, bilirubin, alkaline phosphatases, alanine aminotransferase and μ-globuline for patients known to be suffering from a hepatobiliary disease and considered which test or tests should be used to make a distinction between diagnostic categories such as cirrhosis and hepatic tumour. As an example of the results obtained, they found that alanine aminotransferase contributes most to the allocation when only three main diagnostic categories are considered and that in this instance no significantly better results were obtained when a second test was added. When eleven categories were considered, it was found that four tests contributed significantly to the discrimination. The two other tests can be considered to be redundant in this context. Other interesting examples concern the evaluation of phosphate clearance tests in the diagnosis of hyperparathyroidism (Amenta and Harkins, 1971) and thyroid function tests (Barnett et al., 1973).

Stepwise linear discriminant analysis (see section 20.4.1) was applied by Powers and Keith (1968) in food analysis. They classified coffee into four flavour categories by using ratios of the peak heights in GLC chromatograms as the parameters and used the statistical technique for the selection of the most meaningful of these ratios. The same was done for potato chips using headspace volatiles.

Kawahara and Young (1976) and Mattson et al. (1977) employed analogous procedures for the classification of petroleum pollutants using infrared spectral patterns. The emphásis in both papers, however, was on the classification and not on the feature selection problem.

20.3. NON-PARAMETRIC METHODS

20.3.1. The learning machine and related methods

Consider again Fig. 20.1. One wishes to find a decision line that separates the two classes K and L. In the d-dimensional case a decision surface must be found with a lower dimensionality than the pattern space. For reasons of computational convenience that will become clear later, it is preferable that the decision surface should be linear and pass through the origin of the pattern space. This is not possible in Fig. 20.1. It becomes possible, however, if the two-dimensional space is augmented by the addition of a third dimension, z. This extra dimension is usually given a value of unity and it is added to all pattern vectors. The vector $\vec{x} = (x_1, x_2)$ (see section 20.1) now becomes $\vec{x} = (x_1, x_2, z)$. It is now possible to let a linear decision surface (here a plane) separate the two classes (see Fig. 20.4). Class K falls above the plane and class L below.

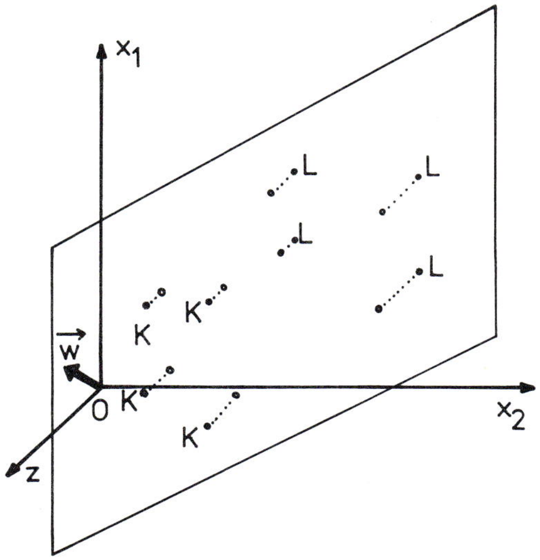

Fig. 20.4. Separation of two classes by a plane through the origin.

This plane can now conveniently and unambiguously be described by an orthogonal or normal vector, called the weight vector and represented by $\vec{w}$.

This can be generalized to the d-dimensional case. To all pattern vectors a (d + 1)th component is added so that they are now given by $\vec{x} = (x_1, x_2, \ldots, x_d, z)$ and a linear decision surface (also called a linear discriminant function) can be sought that will be represented by its normal vector.

The normal vector not only permits a specification of the surface but also allows easy classification. The scalar product of the normal vector and the pattern vector is given by

$$\vec{w} \cdot \vec{x} = |w| \quad |x| \cos \theta \tag{20.9}$$

where $|w|$ and $|x|$ are the magnitudes of vectors $\vec{w}$ and $\vec{x}$ and θ is the angle between the two. When $\vec{w}$ and the pattern vector lie on the same side of the plane, $270° < \theta < 90°$. Therefore, $\cos \theta > 0$ and, as $|w|$ and $|x|$ are positive quantities, $\vec{w} \cdot \vec{x}$ must be positive. When the pattern vector of a sample of class L is considered, $\vec{x}$ and $\vec{w}$ do not lie on the same side of the plane and therefore $90° < \theta < 270°$, so that $\cos \theta < 0$. The scalar product of $\vec{w}$ and $\vec{x}$ is now negative.

In other words, one determines the sign of the scalar product to decide whether the sample is part of K ; when it is negative it should be classified in L.

In practice, the calculations are not carried out by using eqn. 20.9. In the same way as $\vec{x}$ is represented by its component along the axes, $\vec{w}$ can also be decomposed in its components w_1, w_2, ..., w_d, w_{d+1} along the same axes. The scalar product $\vec{w} \cdot \vec{x}$ is then equal to the sum of the products of the components

$$\vec{w} \cdot \vec{x} = w_1x_1 + w_2x_2 + \ldots + w_dx_d + w_{d+1} \cdot z \qquad (20.10)$$

The coefficients w_1, ... can be considered as weights of the variables x_1, ..., which is why $\vec{w}$ is called the weight vector.

The determination of $\vec{w}$ leads to a simple classification rule. Before a classification is possible it is necessary, however, to find a decision surface (or its associated weight vector) that permits the separation of classes K and L. This is accomplished during the training or learning step, using an iterative procedure. It is initiated by selecting, sometimes arbitrarily, an initial weighting vector and investigating if the pattern vectors fall on the correct side of the associated surface. When a pattern $\vec{x}_j$ is found to be misclassified because the product

$$\vec{w} \cdot \vec{x}_j = s \qquad (20.11)$$

produces the wrong sign, a new decision surface is obtained by reflecting it about the misclassified point. This means that one determines $\vec{w}'$, so that

$$\vec{w}' \cdot \vec{x}_j = -s \qquad (20.12)$$

This process is repeated whenever a misclassified sample is found until a completely succesful value of $\vec{w}$ is obtained. In this case the process is said to converge. For mathematical details, one should consult the literature (for example, Nilsson, 1965).

The procedure gradually "learns" the correct answer to the task of finding a correct decision surface and is therefore called the learning machine. The classification and prediction performance is of lesser importance in this book, where the emphasis is on feature selection (see Chapter 16). A general book on the learning machine has been written by Nilsson (1965). Computer packages such as ARTHUR (Kowalski, 1975) include this and many other pattern recognition methods. Many examples of the application of the learning machine in analytical chemistry are to be found in the book by Jurs and Isenhour (1975). Most examples come from the fields of mass and infrared spectrometry. Electrochemical examples are also cited. Gas chromatographic data (Clark and Jurs, 1975) and trace element concentrations (Kowalski et al., 1972) have been used.

Vandeginste (1977) proposed a pattern recognition procedure to select an analytical method from various alternatives. This study is based on a proposal by Kaiser (1970). According to Kaiser, a more systematic approach to problem solving in analytical chemistry could be achieved by describing an analytical problem with the help of a set of parameters such as the element to be analysed, the amount of sample available and desired values of performance characteristics such as precision and cost. In this way, the analytical problem can be represented by a pattern vector (or a point in multi-dimensional space). Problems solved by the same analytical method should lie closely together in the hyperspace and therefore the method selection problem should be transformed into a pattern recognition problem. Vandeginste (1977) applied this to the choice between atomic-absorption spectroscopy and spectrophotometry for a number of analytical problems and obtained prediction abilities from 75 to 92%, depending on the set of problems investigated.

The learning machine as explained here is the simplest of a large class of methods called threshold logic unit (TLU) methods. In the method discussed here, s is compared to zero (the threshold) to make a binary decision (K or L). Non-zero TLU methods also exist and the learning machine can be adapted for multi-category decisions by splitting them into sequences of binary decisions. An introduction to these methods can be found in the book by Jurs and Isenhour

(1975), which was written particularly for chemists, most of the applications being taken from analytical chemistry.

20.3.2. The nearest neighbour method

A mathematically very simple classification procedure is the nearest neighbour method. In this multi-category method, one computes the distance between an unknown, represented by its pattern vector, and each of the pattern vectors of the training set. Usually one employs the euclidian distance (eqn. 18.2). If the training sets consist of a total of n samples, then n distances are calculated and one selects the lowest of these. If this is Δ_{ul} , where u represents the unknown and l a sample from learning group L, then one classifies u in group L. In a more sophisticated version of this technique, called the k-nearest neighbour method (often abbreviated to the KNN method), one selects the k nearest samples to u and classifies u in the group to which the majority of the k samples belong.

The mathematical simplicity of this method does not prevent it from yielding results as good as and often better than the much more complex TLU methods discussed in the preceding section, provided that the training set is sufficiently large. It also has the advantage of being a multi-category method whereas most TLU methods are fundamentally binary decision methods. Its most important disadvantage is that it requires the computation of n distances for each classification decision. It is possible, however, to represent each class in the training set in a first stage of the calculation by a few representative patterns (Gates, 1972) and, according to the present authors, these representative patterns could be chosen in an optimal way using the branch and bound algorithm of section 22.2.

An application of the nearest neighbour method of interest in the context of this book was proposed by Leary et al. (1973). It concerns the GLC classification problem, which was called GLC example II in Chapter 16 and was solved by using numerical taxonomy in Chapter 18. Let us recall briefly that one wants to classify 226 GLC phases according to the retention indices of 10 probes obtained for each of these phases.

In a first step, Leary et al. select more or less subjectively 12 "preferred" phases from the 226. These 12 are well tested phases and are known to have different behaviour towards the 10 probes. Each of these phases constitutes a learning group. Learning groups of one sample are not, of course, to be recommended in most classification problems, but in this particular instance it is unavoidable. The distance between the 214 other phases and each of the 12 learning group phases is then calculated. Eqn. 18.2 becomes

$$\Delta_{ul} = \left(\sum_{i=1}^{10} (\Delta RI_{iu} - \Delta RI_{il})^2 \right)^{1/2} \qquad (20.13)$$

where ΔRI_{iu} is the retention index relative to squalane for probe i on unclassified phase u and ΔRI_{il} is the same characteristic for preferred phase l, while Δ_{ul} is the distance between u and l.

Owing to the a priori choice of 12 preferred phases, this classification method does not allow one, for example, to select "abnormal" phases as is possible with some of the other metods used to solve the same classification problem and discussed in Part III of this book. A more objective choice of the preferred phases would have been possible by using a method such as that described in sections 22.2.1 and 22.2.3. On the other hand, it was the first application of a mathematical classification procedure proposed in this domain and it does allow one to establish those phases with very similar characteristics. This method has also been applied by Lowry et al. (1974) and Haken et al. (1976), both groups also considering GLC selection problems.

20.3.3. SIMCA

Discriminant analysis and the learning machine are aimed at the quantification of differences between classes. The SIMCA (simple modelling of class analogy) method instead aims at finding the similarities within each class in terms of the multivariate data given for the objects in the classes. Thus, in SIMCA a separate mathematical description is obtained for each class, completely disjointly and independently from the other classes. If a classification is then desired, this is obtained by comparing each object to be classified with each of the class models (mathematical descriptions) to find the model to which the object is most similar. "Outliers" are also identified readily in this way.

The basis of SIMCA is the ability of a PC model (see Chapter 19) to closely approximate data observed on a group of similar objects. By definition, the objects in the training set in pattern recognition are organized in classes so that each class contains only similar objects to the best of the data analyst's knowledge. It follows that the data vectors observed on objects k within one class K, denoted by $x_{ik}^{(K)}$, $k = 1, 2, \ldots, n_K$; $i = 1, 2, \ldots, d$), can be closely approximated by the PC model (19.2) with a small number of product terms, r_K

$$x_{ik}^{(K)} = \bar{x}_i^{(K)} + \sum_{p=1}^{r_K} b_{ip}^{(K)} u_{pk}^{(K)} + e_{ik}^{(K)} \qquad (20.14)$$

In this book, we have always tried to present measurement results as y and concentrations as x. Both can be used here, so that x can be replaced with y in eqn. 20.14, yielding equations resembling those in Chapter 19.

Remembering that a PC model is geometrically represented as an r_K-dimensional plane in the d-dimensional measurement space, we have the geometrical interpretation of the SIMCA method shown in Fig. 20.5.

The SIMCA method works as follows. In phase I, the training phase, data observed on objects "known" to belong to the classes - the training set - are used to estimate parameters in the separate PC models, one for each class. The "modelling power" of the variables is calculated and variables with much noise (see below) and also "outliers" in the training set are deleted. The PC models of the classes are then recalculated from the reduced data set.

In phase 2, the classification of the objects in the test set is effected by fitting each of the corresponding data vectors to each of the PC-class models by means of multiple regression as described in Chapter 19 (eqn. 19.25).

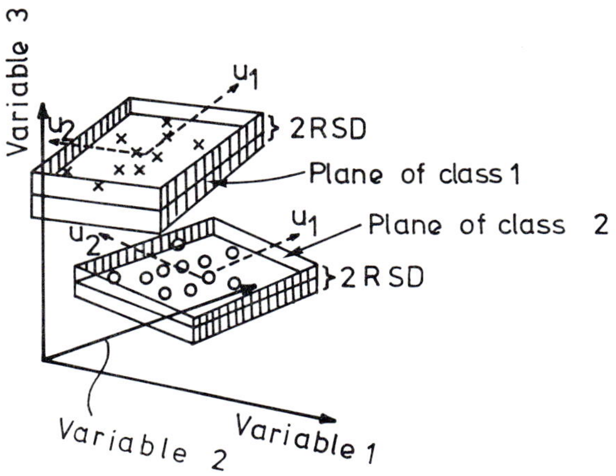

Fig. 20.5. The SIMCA method describes the structure of each class (two above) by means of separate PC models (planes above) and "confidence boxes" based on the residual standard deviations of the class and the component values $\vec{u}$ of the class.

Depending on the ambition level of the data analysis, objects in the test set are classified either (i) in the class to which model hyperplane they are closest (smallest residual standard deviation) or (ii) in the class to which model hyperplane they are sufficiently close, i.e., within the confidence region of the PC model (see fourth paragraph in section 19.4). If an object is not inside the confidence region of any class, it is an "outlier", an object of a new type.

The following information is obtained by a SIMCA analysis :

(1) A description of the "regularities" in each class K by means of a separate PC model with the parameters r_K, $\bar{x}_i^{(K)}$, $b_{ip}^{(K)}$ and $u_{pk}^{(K)}$. The values of the parameters $u_{pk}^{(K)}$ can be used to construct a confidence region within which the u values of a new object shall fall in order for the object to be considered a member of the class.

(2) The "noise" of each variable in each class, $s_i^{(K)}$, and over all classes s_i

$$s_i^{(K)2} = \sum_{k=1}^{n_K} e_{ik}^{(K)2} / (n_K - r_K - 1) \qquad (20.15)$$

$$s_i^2 = \sum_{K=1}^{Q} s_i^{(K)2} / Q \tag{20.16}$$

where n_K is the number of objects in class K, r_K the number of product terms for class K and Q the number of classes. These variances relate to the "modelling power" of each variable, i.e., to the extent to which they participate in the description of the classes (see Chapter 19).

(3) The "noise" for each class, $s_0^{(K)}$, as based on the training set

$$s_0^{(K)2} = \sum_{i=1}^{d} \sum_{k=1}^{n_K} e_{ik}^{(K)2} / (d-r_K)(n_K - r_K - 1) \tag{20.17}$$

This class standard deviation, $s_0^{(K)}$, can be used to construct a confidence region around the hyperplane of the class model within which an object should fall to belong to the class with a given probability, i.e., its residual standard deviation (eqn. 19.14) should be smaller than $s_{lim}^{(K)}$ (see eqn. 19.15)

$$s_{lim}^{(K)2} = F_{crit} \cdot s_0^{(K)2} \tag{20.18}$$

where F_{crit} is chosen from an F-distribution with the appropriate probability and degrees of freedom.

(4) By fitting all objects in the training set to all PC models except the one to which the object "belongs", one obtains information about the separation between the classes, both totally and per variable. Thus, by comparing for each variable i, the residual standard deviation (RSD) when all objects in the training set are fitted to all "other classes" to the RSD when the same objects are fitted to their "own classes", one obtains information on the discriminating power of the variable. Similarly, a measure of the distance between two classes, say K and L, is obtained when objects in class K are fitted to the class model of class L and *vice versa* and the resulting RSDs are compared with the "own" RSDs of the two classes. Let $s^{(K,L)}$ denote the RSD obtained when all objects in class K are fitted to the class model of class L, then we have the distance D_{KL} :

$$D^2_{KL} = (s^{(K,L)2} + s^{(L,K)2}) / (s^{(K,K)2} + s^{(L,L)2}) \qquad (20.19)$$

It can be seen that the scope of SIMCA goes beyond that of mere classification to give also a model of each class in terms of a PCF model. In the context of this book, the most important advantage of SIMCA is the relative ease with which it provides measures of the relevance of each variable, measures describing both how important a variable is for describing the within class similarity and how important a variable is to distinguish between classes.

Finally, as SIMCA is not conditioned to maximize the separation between the classes, the method gives a fairly unbiased measure of the real "distance" between classes - if one finds a definite difference between two classes it probably is real. If one finds no difference, one can be confident that the data involved show no difference between the classes. SIMCA has been applied to a variety of problems in analytical chemistry. Duewer et al. (1975) used SIMCA and other methods to classify simulated oil spills according to their source on the basis of trace elements measured on each "oil spill". They found SIMCA to compare favourably with the other methods. Wold and Sjöström (1977) described the use of SIMCA in the structure determination of unsaturated ketones based on their IR and UV spectra. Sjöström and Edlund (1977) applied SIMCA in the analysis of C-13 NMR data of *exo*- and *endo*-norbornanes. Sjöström and Kowalski (1978) make a comparison of SIMCA with other standard methods on a number of sets of real data. Ulfvarsson and Wold (1977) analysed possible differences in trace element concentrations in blood between welders and controls. Christie and Alfsen (1978) used SIMCA to classify archeological artefacts on the basis of trace element concentrations. Other applications can be found in papers by Strouf and Wold (1977) and Dunn and Wold (1978).

20.4. FEATURE SELECTION

20.4.1. General aspects

One general way of selecting features is to compare the means and the variance of the different variables before pattern recognition is applied. Intuitively, a variable for which the mean is the same for each class is of little use for discriminating the classes in question. In the same way, variables with widely different means for the classes and small intraclass variance should be of value. This can be evaluated, for example, by variance-weighting (Kowalski and Bender, 1972). Variance-weighting permits weights to be given to the variables on the basis of their power to discriminate between the training sets. These weights are measures of the ratio of between-class variance to within-class variance for the learning groups. It should be noted that, as the correlation between variables is not taken into account, one selects in this way the best individual variables, but not necessarily the best combination of variables. In section 17.5 it was noted that, owing to correlation, the best set of two or three GLC phases does not necessarily contain the individually best phase, and in Chapter 18 methods were proposed to arrive at the selection of best sets when correlation is an important factor. The same problem is discussed further in this section.

For two classes, K and L, the weights are obtained by using the equation

$$w_i = \frac{\dfrac{n_K \cdot n_L}{n^2} \sum_{k=1}^{n_K} \sum_{l=1}^{n_L} (x_{ik} - x_{il})^2}{\dfrac{n_K}{n} \sum_{k=1}^{n_K} \sum_{k'=1}^{n_K} (x_{ik} - x_{ik'})^2 + \dfrac{n_L}{n} \sum_{l=1}^{n_L} \sum_{l'=1}^{n_L} (x_{il} - x_{il'})^2} \qquad (20.20)$$

where n_K and n_L are the number of individuals that are members of classes K and L ($n_K + n_L = n$), x_{ik} is the concentration of the ith variable or the measurement value of this variable (k and k' are part of K, l and l' of L) and w_i is the weight of the ith variable. The general equation for more than two classes was given by Kowalski and Bender (1972).

Another general procedure is the stepwise procedure, which is applied mostly in statistical linear discriminant analysis. One starts by selecting the test giving the best discrimination, then one adds at each step the parameter that yields the highest increase in discrimination. In this way, one ranks the parameters according to utility and is able to develop the best set of measurements. This was applied, for instance, by Winkel et al. (1975) in studies on the diagnostic utility of liver tests.

Gray (1976) showed that if the number of dimensions, d, is relatively large compared with the sample size of the training set, n, unsignificant classifications are obtained. He was able to separate with 100% success two sets of numbers obtained from a random number generator and for which there was therefore no genuine reason for separability. The number of dimensions was 30 and the sample size 50. In general, n/d should be > 3 (Bender et al., 1973). These considerations lead us to discuss some fundamental problems in connection with feature selection. One can distinguish two situations. In the first, the number of objects in the training set, n, is fairly large compared with the number of variables, d. It is then possible to select variables that contribute most to the separation of the classes. Such methods were discussed above and are further discussed below. They can be called methods conditioned on separation. In the second situation, when the number of variables approaches the number of objects in the training set, methods of feature selection that are conditioned to find variables that discriminate between classes will not work. By experience, one has found that when the initial number of variables, d, exceeds a third of the number of objects in the training set, n/3, this chance of "pathological" selection becomes unsatisfactorily large. One must then use a feature selection method that is not conditioned on separation of classes. For intermediate situations, when d is not much larger than n, say d < 2n, one can select variables according to their modelling power in SIMCA. When d is even larger, however, radically different methods must be used.

The only strategy that seems feasible at present in this situation when $d > 2n$ is to make a cluster analysis of the variables over all objects including those in the test set. Thus, the variables are treated as objects and vice versa, and one looks for a grouping of the former that indicates some kind of similarity in their behaviour over the data set. Thus, one hopefully finds that the variables cluster into a "small" number of groups. One can then take one variable from each group and proceed with the selection among this reduced set, using the separation conditioned methods described below. This is, in fact, the same philosophy as was applied to the selection of optimal combinations of tests (thin-layer chromatographic systems or GLC columns) in section 17.4.5 and Chapter 18.

An important aspect of feature selection is that it is often found that a few irrelevant variables introduce so much noise that a good classification cannot be obtained. When these irrelevant variables are deleted, however, a clear and well separated class structure is often found. The deletion of irrelevant variables is therefore an important aim of feature selection.

20.4.2. Methods conditioned on separation

The methods used with linear discriminant analysis are generally stepwise methods, as described above (however, see also section 20.2.1). This is often the case also with SIMCA, where feature selection can be made directly on the basis of the calculated relevance of each variable. One can then delete variables that have both low modelling and discriminating power and re-make the SIMCA analysis with the reduced data set. Therefore, we shall discuss in this section mainly methods that are applied in procedures of the learning machine type.
The simplest procedure consists in the elimination of those features with the smallest weights (Kowalski et al., 1969 ; Jurs et al., 1969a, b ; Sybrandt and Perone, 1972). In the same way as described for statistical linear discriminant analysis, it is reasoned that features with small weights must be considered as less important for classification and should therefore be the first to be

eliminated, if feature selection is wanted. After discarding some of the features, one usually carries out again the training procedure and/or the prediction with the reduced set of variables to observe whether there is still linear separability and/or the correct prediction rate does not decrease too much. This method is called weight-magnitude feature selection. Other methods such as the weight-sign feature selection, the distance metric feature selection (Preuss and Jurs, 1974) and weight-variance feature selection (Zander et al., 1975) have also been described.

20.4.3. Some applications

Although feature selection is considered by most workers in this field to be a very important part of pattern recognition, it is usually not carried out with the special purpose of constituting optimal sets of tests, but rather to simplify the classification step or even to make it more significant, as discussed in section 20.4.1. In this section, we shall limit the discussion to a few analytical applications where feature selection is an important aspect. This is the case, for example, in the paper by Duewer et al. (1975), who investigated the source identification of oil spills by pattern recognition analysis using elemental composition data. Table 20.I gives the variance weights for the 22 elements used to separate 40 categories of 10 samples. One observes that the highest weight is found for V and the second highest for Ni or S. These should then be the three most significant elements. This result is not unexpected, as the oil industry uses Ni/V ratios and S concentrations to characterize oil samples. It does, however, prove the validity of the feature selection conclusions obtained by Duewer et al. Remembering that a weight of 1.00 means that no discriminating information is obtained, it is clear that elements with very low weights, such as Sr, Mo, Sb, Sm and Au, are of very doubtful utility. The authors eliminated these and a few other elements such as Cl because they are highly correlated with other elements.

Table 20.I

Variance weights for 22 elements (from Duewer et al., 1975)

	AUTO a)	LNAUTO b)
Na	2.97	4.15
Al	1.77	1.98
S	8.61	10.51
Cl	2.23	2.71
V	51.37	156.34
Mn	7.46	7.85
Co	2.18	2.70
Ni	7.81	23.47
Zn	2.42	3.44
Ga	3.72	4.05
As	4.32	4.64
Br	2.14	2.30
Sr	1.02	1.04
Mo	1.26	1.39
In	1.81	1.81
Sn	1.93	2.27
Sb	1.07	1.07
I	3.47	4.08
Ba	4.65	5.60
La	1.50	1.54
Sm	1.33	1.33
Au	1.05	1.05

a) data transformed by scaling to obtain mean of zero and variance of unity
b) data transformed by computing ln(1.0 + x) and scaling as in a.

In fact, this comes very close to the classification approach discussed in Chapter 18, where groups of tests (in Chapter 18, TLC systems) are classified (and the correlation coefficient is one possible similarity criterion) and where in each resulting class the best test is chosen. A weight-sign procedure was used by Clark and Jurs (1975) in a problem similar to the milk problem (see section 20.2.2), as it concerns the classification of petroleum sample types according to their GLC spectra. The feature selection showed that sufficient information was present in a small fraction of the original 19 characteristics to carry out the classification. Both the weight-magnitude and the weight-sign procedures were applied to mass spectrometric problems (Kowalski et al., 1969 ; Jurs, 1970). A combination of both was used by Sybrandt and Perone (1972) in the classification of overlapping peaks in polarography. In this instance, the original set of 132 parameters was reduced to 22. The weight-variance procedure was employed by Zander et al. (1975) to select features from mass spectrometric spectra.

Applications of feature selection in statistical linear discriminant analysis are discussed in section 20.2.2.

REFERENCES

J.S. Amenta and M.L. Harkins, Amer. J. Clin. Path., 55 (1971) 330.
H.C. Andrews, Introduction to mathematical techniques in pattern recognition, Wiley-Interscience, New York, 1972.
D.B. Barnett, A.A. Greenfield, P.J. Howlett, J.C. Hudson and R.M. Smith, Brit. Med. J., 2 (1973) 144.
C.F. Bender, H.D. Shepherd and B.R. Kowalski, Anal. Chem., 45 (1973) 617.
O. Christie and B. Alfsen, 1978, in preparation.
H.A. Clark and P.C. Jurs, Anal. Chem., 47 (1975) 374.
D. Coomans, I. Broeckaert and D.L. Massart, 1978, unpublished results.
R.O. Duda and P.E. Hart, Pattern classification and scene analysis, Wiley, New York, 1973.
D.L. Duewer, B.R. Kowalski and T.F. Schatzki, Anal. Chem., 47 (1975) 1573.
W.J. Dunn, III and S. Wold, J. Med. Chem., (1978) in press.
G.W. Gates, IEEE Transactions, IT-18 (1972) 431.
N.A.B. Gray, Anal. Chem., 48 (1976) 2268.
J.K. Haken, M.S. Wainwright and N. Do Phuong, J. Chromatogr., 117 (1976) 23.
T.L. Isenhour, B.R. Kowalski and P.C. Jurs, Crit. Rev. Anal. Chem., 4 (1974) 1.
P.C. Jurs, Anal. Chem., 42 (1970) 1633.
P.C. Jurs and T.L. Isenhour, Chemical application of pattern recognition, Wiley-Interscience, New York, 1975.
P.C. Jurs, B.R. Kowalski, T.L. Isenhour and C.N. Reilley, Anal. Chem., 41 (1969a) 690.
P.C. Jurs, B.R. Kowalski, T.L. Isenhour and C.N. Reilley, Anal. Chem., 41 (1969b) 1949.
H. Kaiser, Anal. Chem., 42 (1970) 24A.
F.K. Kawahara and Y.Y. Yang, Anal. Chem., 48 (1976) 651.
M. Kendall, Multivariate analysis, Charles Griffin, London and High Wycombe, 1975.
B.R. Kowalski, Anal. Chem., 47 (1975) 1153A.
B.R. Kowalski and C.F. Bender, J. Amer. Chem. Soc., 94 (1972) 5633.
B.R. Kowalski, P.C. Jurs, T.L. Isenhour and C.N. Reilley, Anal. Chem., 41 (1969) 695.
B.R. Kowalski, T.F. Schatzki and F.H. Stross, Anal. Chem., 44 (1972) 2176.
J.J. Leary, J.B. Justice, S. Tsuge, S.R. Lowry and T.L. Isenhour, J. Chromatogr. Sci., 11 (1973) 201.
S.R. Lowry, S. Tsuge, J.J. Leary and T.L. Isenhour, J. Chromatogr. Sci., 12 (1974) 124.
J.S. Mattson, C.S. Mattson, M.J. Spencer and F.W. Spencer, Anal. Chem., 49 (1977) 500.
N.H. Nie, C.H. Hull, J.G. Jenkins, K. Steinbrenner and D.H. Bent, SPSS - Statistical package for the social sciences, McGraw-Hill, New York, 1975.
N.J. Nilsson, Learning machines - Foundations of trainable pattern-classifying systems, McGraw-Hill, New York, 1965.
J.J. Powers and E.S. Keith, J. Food Sci., 33 (1968) 207.
D.R. Preuss and P.C. Jurs, Anal. Chem., 46 (1974) 520.
S. Radhakrishna, Statistician, 14 (1964) 147.
J.M. Romeder, Méthodes et programmes d'analyse discriminante, Dunod, Paris, 1973.
M. Sjöström and U. Edlund, J. Magn. Reson., 25 (1977) 285.

M. Sjöström and B.R. Kowalski, 1978, in preparation.
J. Smeyers-Verbeke, D.L. Massart and D. Coomans, J. Ass. Offic. Anal. Chem., 60 (1977) 1382.
O. Strouf and S. Wold, Acta Chem. Scand., A 31 (1977) 391.
L.B. Sybrandt and S.P. Perone, Anal. Chem., 44 (1972) 2331.
U. Ulfvarsson and S. Wold; Scand. J. Environm. Work Health, 3 (1977) 183.
B.G.M. Vandeginste, Anal. Lett., 10 (1977) 661.
M. Werner, S.H. Brooks and G. Cohen, Clin. Chem., 18 (1972) 116.
P. Winkel and E. Juhl, Lancet, (1971) 435.
P. Winkel, R. Ramse, J. Lyngbye and N. Tygstrup, Clin. Chem., 21 (1975) 71.
S. Wold, Pattern Recognition, 8 (1976) 127.
S. Wold and M. Sjöström, in B.R. Kowalski (Editor), ACS Symposium Series 52, American Chemical Society, Washington, D.C., 1977, page 243.
T.Y. Young and T.W. Calvert, Classification, estimation and pattern recognition, Elsevier, Amsterdam, 1974.
G.S. Zander, A.J. Stuper and P.C. Jurs, Anal. Chem., 47 (1975) 1085.

Chapter 21

OPERATIONAL RESEARCH : LINEAR PROGRAMMING, QUEUEING THEORY AND SOME RELATED METHODS *

21.1. LINEAR PROGRAMMING

Linear programming is used when one tries to maximize (or minimize) a linear function of several variables and when these variables are subject to constraints. Perhaps the best known example is the diet problem. The price of a number of ingredients being known, one seeks the composition of the cheapest diet, so that certain requirements (maximal or minimal utilization rate of the ingredients, number of calories required, etc.) are fulfilled. Linear programming is one of the methods used to solve such allocation problems, i.e., problems in which resources (the ingredients in the diet problem) have to be distributed in the most efficient manner possible. As far as is known to the authors, linear programming has not been used in analytical chemistry. However, as it is one of the oldest methods of O.R., a short account of the method is given here, using as an example a simple situation which might occur in an analytical chemical laboratory. A laboratory must carry out routine determinations of a substance P and uses two methods, A and B, to do this. With A, one technician can carry out ten determinations per day, with method B twenty determinations per day. There are only three apparatuses available for method B and there are five technicians in the laboratory.

The first method, although it needs more man-hours, is cheaper and costs 100 units per determination, method B costs 300 units per determination and the available daily budget is 14,000 units. How should the technicians be divided over the two available methods, so that as many determinations as possible are

---

* Sections 21.3 and 21.4 were written by B.G.M. Vandeginste, Department of Analytical Chemistry, Catholic University of Nijmegen, The Netherlands.

carried out ? Let the number of technicians working with method A be a and with method B b, and the total number of determinations z ; then, the economic function (or objective function) is given by

$$z = 10\,a + 20\,b \tag{21.1}$$

The following restrictions (constraints) must be taken into account

$$b \leqslant 3 \text{ (apparatus restriction)} \tag{21.2}$$

$$a + b \leqslant 5 \text{ (personnel restriction)} \tag{21.3}$$

$$(10 \times 100)\,a + (20 \times 300)\,b \leqslant 14{,}000 \text{ (budget restriction)} \tag{21.4}$$

The problem is to find values of a and b that maximize z.

In Fig. 21.1, the restrictions are shown graphically. All of the solutions (combinations of an a and a b value) outside these limits violate one of the restrictions and are therefore impossible (shaded area). The remaining unshaded area defines the so-called feasible region.

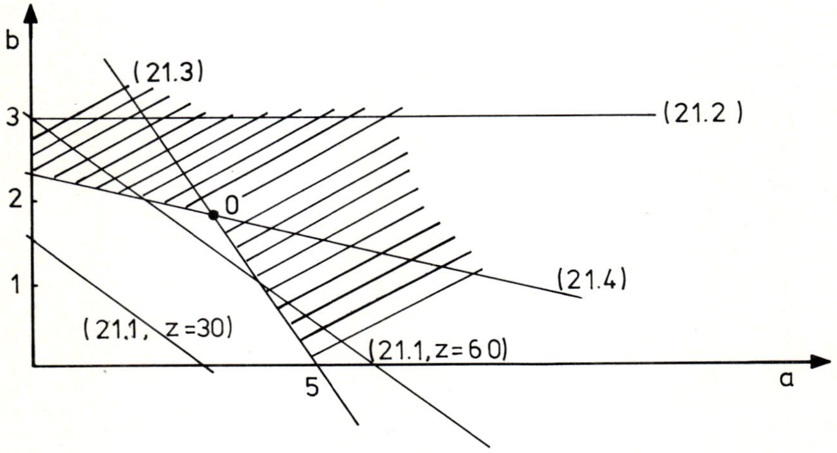

Fig. 21.1. An example of linear programming.

The solutions must satisfy eqn. 21.1. All solutions possible for one particular value of z fall on a straight line. The lines for z = 30 and z = 60 are shown in the figure. One observes that these lines are parallel. By moving the line upwards, better solutions are obtained. One can now restate the optimization problem as follows : find a point in the feasible region, situated on a line given by eqn. 21.1 and situated as far from the origin as possible. In the example studied here, this is given by point 0 (b = 1.8, a = 3.2, z = 68). One can show that in this way, one always selects one of the corner points of the feasible region. As linear programming has not been used in analytical chemistry, we shall not go into the details of the solution methods for more complex problems. The solution method is called the Simplex method and it consists in moving over the polyhedron formed by the constraints, from corner point to corner point, in such a way that the value of the economic function increases until the optimum is reached.

One should observe that the results obtained in this example imply that 1.8 technicians should work with the automatic apparatus. This can be solved by having one technician working full time and another four days out of five with this apparatus. When this is not permissible, the solution is not feasible. This is not a rare occurrence and one should then use a method called integer programming in which only solutions with integer numbers are permitted (see Chapter 22). In addition, one should observe that only linear economic functions or constraints are considered. This severely limits the possibilities for application in analytical chemistry. A generalization to non-linear programming is possible, but mathematically much more sophisticated and complicated.

Many books on linear programming have been written, and that by Hadley (1962) can be recommended. One should also consult general books on operational research such as those by Ackof and Sasieni (1968), Hillier and Lieberman (1974) and Wagner (1972). This is true for each of the sections in Chapters 21, 22 and 23.

21.2. GAME THEORY

21.2.1. Some examples

Consider the following situation. An analyst has to decide which of five possible species is present in a solution. To do this, he uses three spot tests. He knows for example that substance 5 will be detected unambiguously and with certainty by test 2, while test 1 will fail. He estimates also that test 1 will detect substance 1 with a probability of 0.3 (because, for example, only sufficiently large amounts allow identification or because the test works only in the absence of certain anionic substances). He knows from past experience that the *a priori* probability of occurrence of the five substances is the same. From these kinds of considerations, he is able to construct the following probability matrix

		Cationic species				
		(1)	(2)	(3)	(4)	(5)
Tests	(1)	0.3	0.4	0.5	1	0
	(2)	0.2	0.3	0.6	0	1
	(3)	0.1	0.5	0.3	0.1	0

This situation can be described as a game against nature *. This terminology is explained in the mathematical section, which discusses game theory in a more systematic way than is possible in this introduction. We shall, however, solve this problem here to demonstrate the philosophy of game theory. If one considers carefully the matrix, one observes that species (1) is always more difficult to detect than species (2) and (3). In elaborating a strategy one should therefore concentrate on (1) and one may eliminate columns (2) and (3). Species (1) is said to dominate species (2) and (3).

* The game described in this section is essentially the same problem as that treated by Kaufmann (1968) concerning the choice from three antibiotics to fight five possible microorganisms.

The matrix now reduces to

		Cationic species		
		(1)	(4)	(5)
Tests	(1)	0.3	1	0
	(2)	0.2	0	1
	(3)	0.1	0.1	0

In the same way, test (1) is always better than test (3) as the probability for succesful application is higher with species (1) and (4) and the same for species (5). Clearly, it would not be intelligent to select test (3). Test (1) is said to dominate test (3). The latter may be eliminated so that the following matrix is now obtained

		Cationic species		
		(1)	(4)	(5)
Tests	(1)	0.3	1	0
	(2)	0.2	0	1

What strategy should now be chosen ? This is, of course, a question of criteria. In games against nature this is a particularly difficult question and several criteria (minimax, Laplace, Wald, Hurwicz, etc.) have been proposed. The most classical criterion is the minimax criterion. The optimal strategy is said to be the one that maximizes the smallest probability whatever the state of nature. For both pure (see section 21.2.2) strategies this is zero. For example, if nature is in state (5), the detection probability with test (1) is 0. If a mixed strategy is employed, so that test (1) is used with a probability of 8/11 and test (2) with a probability 3/11, then the probability of detecting species (1) is

$$0.3 \times \frac{8}{11} + 0.2 \times \frac{3}{11} = 0.273$$

For the four other species it is 0.373, 0.527, 0.727 and 0.273, respectively. This means that if the analyst carries out his selection by putting eight cards with the text : "test (1) must be chosen", and three cards directing him to choose test (2) in a box and by obeying the card he chooses at random from the box, that his probability of identifying the species contained in the solution is at least 0.273. This is, of course, a rather surprising way of making a decision and it is probable that the reader of this book will view this result with some scepticism. The example is clearly a very simple one but the structure of the problem it poses is a very common one.

Another also very simple example is the following. An element A should be determined with a certain precision. If it is present in a concentration > 1 ppm, one will be able to do this by flame AAS, if its concentration is higher than 0.1 ppm by atomic absorption after extraction and if it is higher than 0.01 ppm by neutron-activation analysis. The times for carrying out these techniques are 8, 10 and 12 units, respectively. If one carries out the flame AAS method and the concentration is found to be too low, one can extract the solution, use the already prepared standards and measure again. This will take an additional 2.2 units. In the same way, one is able to calculate the time necessary with each method and for each hypothesis which yields the following matrix

	First method tried		
	AAS	Extraction-AAS	Neutron activation
> 1 ppm	8	10	12
< 1 ppm, > 0.1 ppm	10.2	10	12
< 0.1 ppm, > 0.01 ppm	12.8	12.4	12

What is a prudent and intelligent strategy ? In this instance (which is again a rewording of a problem given by Kaufmann) the matrix contains a saddle point (see section 21.2.2) at the intersection of the third row and third column, which yields the optimal strategy. Neutron activation should be chosen. It is evident that many analytical optimization problems can be formulated as games against nature. Whether game theory will be able to give meaningful answers is a matter for speculation.

Although much more complex games than those described here have been solved, practical game problems are still more complex. It is also possible that the difficulty of selecting a relevant criterion which exists in all applications of game theory, precludes a meaningful use in analytical chemistry. Nevertheless, we think that research in this direction should be carried out even if no immediate rewards are to be expected.

21.2.2. Mathematical

A game can be defined as a sequence of moves, each of which is made by one of the players of the game. At each move the player making the move chooses amongst several possibilities. The outcome of a move is to change the position of the game. The knowledge of the position is an important factor of the game. In many games such as chess the position is completely known by each player at the time of his move, whereas in other games such as bridge this knowledge is incomplete. At the end of a game there is usually some sort of payoff. Here the difference can be made between zero-sum games, in which the total of the sums won and lost amounts to zero and games in which this is not the case.

A strategy for a player of a game is defined as a function which assigns a move to each possible situation with which the player can be confronted while playing the game.

In practice the decision to make a given move is made during the game but when studying a game it can be assumed that all strategies are enumerated beforehand. Let us now limit ourselves to games with two players, A and B, with zero-sum. By this we mean that any sum won by one of the players must be lost by the other. Let us denote by the set $I = \{1, 2, \ldots, n\}$ the set of all strategies of player A and by the set $J = \{1, 2, \ldots, m\}$ the set of all strategies of player B. If player A chooses strategy i ($i \in I$) and player B chooses strategy j ($j \in J$), this will lead to a specific outcome of the game and to a payoff. We shall call a_{ij} the sum player A wins and player B loses in this situation. The values a_{ij} can be represented by a matrix. The rows of the matrix represent the strategies of player A and the columns the strategies of player B.

			B		
		1	2	. . .	m
	1	a_{11}	a_{12}	. . .	a_{1m}
	2	a_{21}	a_{22}	. . .	a_{2m}
A	.				.
	.				.
	.				.
	n	a_{n1}	a_{n2}	. . .	a_{nm}

When examining this matrix from the point of view of player A, it can be seen that if he chooses strategy i he will at least obtain $\underset{j}{\text{Min}}\ a_{ij}$ from the game. By examining his various strategies with the criterion of his minimal gain, he will choose the strategy i for which he maximizes this value

$$\underset{i}{\text{Max}}\ (\underset{j}{\text{Min}}\ a_{ij})$$

Likewise player B, if he selects strategy j, will in the worst case lose $\underset{i}{\text{Max}}\ a_{ij}$. Therefore, he will try to minimize this value and choose strategy j, which minimizes this worst loss

$$\underset{j}{\text{Min}}\ (\underset{i}{\text{Max}}\ a_{ij})$$

It can be proved mathematically that

$$\underset{j}{\text{Min}}\ (\underset{i}{\text{Max}}\ a_{ij}) \geqslant \underset{i}{\text{Max}}\ (\underset{j}{\text{Min}}\ a_{ij})$$

If the matrix of a zero-sum two-person game is such that

$$\underset{j}{\text{Min}}\ (\underset{i}{\text{Max}}\ a_{ij}) = \underset{i}{\text{Max}}\ (\underset{j}{\text{Min}}\ a_{ij}) = v$$

the game is said to have a point of equilibrium or a saddle point. The quantity v is then called the value of the game.

From the definition, it follows that if a saddle point is chosen by both players it must be the smallest element in its row and the largest in its column. This choice of a row of the matrix by player A or of a column by player B is usually refered to as a pure strategy. By this it is meant that the choice by a player is made once and for all. If both players choose a pure strategy corresponding to a saddle point of the game, the following interesting fact can be observed : if the opponent maintains his strategy, changing one's own strategy can only lead to loss.

If no saddle points exists in a game, it is possible to introduce a concept of equilibrium in another way. For this we need the following definition : a mixed strategy is a probability distribution of the set of pure strategies. A mixed strategy for player A can be obtained by assigning probabilities $p_1, p_2, \ldots, p_i, \ldots, p_n$ to each of his pure strategies, such that

$$p_i \geqslant 0 \qquad i = 1, \ldots, n \quad \text{and} \quad \sum_{i=1}^{n} p_i = 1$$

Likewise, a mixed strategy for player B is found by assigning probabilities $q_1, q_2, \ldots, q_j, \ldots, q_m$ to each pure strategy of B, such that

$$q_j \geqslant 0 \qquad j = 1, 2, \ldots, m \quad \text{and} \quad \sum_{j=1}^{m} q_j = 1$$

Of course, if a game is played only once, one of the pure strategies must be chosen by each player. A mixed strategy can be chosen if the player draws a pure strategy at random using the probabilities p_i or q_j. If the game is played N times the mixed strategy is found by playing pure strategy 1 $N.p_1$ times, pure strategy 2 $N.p_2$ times, etc. Of course, the order of playing these pure strategies must be either random or kept secret from the opponent who could make use of this information.

To simplify a game one can often use the property of domination. To illustrate this process, consider a game with the following payoff matrix

		B 1	B 2
A	1	5	4
A	2	2	6
A	3	2	3

Consider the pure strategies of player A ; it can be observed that whatever the strategy of player B it is always better for player A to choose strategy 1 than strategy 3. This action will lead to a payoff at least as good. Therefore, strategy 1 is said to dominate strategy 3. As A will never select strategy 3, this game is in fact equivalent to a game in which strategy 3 has been removed

		B 1	B 2
A	1	5	4
A	2	2	6

Let us consider now a game in which one of the players is nature and the payoff obtained by the other player is influenced by the state in which nature is.

An element a_{ij} of the payoff matrix is defined as the payoff obtained by the player if he selects strategy i and if nature is in state j. Consider a farmer confronted with an investment problem. If he decides to invest, his returns are strongly influenced by the weather. In the simplified case in which only good or bad weather is considered, the payoff matrix is given by

		Nature: Good weather (1)	Nature: Bad weather (2)
Farmer	Investment (1)	100.000	- 20
Farmer	No investment (2)	- 10	- 10

Let us apply the minimax strategy to this game. If he chooses to invest he will in the worst case loose 20 and if he does not invest he will loose 10

$$\operatorname*{Max}_{i} \operatorname*{Min}_{j} = -10$$

He will therefore choose not to invest. It is obvious that it would be wiser for the farmer to take the risk of loosing 20 instead of 10 in the hope of winning 100,000. To express this type of reasoning several alternative criteria have been suggested for solving games against nature. Two of these will be examined in this section.

Hurwicz suggested defining the optimism of the player by a number k, with

$$0 \leqslant k \leqslant 1$$

Then, for each strategy i the worst and best possible outcomes are calculated

$$a_i = \operatorname*{Min}_j a_{ij}$$

$$A_i = \operatorname*{Max}_j a_{ij}$$

If k is the optimism of the player he expects to gain

$$P_i = k A_i + (1-k) a_i$$

from the game, when choosing strategy i.

He then chooses the strategy i which maximizes his expected return

$$\operatorname*{Max}_i P_i$$

Arbitrarily fixing the optimism of the player at k = 0,1 this gives the following values for the game described above

$$a_1 = -20 \quad A_1 = 100{,}000 \quad P_1 = 0.1 \times 100{,}000 + 0.9 \times (-20) = 9982$$

$$a_2 = -10 \quad A_2 = -10 \quad P_2 = 0.1 \times (-10) + 0.9 \times (-10) = -10$$

as $P_1 > P_2$, the player chooses to invest.

An alternative is Laplace's criterion. If the reaction of nature is unknown it can be assumed that the states of nature occur with equal probabilities. In this instance the expected return for strategy i is given by

$$P_i = \frac{1}{n}(a_{i1} + a_{i2} + \ldots + a_{in})$$

and the strategy is chosen for which the expected return is maximized

$$\max_i P_i$$

In the example, the P_i have the following values

$$P_1 = \frac{1}{2}\left[100{,}000 + (-20)\right] = 49990$$

$$P_2 = \frac{1}{2}\left[(-10) + (-10)\right] = -10$$

Again, strategy 1 is chosen.

If probabilities are known for the different states of nature, this criterion can be generalized by giving weights to the payoffs, equal to the known probabilities.

The basic reference for this section is the book by Von Neumann and Morgenstern (1953).

21.3. QUEUEING

21.3.1. Models and assumptions

In Chapter 9 the time taken for an analysis was mentioned as one of the performance characteristics of an analytical procedure. In the same way, the time between the arrival of a sample and the communication of a result is a performance characteristic for an analytical laboratory. If one considers how this time is composed in practice one often finds that a large part of it is spent waiting. In fact, there are two waiting times that are of importance to a laboratory : (a) the time a sample waits before being processed because the apparatus or personnel are occupied with previous samples and (b) the time that apparatus or personnel wait because no samples are available. If these waiting times are too long, there is clearly insufficient agreement between the proposed work load and the analysis capacity and it is therefore obviously of interest to investigate the relationships between these two quantities. This is done with the use of queueing theory.

The waiting time for a determination depends on :

- the mean rate of arrival of samples, λ
- the mean rate of analysis or, to use queueing theory language, the mean service rate per channel, μ. $\mu = 1/\bar{t}_a$, where $\bar{t}_a$ is the average service time
- the number, m, of apparatus (or technicians) that are available to carry out the analysis ; in queueing theory language, the number of channels.

The objective of the queueing analysis is to determine parameters such as $\bar{w}$, the mean waiting time before commencement of the actual determination (queueing theory language : mean waiting time in the queue), or $\bar{n}_q$, the mean queue length. The analysis is primarily of interest when $\lambda/m < \mu$, i.e., when the mean number of samples submitted for analysis is smaller than the analysis capacity. If this is not the case the queue will grow indefinitely.

The quantity $\lambda/m\mu = \rho$ is called the traffic intensity or utilization factor and plays an important role in computations, as is shown later. When $\rho < 1$, queueing analysis can be carried out and it has been shown that in this instance a steady state is reached. This means that after a certain initial time, needed to establish the steady state, a situation is reached in which one is able to predict the values of the parameters $\bar{t}_a$ and $\bar{n}_q$. For the calculation of $\bar{t}_a$ and $\bar{n}_q$, certain assumptions about the distributions are necessary, however.

The following distributions of interarrival time and service time can be distinguished : exponential, (M) ; r-stage Erlangian, (E_r) ; R-stage hyperexponential, (H_R) ; deterministic, (D) ; general, (G). The exponential distribution of the service time, for instance, is given by

$$p(t_a) = \mu\, e^{-\mu t_a}$$

Queues are described with a shorthand notation such as A/B/m, where A, B and m represent the distributions for interarrival time and service time and the number of channels. For example, in the M/M/1 system both the interarrival time and the service time are exponentially distributed and there is only one service channel. This system is the simplest one and is the one that is usually discussed in introductory texts on queueing theory. The average waiting time and distribution of the waiting time and the influence of priority rules are easily calculated.

The mean waiting times for G/M/1, M/G/1 and G/M/m systems are also easily assessed. Problems arise, however, for G/G/1, G/G/m and M/G/m systems for which one has only approximating formulae.

Clearly, the applicability of queueing theory depends strongly on the distribution functions of interarrival time and analysis time. It also depends on the complexity of priority rules and of the laboratory model (see Chapter 30 for a discussion of laboratory models).

21.3.2. The M/M/1 and M/M/n systems

In the simplest model (M/M/1), the following assumptions are made concerning the distribution of the arrival rate and the service times.

(a) The number of arrivals during an interval has a Poisson distribution. This hypothesis implies that the probability of n arrivals in an interval (0,t) is given by

$$P_n(t) = \frac{e^{-\lambda t}\,(\lambda t)^n}{n!} \qquad (21.5)$$

where λt is the average number of arrivals during this interval.

The result for the probability of n arrivals in a Poisson distribution can be obtained from the three basic assumptions of this distribution :

(i) The probability of a single arrival during a small time interval, Δt, is proportional to the length of the interval. It is given by $\lambda \Delta t + o(\Delta t)$ where λ is the parameter of the Poisson distribution and $o(\Delta t)$ is a small value which becomes negligible for small Δt. In terms of the probability, $P_1(t)$, it can be seen that eqn. 21.5 implies $\lim_{\Delta t \to 0} \frac{P_1(\Delta t)}{\Delta t} = \lambda$.

(ii) The probability of more than one arrival during a small interval Δt is negligible for small t. As a result of these two assumptions,

$$\lim_{\Delta t \to 0} \left[P_1(\Delta t) + P_0(\Delta t) \right] = 1$$

(iii) The numbers of arrivals during non-overlapping time intervals are statistically independent.

(b) The service time has an exponential distribution. This hypothesis means that the probability that the service time equals t is given by $\mu e^{-\mu t}$, where μ is the parameter of the exponential distribution. Further, it is assumed that times between arrivals and service times are independent.

To describe the state of the system at a given time t, we shall use the following definitions.

The number of elements present in the system at an instant t is called N(t). The probability that there are n elements at an instant t is called $p_n(t)$. The study of queueing systems is mainly concerned with the behaviour of the system in a state of equilibrium. At this point, the probabilities $p_n(t)$ do not depend on t and are called p_n. In the single-server queue, the condition for reaching such a state is given by

$$\rho = \frac{\lambda}{\mu} < 1$$

The main results which can be obtained from these assumptions concern a number of parameters which describe the expected way the system will behave. For the simple model described above, we shall give formulae for the following parameters :

p_n = probability for n customers to be present in the system ;

$\overline{n}$ = mean number of customers present in the system ;

$\overline{n}_q$ = mean number of customers present in the queue ;

$\overline{w}$ = mean waiting time ;

$p(W \leqslant A)$ = probability that the waiting time is smaller than or equal to A ;

$$p_n = \rho^n(1-\rho) \qquad n = 0, 1, 2, \ldots \tag{21.6}$$

$$\overline{n} = \frac{\rho}{1-\rho} \qquad \overline{n}_q = \frac{\rho^2}{1-\rho} \qquad \overline{w} = \frac{\rho}{\mu(1-\rho)} \tag{21.7}$$

$$P(W \leqslant A) = 1 - \rho\, e^{-\mu A(1-\rho)} \tag{21.8}$$

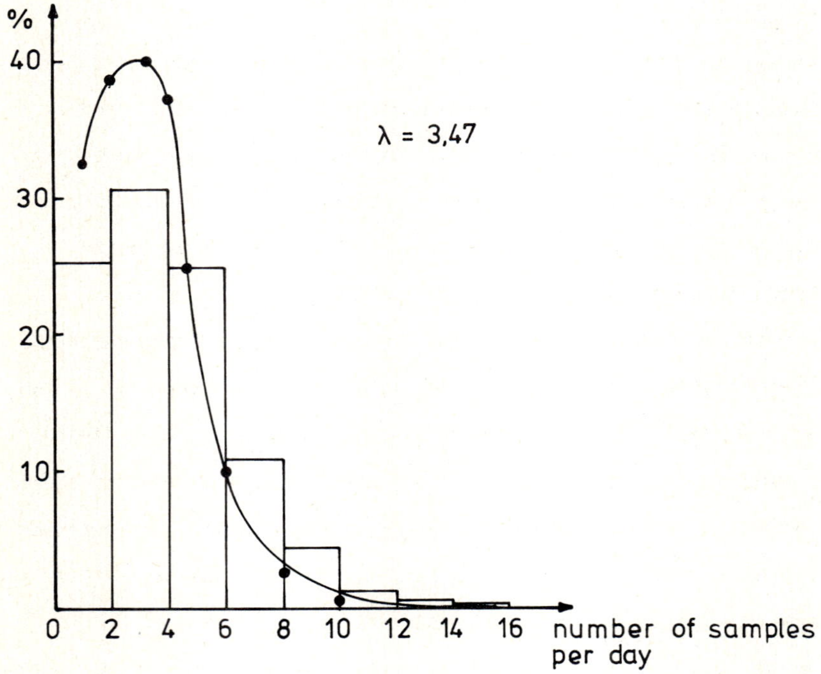

Fig. 21.2. Distribution of the number of samples per day obtained in a laboratory for structural analysis over a period of 250 days compared with a theoretical Poisson distribution with $\lambda = 3.47$ (Vandeginste, 1978).

A laboratory usually consists of several service posts. For example, the samples have to be centrifuged (service post 1) before being distributed over several apparatuses for determination of the concentration (service posts 2, ..., m). Jackson (1957) demonstrated that if some service node in a network of service points receives samples from various sources (i), each with a Poisson input rate λ_i, in general the total input will not be a Poisson process, but this node still behaves as if it is an M/M/n system with input rate $\Sigma\lambda_i$. Therefore, many systems can be investigated using introductory queueing theory.

It remains to be shown that the rate of arrival of samples in an analytical laboratory follows a Poisson distribution. Vandeginste (1978) has shown that this is the case in at least some laboratories. Fig. 21.2 gives the distribution of arrival rates in a laboratory for structural analysis. The actual distribution can be represented by a Poisson distribution with $\lambda = 3.47$.

21.3.3. Applications in analytical chemistry

The literature on applications of queueing theory in analytical chemistry is restricted to a rather general introduction by Adeberg and Doerffel (1975). In this section, the more important conclusions of Vandeginste (1978) are given.

One should observe first that not all laboratories can be studied with queueing theory. The sample input of some laboratories, such as clinical and some industrial control laboratories, is time dependent, and the time elapsed between sampling and producing the analytical result may vary from a few hours to a day. In the early morning the laboratory is almost empty, and by the evening all samples received are completed. In terms of queueing theory, such laboratories never reach a "steady state". The solution of non-stationary queues requires complex mathematics, and to our knowledge, up to now, queues in such kinds of laboratories have never been calculated.

Other laboratories, for instance analytical departments in research laboratories, receive a more or less continuous flow of samples. This flow can be described in terms of statistical parameters such as the mean and the variance of the interarrival time of the samples. If these parameters, together with the parameters describing the statistical behaviour of the analysis time, remain constant for a sufficiently long period, a steady state will be observed, allowing the application of queueing theory. However, this is only true for fairly simple systems and it is obvious that such complex systems as real analytical laboratories cannot be described easily by the rather simple models on which queueing theory is based. Therefore, digital simulation which will be discussed in section 21.4 can be considered as an alternative method for handling more complex models. Nevertheless, queueing theory provides some interesting conclusions about delay times in analytical laboratories, and some examples of general interest will be discussed.

(a) Fluctuations of the analysis time

From eqn. 21.8, one concludes that, for exponentially distributed interarrival times of the samples and analysis times, the delay time shows an exponential distribution. Other distributions of interarrival and analysis time also give rise to an exponential distribution. This means that the waiting time for the results of some samples is much longer than the average waiting time. The statistical nature of both times causes the delay of some samples to be much longer than the average delay.

One of the most powerful means of controlling the average delay is to control the probability distribution of the analysis time. Kleinrock (1975 a) demonstrated that systems without statistical fluctuations of the analysis time show half the waiting time of a M/M/1 system. The magnitude of the fluctuations of the analysis time can be expressed as the coefficient of variation, C_b, which is the ratio of the standard deviation and the mean of the probability distribution function. The influence of this coefficient on the waiting time is given by the well known Pollaczek-Khinchin mean-value equations :

$$\frac{\overline{w}}{\overline{t}_a} = \frac{\rho(1+C_b^2)}{2(1-\rho)} \tag{21.9}$$

$$\frac{\overline{T}}{\overline{t}_a} = 1 + \frac{\rho(1+C_b^2)}{2(1-\rho)} \tag{21.10}$$

where $\overline{T}$ is the delay time ($\overline{T} = \overline{w} + \overline{t}_a$).

Statistical fluctuations in the analysis time can be eliminated or reduced by standardizing the manipulations or by introducing automated procedures. The effect of doing this can therefore be predicted by using eqns. 21.9 and 21.10.

(b) Influence of the analysis time

It is obvious from eqn. 21.7 that the utilization factor has an important effect on the mean delay in the laboratory. As ρ approaches unity, the average time in the laboratory grows in an unbounded fashion. In some laboratories it is common practice to analyse all samples twice, in order to detect outliers. Assuming that a duplicate analysis requires double the time of a single analysis, a considerable decrease in the average delay is observed if the second analysis is omitted. For example, for a system with $\rho = 0.9$ and an average analysis time of 0.5 h, the average delay time decreases from 10 to 0.9 h. The analyst should then decide whether the decrease in the average delay is worth the increased probability of obtaining outliers.

(c) The number of channels of the apparatus

Consider an instrument with m service channels. When the instrument becomes available, it will accept m samples from the queue and analyse them simultaneously. If the analyst finds less than m samples in the queue, then this number of samples will be analysed. For this particular instance, the change in delay time on increasing the capacity of the instrument, assuming a constant analysis time, can be calculated from the equations given by Kleinrock (1975 b). Fig. 21.3 demonstrates the effect of increasing the instrument capacity for a system with $\rho = 0.9$.

(d) The number of analysts

Suppose that two analysts perform the same analytical procedure. From the equations of an M/M/m queueing system (Kleinrock, 1975, c), the effect of admitting a third analyst can be calculated as a function of ρ. From Fig. 21.4, one can see that for $\rho = 1.4$, that is both analysts are 70% employed, the admittance of a third analyst reduces the delay time to 47% of the initial value.

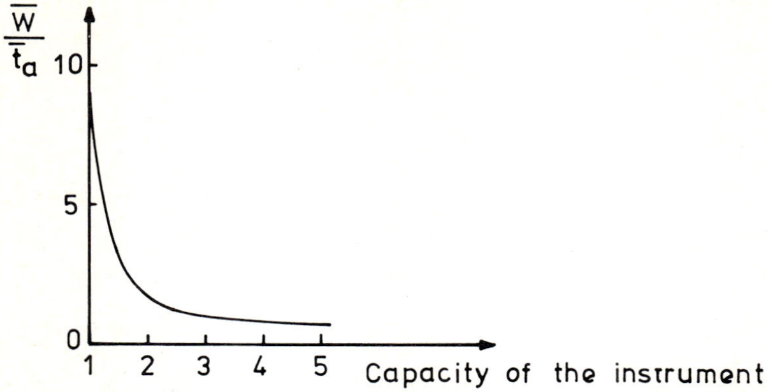

Fig. 21.3. Waiting time in units of average analysis time as a function of the capacity of the instrument (ρ = 0.9).

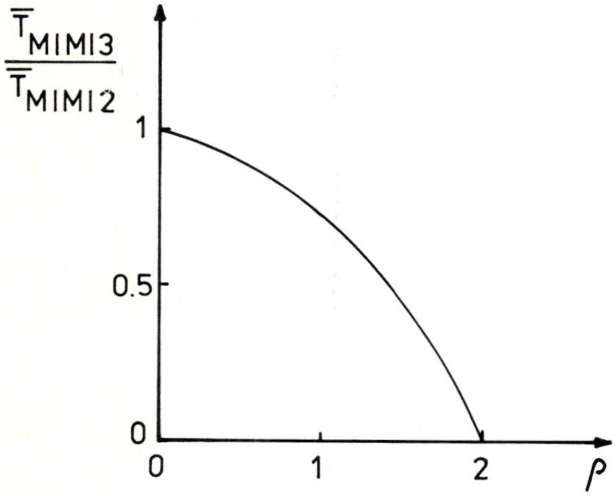

Fig. 21.4. Relative decrease in the average delay time by increasing the number of analysts from 2 to 3 as a function of the utilization factor ρ.

(e) Priorities

The samples submitted to a laboratory do not necessarily have the same priority. For example, urgent samples in a clinical laboratory are positioned at the top of the queues and are analysed before all samples of lower priority, irrespective of their delay time.

For equal average analysis times of both priority classes, the total average delay time of the samples is not affected by a priority rule, but a great difference in average delay may be observed between the classes. It can be derived (Kleinrock, 1976) that for absolute priorities in an M/M/1 system the delay time of the high- and low-priority classes are $(\rho_1 \bar{t}_{a1} + \rho_2 \bar{t}_{a2}) / (1 - \rho_1)$ and $(\rho_1 \bar{t}_{a1} + \rho_2 \bar{t}_{a2}) / (1 - \rho_1)(1 - \rho)$, respectively, where subscript 1 indicates the high priority class and $\rho = \rho_1 + \rho_2$. From Fig. 21.5, it can easily be seen that an increased delivery of samples of the low-priority class has a very small effect on the delay time of the high-priority samples but has a strong effect on the delay time of the low-priority samples. It can also be shown that an increased delivery of high-priority samples affects the delay time of both classes.

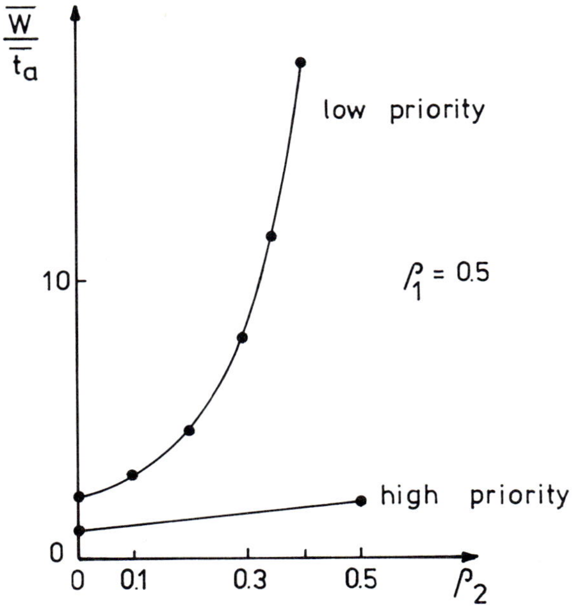

Fig. 21.5. Waiting time in units of average analysis time for two priority classes under absolute priority discipline as a function of the utilization factor ρ_2 for the samples of low priority.

All of the examples presented above are calculated for open systems. These are systems for which the interarrival times of the samples do not depend on the delay times. However, analytical laboratories often interact with their environment and form a closed system with it. When the investigator receives the analytical results, he starts new experiments and sends new samples to the analytical laboratory. The time lag between two samples (or sample series) consists of the delay time of the result and the time needed to react on the result. Decreasing the delay time of the result in a closed system diminishes the average interarrival time of the samples. As a consequence, the expected reduction in the delay times is not obtained. However, the throughput of the laboratory is increased.

Delay times can be considered to be performance characteristics of the analytical laboratory. However, this criterion is interrelated with another criterion, namely cost. Enhancement of the equipment (technicians and instruments) decreases the delay times of the samples. As a result the operation of the laboratory is more expensive, but the gross profit increases. A maximal net profit is then obtained for some mean delay time (Fig. 21.6).

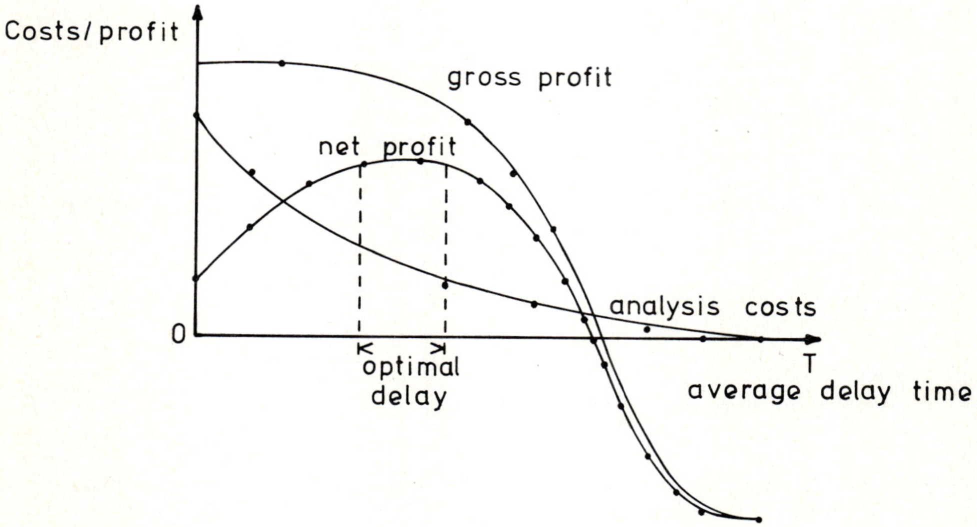

Fig. 21.6. Cost/profit relationships.

21.4. SIMULATION TECHNIQUES

21.4.1. Introduction

We mentioned above (section 21.3) that for complex queueing models with general distributions of the interarrival time and analysis time, analytical solutions cannot be obtained or can be obtained only with difficulty by the application of queueing theory. It is obvious that the alternative way, i.e. performing experiments with the real laboratory system, cannot be considered because this would be too expensive and time consuming and might even lead to chaos. In order to study the behaviour of the system, one can simulate it with a model, which may be physical, verbal, pictorial or mathematical. For simulation on a digital computer a mathematical model is required. The behaviour of the laboratory system is then simulated over a long period under stochastic and/or dynamic circumstances. Dynamic stochastic models are used because the interarrival time and analysis time are defined by their distribution function and because the interactions between variables are time dependent, e.g., the efficiency of the analyst may depend on the number of samples waiting in the laboratory.

There are other reasons for using simulation. First, the model of the system can be altered in order to investigate alternative systems and the effects of the alterations. For example, in a spectroscopic laboratory (IR, NMR and mass spectrometry) the analyst may be responsible for both the acquisition and the interpretation of the spectra or, in an alternative model, operators may run the spectra, which are interpreted by specialists. Secondly, the detailed observation of the system, which is necessary to construct the model, provides in itself a better understanding of the system. Valuable suggestions for improving the system organization can often be made, even before simulation with the model has been carried out. Therefore, the stage of constructing the model is as important as the execution of the experiments themselves.

Building a computer model of a system and performing meaningful experiments with it is not easy. It requires a knowledge of computer programming, statistics, probability theory and experimental optimization techniques. Furthermore, research entirely by experimental methods is a slow and difficult process, even under the ideal conditions of control which simulation provides. Because in a simulation model a great number of variables is involved, a good experimental optimization method is very important in order to obtain the desired information.

Computer simulation experiments usually consist of the following stages (Naylor et al., 1966).

(1) Formulation of the problem. This consists in the formulation of the questions to be answered, the hypothesis to be tested or the effects to be calculated. For example, should one add an instrument or an analyst when the number of samples increases by a certain extent ? What is the increase in the throughput of samples on increasing the number of technicians ? What is the effect of automated data processing ?

(2) Collection and processing of laboratory data. In order to formulate the problem exactly, some primary observations of the system have to be made. From detailed observations, the value of a number of parameters and variables must be determined, such as the probability function of the analysis time and interarrival time of the samples, the mean down-time of the instruments and the mean time between instrument failure. Some of these observations, such as those concerning the arrival of samples and the delivery of analytical results, can be obtained from the administration of the laboratory. Other data can be obtained from interviews with the analysts, e.g., to find out which priority policies are used to choose an analytical method in a laboratory.

(3) Formulation of the mathematical model. This is the most difficult and time-consuming stage of computer simulation because here all variables to be included in the model must be defined. The variables are selected on the basis of an estimate of their relative importance. If one or more important variables are missed, the simulation results are inaccurate. On the other hand the inclusion of too many variables renders the computer simulation needlessly complex. Furthermore, it is necessary to build the model as efficiently as possible in order to obtain accurate results with a minimum of effort. Therefore, various computer languages have been developed especially for programming simulation models such as GPSS (1970), Simscript (Markovitz et al., 1962) and GASP (Kiviat, 1963). GPSS and GASP are typically languages used for the simulation of queueing and scheduling systems and are therefore suitable for the simulation of laboratory systems.

(4) Estimation of the parameters. The moments of the distributions of the analysis time and interarrival time must be estimated. Furthermore, the correlation between successive interarrival times must be investigated by calculating the autocorrelation function. After the inclusion of the estimates in the model, the resulting theoretical distributions must be compared with the experimental values from the laboratory. To do this, various statistical tests (such as the χ^2 test, Chapter 8) can be used.

(5) Validation of the model. One must decide whether the results obtained by simulation are accurate. Some assurance of validity would be provided by a demonstration that for at least one alternative version of the simulated system and one set of conditions the model produces results that are consistent with the known performance of the laboratory. The simulation experiments must be designed in such a way that the fluctuations of the results are minimal, e.g., by using variance reduction techniques (Mitchell, 1973). The statistical analysis of simulated data is often more difficult than for real data, because of the large number of parameters and variables and the fact that the results are merely correlated, non-stationary time series.

From all this, it is clear that the simulation of laboratories should not be undertaken lightly. Moreover, simulation is a slow and difficult process, which can succeed only if there is enough multidisciplinary knowledge available. A survey providing a theoretical background of digital simulation was given by Naylor et al. (1968).

21.4.2. Applications in analytical chemistry

Clearly, the quantitative study of waiting-line situations in analytical laboratories should permit a better use of the capacity of laboratories and the reduction of delays. However, only a few studies have been published on laboratory activities using simulation methods. To date, reports of simulation studies with particular reference to laboratory activities are only known from Schmidt (1976, 1977), Delon and Smalley (1969), Rath et al. (1970) and Väänänen et al.

(1974). The last author published the results of the simulation of a clinical laboratory, including 64 different analytical methods and 37 different instruments. For some of the analytical procedures more than one instrument was available. Analysis could be carried out in a single batch, with a variable number of samples or several batches had to be started at different times. The study by Väänänen et al. was concerned essentially with the effect on the delay time of two key factors, namely the number of laboratory technicians and the number of specimens analysed. They calculated the effect of an increase in the number of analysts and found that if this bottleneck is eliminated the number of instruments determines the waiting time for specimen batches. The computer program was written in GPSS and proved to be generally useful in its application to other laboratories. The results of the simulation study have been applied in a rationalization of the work in the laboratory in which this study was performed.

The results of Vandeginste (1978), who designed a simulation model of a laboratory for structural analysis (with IR, NMR and mass spectrometry) of a pharmaceutical industry, confirm that a detailed observation of the system makes it possible to propose modifications to the system, yielding a distinct increase in the throughput.

REFERENCES

R.L. Ackof and M.W. Sasieni, Fundamentals of Operations Research, Wiley, New York, 1968.

V. Adeberg and K. Doerffel, Communication presented at the Euroanalysis Congress, Budapest, 1975.

G.L. Delon and H.E. Smalley, Health Serv. Res., 4 (1969) 53.

GPSS, General Purposes Simulation System - 360 (user manual), GH20-326-3, 4th Edition, International Business Machines Corporation, New York, 1970.

G. Hadley, Linear Programming, Addison-Wesley, Reading, 1962.

F.S. Hillier and G.J. Lieberman, Introduction to Operations Research, Holden-Day, San Francisco, 2nd ed., 1974.

J.R. Jackson, Networks of Waiting lines, Operations Res., 5 (1957) 518.

A. Kaufmann, Graphs, Dynamic Programming and Finite games, Academic Press, New York, 1967.

A. Kaufmann, Méthodes et Modèles de la Recherche Opérationelle, Tome II, Dunod, Paris, 2nd ed., 1968.

P.J. Kiviat, GASP - A general activity simulation program, Proj No 90, 17-019 (2), Applied Research Laboratory, United States Steel, Monroeville, Pa. July 8, 1963.

L. Kleinrock, Queueing Systems, Vol I, p 180, Wiley, New York, 1975 a.

L. Kleinrock, Queueing Systems, Vol I, p 137, Wiley, New York, 1975 b.

L. Kleinrock, Queueing Systems, Vol I, p 102, Wiley, New York, 1975 c.

L. Kleinrock, Queueing Systems, Vol II, p 119-126, Wiley, New York, 1976.

H.M. Markowitz, B. Hausner and H.W. Karr, Simscript : A simulation programming language, The RAND Corporation, R.M.-3310, 1962.
B. Mitchell, Operations Res., 21 (1973) 988.
T.H. Naylor, J.L. Balintfy, D.S. Burdick and K. Chu, Computer Simulation Techniques, Wiley, New York, 1966.
G. Owen, Game Theory, W.B. Saunders Company, Philadelphia, Pa. 1968.
G.J. Rath, J.M.A. Balbas, T. Ikeda and O.G. Kennedy, Health Serv. Res., 5 (1970) 25.
B. Schmidt, GIT Fachz.20, (11), (1976) 1167.
B. Schmidt, Z. anal. Chem., 287 (1977) 157.
I.K. Väänänen, S. Kivirikko, J. Koskenniemi, J. Koskimies and A. Relander, Meth. Inform. med., 13 (3) (1974) 158.
B.G.M. Vandeginste, to be published, 1978.
J. Von Neumann and O. Morgenstern, Theory of Games and Economic Behavior, Princeton University Press, Princeton, 1953.
H.M. Wagner, Principles of Operations Research, Prentice Hall, London, 1972.

Chapter 22

PARTIAL ENUMERATION METHODS

22.1. THE OPTIMAL CONFIGURATION OF APPARATUS IN A (CLINICAL) LABORATORY

Routine laboratories such as clinical laboratories are often faced with the problem of the selection of an optimal "configuration" of apparatus. This problem was studied in detail by De Vries (1974), using operational research methods, and was introduced in Chapter 16. By configuration is meant a particular combination of apparatus. Let us consider the simple example of a laboratory that has to carry out only two types of determinations. At least two configurations are then possible : one can choose either two different instruments (or manual methods), one for each type of determination, or a two channel apparatus that carries out both. Other configurations can be introduced if one takes into account, for example, that the two channel apparatus can be completed with a printer or not. The problem is to decide which configuration is cheapest for a particular work load. It is a relatively simple problem to solve if the necessary data (costs, number of analyses per day, etc.) are known, and the number of different determinations is small.

This is no longer true when it is large, as in De Vries' work where this number, n = 18. De Vries (1974) included in his model initially 77, later 59, apparatuses, performing one or different combinations of 2, 3, 4, 6 or 12 different determinations. The number of possible configurations now exceeds 10000, so that one is clearly faced with a problem of a combinatorial nature.

Each apparatus j can be represented by a vector $\vec{m}_j$, $\vec{m}_j = (m_{1j}, \ldots, m_{Ij})$. The elements of this vector take the value 1 when apparatus j can perform determination i and 0 when this is not the case. In this way, a matrix $M = (m_{ij})$ is obtained. An example of such a matrix is shown in Table 22.I.

Table 22.I

M-matrix representing possible apparatus, from De Vries (1974) (only about half of the 59 possibilities given in the original table are reproduced)

Determination i	m_{ij}																																	
1	0	0	0	0	0	0	0	0	0	0	0	1	0	0	0	0	0	0	0	0	0	0	0	0	0	0	0	0	0	0	1	0	0	0	0
2	1	1	1	1	1	1	1	1	1	1	0	0	0	0	0	0	0	0	0	0	0	0	0	0	0	0	0	0	0	0	0	0	0	0	0
3	1	1	1	1	1	1	1	1	0	0	0	0	0	0	0	0	0	0	1	0	0	0	0	0	0	0	0	0	0	0	0	0	0	0	0
4	1	1	0	1	0	1	1	0	1	0	1	0	0	0	0	0	0	0	0	1	0	0	0	0	0	0	0	0	0	0	0	0	0	0	0
5	1	1	1	1	0	1	0	0	0	0	1	0	0	0	0	0	0	0	0	0	1	0	0	0	0	0	0	0	0	0	0	0	0	0	0
6	1	0	0	0	1	0	0	0	0	0	1	0	0	0	1	0	0	0	0	0	0	1	0	0	0	0	0	0	0	0	0	0	0	0	0
7	1	0	0	0	1	0	0	0	0	0	0	0	0	0	1	0	0	0	0	0	0	0	1	0	0	0	0	0	0	0	0	0	0	0	0
8	1	0	0	0	1	0	0	0	0	0	1	0	0	1	1	0	0	0	0	0	0	0	0	1	0	0	0	0	0	0	0	0	0	0	0
9	1	0	0	1	0	0	0	0	0	0	1	0	0	1	0	0	0	0	0	0	0	0	0	0	1	0	0	0	0	0	0	0	0	0	0
10	1	0	1	0	0	0	0	0	0	0	1	0	0	1	0	1	0	0	0	0	0	0	0	0	0	1	0	0	0	0	0	0	0	0	0
11	1	1	1	1	1	0	0	0	0	0	1	0	0	0	0	1	0	0	0	0	0	0	0	0	0	0	1	0	0	0	0	0	0	0	0
12	1	1	0	1	1	0	0	0	0	0	0	0	0	0	0	0	0	1	0	0	0	0	0	0	0	0	0	1	0	0	0	0	0	0	0
13	1	0	1	1	1	0	0	0	0	0	1	0	0	0	0	0	1	0	0	0	0	0	0	0	0	0	0	0	1	0	0	0	0	0	0
14	0	0	1	0	0	0	0	0	0	0	0	1	0	0	0	0	1	1	0	0	0	0	0	0	0	0	0	0	0	1	0	0	0	0	0
15	0	0	1	0	0	0	0	0	0	0	1	1	0	0	0	0	0	0	0	0	0	0	0	0	0	0	0	0	0	0	0	1	0	0	0
16	0	0	0	0	0	0	0	0	0	0	1	0	1	0	0	0	0	0	0	0	0	0	0	0	0	0	0	0	0	0	0	0	1	0	0
17	0	0	0	0	0	0	0	0	0	0	1	0	1	0	0	0	0	0	0	0	0	0	0	0	0	0	0	0	0	0	0	0	0	1	0
18	0	0	0	0	0	0	0	0	0	0	1	0	1	0	0	0	0	0	0	0	0	0	0	0	0	0	0	0	0	0	0	0	0	0	1
j =	3	4	5	6	7	8	9	10	11	12	13	23	24	25	26	40	41	42	43	44	45	46	47	48	49	50	51	52	53	54	55	56	57	58	59

The following costs are calculated for each apparatus :

C_j : fixed costs (i.e., costs independent of the number of runs, such as depreciation and maintenance costs per day) ;

T_j : labour costs per run ;

R_j : reagent and other variable costs per run.

A run is defined here as the carrying out of one determination of each type that is possible with a particular apparatus.

The daily cost for N_j runs with apparatus j is then $C_j + N_j (T_j + R_j)$. To take into account the fact that the output of a clinical laboratory usually increases from year to year, a proportionality constant, f, is introduced. If the number of requests for analyses doubles, then f = 2. The cost of introducing a particular apparatus j in the configuration is then given by $C_j + f\, N_j (T_j + R_j)$ and the total cost, K, for a configuration is given by

$$K = \sum_{j=1}^{J} y_j \left[(C_j + f\, N_j (T_j + R_j)\right] \qquad (22.1)$$

where J is the total number of apparatuses considered (here 77 or 59) and y_j is a 0 - 1 variable, equal to 1 when apparatus j is part of the configuration considered and 0 when it is not. K is the economic function that has to be minimized. The minimization problem is, however, subject to some constraints, as follows :

(a) each type of determination should be carried out with only one apparatus ; this can be written as $\sum_{j=1}^{J} m_{ij} \; y_j \leqslant 1$ for each i ;

(b) each type of determination must be carried out, which can be written as

$\sum_{j=1}^{J} m_{ij} \; y_j \geqslant 1$ for each i ;

(c) the capacity L_j (i.e., the maximal number of runs per day) of each apparatus in the configuration should not be exceeded or $y_j\, f\, N_j \leqslant y_j\, L_j$.

The minimization problem can now be summarized as: minimize K, subject to the constraints

$$\sum_{j=1}^{J} m_{ij}\, y_j = 1 \qquad i = 1, \ldots, I \qquad (22.2)$$

$$y_j\, f\, N_j \leqslant y_j\, L_j \qquad j = 1, \ldots, J \qquad (22.3)$$

$$y_j \in \{0,1\} \qquad j = 1, \ldots, J \qquad (22.4)$$

A typical result is given in Table 22.II.

Table 22.II

Cheapest configuration as a function of f (from De Vries, 1974)

f	configuration $(j \mid y_j = 1)$
0.2	12,43-59
0.3	9,24,45-56
0.4	7,24,44,45,49,50,54,55,56
0.5	7,23,39,44,45,49,50,54
0.6	7,23,24,44,45,49,50
1.0	6,23,24,26,50
7.4	3,23,24

The most economic configuration for f = 1 is found to consist of apparatus 6, 23, 24, 26, 50. From Table 22.I, one finds that the best configuration is therefore an eight-channel apparatus (j = 6), three three-channel apparatuses (j = 23, 24, 26) and one manual method (j = 50). One obtains the expected result that at low values of f (low work load) only manual or single-channel methods are selected, while for a very high work load a twelve-channel and two three-channel apparatuses are preferred.

The results of De Vries have been subject to controversy in the Dutch literature (Chem. Weekbl., 1975). There are no doubt some shortcomings in the proposed model and in the method of calculating costs but, unsurprisingly, one of the more important arguments against the model is that it does not take into account all of the relevant factors. As was discussed in more detail in Chapter 16, this is due to a misunderstanding between the operational research specialist and the user (here the practising clinical chemist). It is rarely possible to take into account all factors in a model and the results obtained should therefore be considered as guidelines and not as the absolute truth.

22.2. SELECTION OF REPRESENTATIVE PATTERNS

22.2.1. The problem

The selection of representative patterns has rarely been stated in a formal way in analytical chemistry, but it is a common problem. Two examples will show this :

(1) To develop optimal solvents for the high-performance liquid chromatographic (HPLC) separation of a large group of substances such as basic drugs, it is impossible to try out every possible solvent for each basic drug. Possible solvent systems are therefore screened with a few of the drugs, which should be representative of the chromatographic behaviour of the basic drugs. The problem is then the selection of these drugs. The chromatographic properties of a substance can be depicted by a pattern vector whose elements are retention times for a set of solvents. If this vector has d dimensions, then each substance is a point in d-dimensional space. Probably a number of clusters will form and representative substances should then come from different clusters.

(2) As appears from the first example, the selection of representative patterns is a clustering (unsupervized learning) problem. It consists in a search for the centroids of the clusters. Let us consider again the work of Leary et al. (1973), cited in section 20.3.1.2. They selected 12 preferred GLC phases from 226 and classified each of the 214 other phases in one of the groups of which one of the 12 preferred phases is the nucleus. This classification was carried out on the basis of their retention indices with 10 substances (probes). The 12 preferred phases were chosen because they are among the most commonly employed and well tested phases of the 226. A more objective method from the classification point of view would have been to look for the 12 most representative 10-dimensional patterns among the 226, i.e., to separate the 10-dimensional space into 12 clusters, the centroids of which are sought.

An OR method for solving problems of this type is proposed here. To explain it, let us first turn to the non-chemical problem with which this method was introduced (Massart and Kaufman, 1975).

In a region of which Fig. 22.1 is a map, there are 10 villages and one has to locate p supermarkets in this region. The supermarkets must be located in villages and the total distance from the villages to the nearest supermarket should be as small as possible. Clearly, if two supermarkets are wanted (p = 2), these will be located in A and B. The effect is to split up the region in two parts, in each of which five villages are situated (equal numbers are not necessary, however). A so-called 2-median has been found. One can also say that two clusters have been isolated, the centroids of which are A and B.

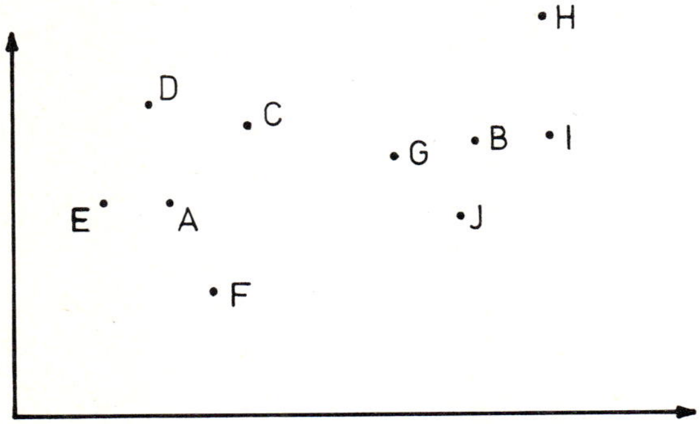

Fig. 22.1. Illustration of location model.
From Massart and Kaufman, 1975. Reprinted with permission. Copyright by the American Chemical Society.

Suppose now that A, B, ..., J are chemical substances and that the map is in fact a graph with the retention times in solvent 1 on the abscissa and those in so̅vent 2 on the ordinate, and that two representative substances should be chosen on the basis of these retention times ; it is evident that these substances should be A and B. It is also evident that it is not so easy to select three representative substances, and a mathematical formulation is then necessary.

22.2.2. The mathematical model

The general non-chemical problem can be stated as follows : "for a finite number of users, whose demands for a given service are known and must be fulfilled and a finite set of possible locations where a given number, p, of service centres may be located, select the locations of the service centres in order to minimize the sum of transportation costs of the users" (De Clercq et al., 1976). The equivalent (chemical or non-chemical) clustering problem is then : "from a finite number n of patterns represented by pattern vectors, select a given number, p, of representative patterns in order to minimize the sum of the distances of the patterns to the representative patterns". Mathematically, the following model is obtained :

Minimize

$$\sum_{i}\sum_{j} d_{ij}\, x_{ij} \qquad (22.5)$$

subject to

$$\sum_{i} x_{ij} = 1 \qquad (22.6)$$

$$x_{ij} \leqslant y_i \qquad (22.7)$$

$$\sum_{i} y_i = p \qquad (22.8)$$

$$y_i \in \{0,1\} \qquad (22.9)$$

$$x_{ij} \in \{0,1\} \qquad (22.10)$$

where

$i = 1, \ldots, n$ and $j = 1, \ldots, n$;

p = number of representative patterns (or probes) ;

d_{ij} = distance between substance j and probe i ;

x_{ij} = a variable that determines which probe is representative of substance j ;

$x_{ij} = 1$ if j is closest to probe i and is therefore represented by i and

$x_{ij} = 0$ when this is not the case ;

y_i = a variable that determines whether a substance is selected as a probe ;

$y_i = 1$ when this is the case and $y_i = 0$ when it is not.

The solution can be obtained by an heuristic method or by a partial enumeration method, called branch and bound.

Both the clinical laboratory model given by eqns. 22.1 - 22.5 and the location model given by eqns. 22.5 - 22.10 are examples of integer programs.

These programs are characterized by the presence of variables that can take only a limited number of values. In the two problems considered, the variables y_j and x_{ij} can take only the values 0 or 1. To find the optimal solution of such a program, it is sufficient to consider all possible combinations of values that can be taken by the different variables. Such a method is called a complete enumeration of the solutions. Unfortunately, for problems of a realistic size, such a method is not feasible. Consider, for example, a problem with 30 0 - 1 variables. For such a problem there are $2^{30} \approx 10^{10}$ different possible combinations, which is a large number even for an electronic calculator.

For this reason a new type of method was introduced for which the enumeration of the set of solutions can be reduced by using some mathematical properties of the economic function to be optimized and of the constraints of the problem. Such methods are called partial enumeration methods. One of these, the branch and bound method, will now be outlined briefly.

Suppose that an objective function is to be minimized and assume that a solution is available (this solution was given, for example, by an heuristic method). Firstly the set of all solutions is divided into several subsets (branch). Then, for each of these subsets a lower bound is computed (i.e., a value at most equal to the smallest value that could be taken by the economic or objective function for the solutions of this subset). If the lower bound of a subset is larger than the value of the best solution already known, this subset is excluded from further consideration. Indeed, none of its solutions can then be better than the solution already known.

Subsequently, one of the remaining subsets is selected for partitioning into smaller subsets. Lower bounds are computed for the subsets and the process is repeated until a subset contains only one element. If its value is smaller than the value of the previous best solution, it replaces this solution. If not, it can also be excluded. When all subsets have been excluded, the best solution so far is the optimal solution.

To explain the branch and bound method, we consider a very simple example, concerning the determination of a pesticide. Starting with a complex mixture (node 0), one can either carry out an extraction, taking 5 units of time (node 1), or a partition chromatography (node $\overline{1}$), which takes 12 time units. After the extraction, one can either back-extract (node 2, 5 time units, 10 cumulative units) or carry out a column clean-up (node $\overline{2}$, 8 time units, 13 cumulative time units). After the partition chromatography, one also has two possibilities, called 2 and $\overline{2}$, taking 11 and 12 time units, respectively. There are now 4 possibilities. Each of these gives rise again to 2 possibilities, etc. The tree depicting all of the possibilities leading to the determination is then given by

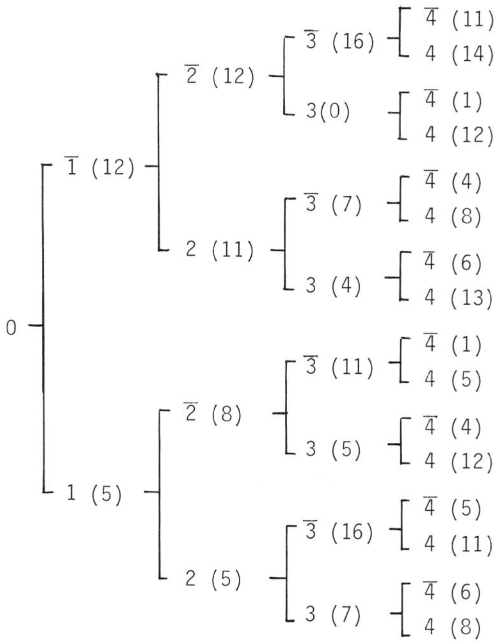

Route 1, 2, 3, 4, for example, depicts a procedure starting with the extraction followed by back-extraction and two other steps. The values in parentheses are the times for each step. For route 1, 2, 3, 4 the total time required is 5 + 5 + 7 + 8 = 25 units. In this simple example, it is easy to verify that the shortest route is represented by 1, $\overline{2}$, 3, $\overline{4}$ (22 time units). Branch and bound methods permit one to arrive at this conclusion without calculating the cumulative time (or cost, etc.) for each node in the tree. This is not necessary for the pesticide example, but it is for the clinical laboratory and GLC probe problems where the number of possible combinations is very high.

There are several ways of applying the branch and bound method to the pesticide example, but we shall consider only one. One first tries to obtain a good (but not necessarily optimal) solution. For example, one can choose the lowest time from the first two alternatives (1, 5 time units), go from there to the node which requires the lowest cumulative time (2, cumulative time 10), from there again to the node with lowest cumulative time (3, cumulative time 17), etc. One then arrives at a good solution (here 1, 2, 3, $\overline{4}$, cumulative time 23), which is considered as a first solution. This is depicted below. The values in parentheses are now cumulative times.

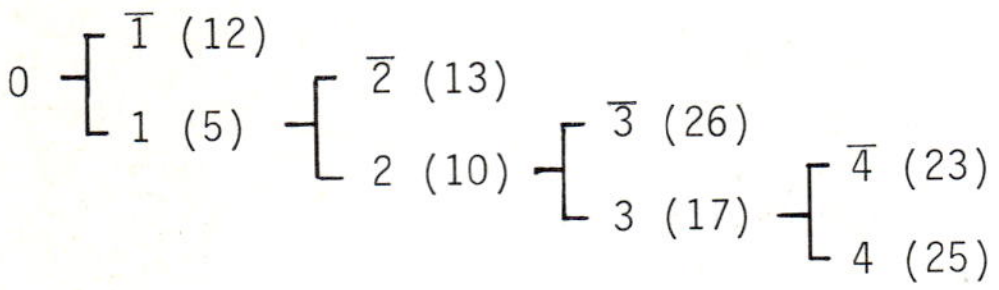

One now works backwards : each node for which the cumulative time equals or exceeds 23 is rejected, and the others are investigated by branching out (from which the name of the technique, branch and bound, is derived). Node 3 (cumulative time 17) leads to the solution functioning as a bound and to node 4, which requires more time. Therefore, one goes back to 2. From there, one can reach $\overline{3}$ with a cumulative time of 26. This is the (lower) bound on the solutions which start with 0 1 2 $\overline{3}$. As this lower bound is larger than the value of the best solution found so far (23), the value of each solution starting with 0 1 2 $\overline{3}$ must exceed 23 and this set of solutions can be discarded. Therefore, one goes back to 1. From there one can branch out to $\overline{2}$, cumulative time 13. This is further investigated, leading to the following situation

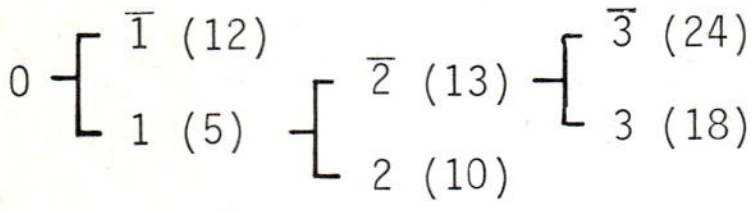

Node $\overline{3}$ exceeds the bound, but node 3 does not, so that one branches out again

0 → $\overline{1}$ (12), 1 (5)
1 (5) → $\overline{2}$ (13), 2 (10)
$\overline{2}$ (13) → $\overline{3}$ (24), 3 (18)
3 (18) → $\overline{4}$ (22), 4 (30)

Route 1, $\overline{2}$, 3, $\overline{4}$ constitutes a new and better solution and is therefore used as a new provisional optimal solution. One now works backwards again. It is not necessary to investigate routes branching out from $\overline{3}$ (which exceeds the bound) and 2 (already investigated), but one must investigate $\overline{1}$. This leads to nodes with costs of 24 and 23, thereby exceeding the bound. No other possibilities can now be investigated and the provisional solution is therefore the optimal one.

One observes that it is possible through the use of this method to eliminate many nodes from consideration. A further discussion of branch and bound methods, especially as they are applied in integer programming, can be found in the book by Zionts (1974).

It should be observed that in analytical chemistry all the times for the steps constituting the decision tree are not necessarily known. However, one can plan the experimental development using the branch and bound principle, i.e. one can first carry out 1 and $\overline{1}$ and determine the times necessary for carrying out these steps. After observing that 1 is the shortest one (or the one with the highest yield or the cheapest), one carries out the alternatives 2 and $\overline{2}$, following 1, etc. in exactly the same sequence as described higher. In this way it will not be necessary to investigate for example route $\overline{1}\ \overline{2}\ \overline{3}\ \overline{4}$.

22.2.3. An application : the selection of GLC probes

In Chapter 18, among others, we have seen that GLC stationary phases are characterized by a pattern vector, the elements of which are the retention times of probe substances. There is a considerable body of literature concerning the choice of these probes, the most succesful being those proposed by Rohrschneider

(1966) and McReynolds (1970). Rohrschneider carried out his selection of five probes from a restricted set of 30 substances and McReynolds selected ten probes from 68 substances. These 30 and 68 substances were themselves, in fact, probes selected from the complete range of known and unknown chemical substances. Rohrschneider and McReynolds carried out their selection on the basis of chemical argumentation. The method described in the preceding section was carried out by De Clercq et al. (1976) to select p = 1, ..., 20 probes from both sets. As the distance between two substances they used 1 - |r|, where r is the correlation coefficient obtained when comparing the 30 retention indices obtained for each of two substances. The results obtained for p = 3, 4 and 5 were compared with those proposed in the literature. Very good concordance of the results was obtained, for example, with the set proposed by Rohrschneider. Rohrschneider proposed ethanol, methyl ethyl ketone, nitromethane, pyridine and benzene while De Clercq et al. proposed ethanol, propionaldehyde, acetonitrile, dioxane and thiophene. Ethanol occurs in both sets of probes and the GLC behaviours of methyl ethyl ketone and propionaldehyde (r = 0.9995), acetonitrile and nitromethane (r = 0.9988), benzene and thiophene (r = 0.9989) and dioxane and pyridine (r = 0.9973) are very similar. This is a very good result for a mathematical procedure in which every form of chemical reasoning is excluded. It should also be noted that in nearly all instances where a comparison could be made, the prediction of retention indices of other substances was better with the probe sets selected by De Clercq than with other sets proposed in the literature.

REFERENCES

Chem. Weekbl., 71 (10) (1975) 31.

H. De Clercq, M. Despontin, L. Kaufman and D.L. Massart, J. Chromatogr., 122 (1976) 535.

T. De Vries, Het Klinisch-Chemisch Laboratorium in Economisch Perspectief, Stenfert-Kroese, Leiden, 1974.

J.J. Leary, J.B. Justice, S.B. Tsuge, S.R. Lowry and T.L. Isenhour, J. Chromatogr. Sci., 11 (1973) 201.

D.L. Massart and L. Kaufman, Anal. Chem., 47 (1975) 1244A.

W.O. McReynolds, J. Chromatogr. Sci., 8 (1970) 685.

L. Rohrschneider, J. Chromatogr., 22 (1966) 6.

S. Zionts, Linear and Integer Programming, Prentice-Hall, Englewood Cliffs, N. J., 1974.

Chapter 23

GRAPH THEORY AND RELATED TECHNIQUES

23.1. GRAPH THEORY : A SHORTEST PATH APPLICATION

A network or graph consists of a set of points (nodes) connected by lines (edges or links). These links can be one-way (one can go from point A to B, but not vice versa) or two-way. When the edges are characterized by values, it is called a weighted graph. In the usual economic problems for which one applies networks, these values are cost, time or distance.

In this section a routing problem is considered, in which one must go from one node (the origin) to another (the terminus). There are many ways by which this is possible and the routing problem consists in finding a path that minimizes (or maximizes) the sum of the values of the edges that constitute the path. This is understood most easily if one supposes that the nodes are towns and the values of the edges between neighbouring towns are the distances along a highway between the towns. The routing problem then consists in choosing the shortest way of going from one town to another distant one. This type of problem is called a shortest or minimal path problem. It has been applied by us to the optimization of chromatographic separation schemes for multicomponent samples (Massart et al., 1972) and more particularly to the ion-exchange separation of samples containing several different ions.

In this type of application, one usually employs more or less rapid and clear-cut separation steps. This can be explained best by considering the simplest possible case, namely the separation of three ions, A, B and C. The original situation is that the three ions have been brought together (not separated) on a chromatographic column and the final situation should be that they are eluted and separated from each other. These two situations constitute the origin and the terminus of the network. They are denoted by ABC// and //A/B/C. The elements which remain on the column are given to the left of symbol

// and the symbol / means that the ions to the left and to the right of it are separated. There are many ways in which one can go from situation ABC// to situation //A/B/C, as shown in the network in Fig. 23.1.

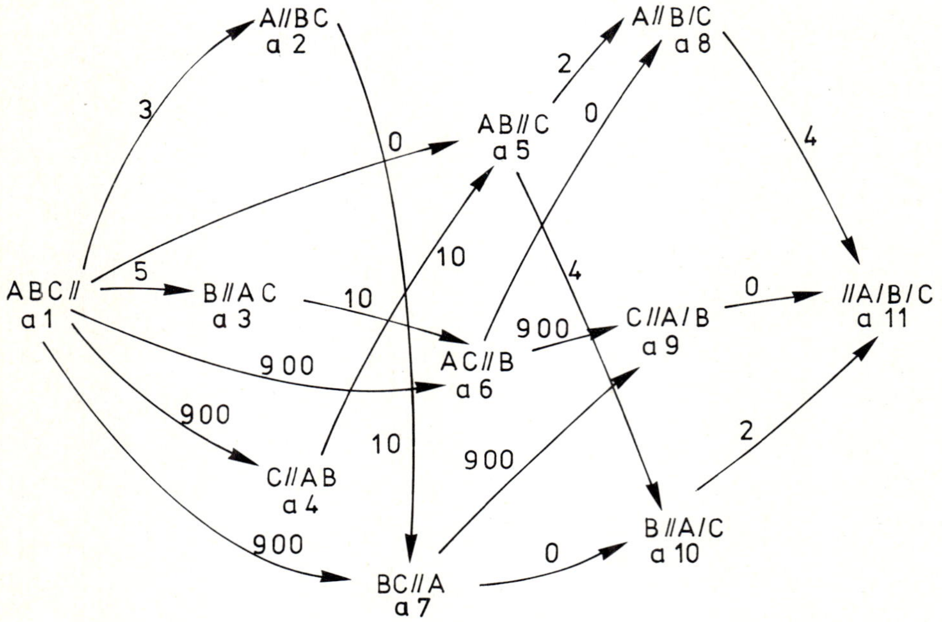

Fig. 23.1. Network describing the separation of three ions, A, B and C (adapted from Massart et al., 1972).

Step 1 : there are two possibilities :

(a) one can elute one element and retain the two others on the column. This leads to nodes AB//C, AC//B, BC//A

(b) one can elute two elements and retain the other. This leads to nodes A//BC, B//AC and C//AB.

Step 2 :

(a) (following step 1a) :

One elutes one of the two remaining ions. For example, if in the first step A was eluted, one now elutes B or C. In step 2a one can reach the situations A//B/C, B//A/C or C//A/B.

(b) (following step 1b) :

Two ions were eluted together and are therefore not separated ; they have to be adsorbed first on another column. In the meantime, one can elute the single ion left on the first column. One then has two elements adsorbed on a column and one eluted. The different possibilities are AB//C, AC//B, BC//A, i.e., the situations reached also after step 1a. From there one proceeds to step 2a.

Step 3 :

Step 3 follows step 2a. Only one ion remains on the column. It is now eluted, so that the terminus is reached.

These different possibilities and their relationships can be depicted as a directed graph (Fig. 23.1) Directed graphs are graphs in which each edge has a specific direction. To find the shortest path, one has to give values to the edges of the graph. As the problem is to find the procedure that permits one to carry out the separation in the shortest time possible, these values should be the times necessary to carry out the steps symbolized by the links in the graph or a variable proportional to the time. We shall not go in detail into the manner in which these times were derived. Essentially, three different possibilities exist :

(a) If the separation depicted by a particular link is possible, the time is considered to be equal to the distribution coefficient of the ion which is the slowest to be eluted in this step (the distribution coefficient as defined in ion exchange chromatography is proportional to the elution time). There is a very large literature on distribution coefficients, particularly for metal ions (it is probable that at least 1000 such coefficients can be found for each ion) and a computer program was used to select the best eluting agent (from nearly 400 possible substances) for each separation step depicted by a link.

(b) If the separation depicted by a particular link is not possible, one gives a very high value to that link.

(c) If the link contains a transfer from one column to another, a high (but not very high) value is given.

One may question whether the application of graph theory is really necessary as no doubt a separation such as that described above can be investigated easily without it. The number of nodes grows very rapidly, however, when more ions are to be separated.

The calculation of the number of nodes is rather complicated (the equations can be found in the original paper). The calculation is simplified in the special case where transfers from one column to another are not allowed (all ions eluted from the column must be completely separated from all of the other ions). In this instance, and for three elements, the following nodes should then be considered : ABC///, AB//C, AC//B, BC//A, A//B/C, B//A/C, C//A/B and //A/B/C. If one considers only the stationary phase (to the left of //), one notes that all the combinations of zero, one, two and three ions out of three are present.

Calling n the total number of ions and p ($0 \leqslant p \leqslant n$) the number of ions in a particular combination taken from these n, this means that the total number of combinations is equal to $\sum_{p=0}^{n} C_n^p$ (where C_n^p is the symbol used for the number of combinations of n elements in sets of p). This can be shown to be equal to 2^n. In this particular instance a separation scheme for eight elements would contain 256 nodes. For the general case (transfers allowed), no less than 17008 nodes would have to be considered ! Even for 4 elements, 38 nodes are obtained and it begins to be difficult to consider all of the possibilities without using graph theory. The shortest (cheapest) path in a graph can be found, for example, with a very simple algorithm due to Ford (Kaufmann, 1968). Let us suppose that one has to construct a highway from a town a_1 to a town a_{11}. There are several possible layouts which are determined by the towns through which one must pass, and these must be selected from a_2 - a_{10}. The values of the links in the resulting graph are given by the estimated costs. The problem is, of course, to find the cheapest route. A value $A_1 = 0$ is assigned to town (node) a_1 and the value of all the nodes a_n directly linked to a_1 is computed by using the equation $A_n = A_1 + l\ (a_1, a_n)$, where $l\ (a_1, a_n)$ is the length of edge (a_1, a_n).

In this way one assigns the values 3, 5, 900, 0, 900 and 900 to the nodes a_2, a_3, a_4, a_5, a_6 and a_7, respectively. One repeats this procedure for the nodes a_m linked directly to one of the nodes a_n by using the equation $A_m = A_n + l(a_n, a_m)$. One continues to do this until a value has been assigned to each of the nodes in the graph. One would, for example, assign the values 15, 1800, 900 and 902 to a_8, a_9, a_{10} and a_{11}, respectively.

In the first stage, one has assigned a possible value to all of the nodes, but not necessarily the lowest possible one. For example, the value A_7 of node a_7 is now 900. The value 900 is artificially high, indicating that for practical reasons it is impossible to go from a_1 to a_7. In the highway example this could mean a mountain ridge and in the ion-exchange case it could be a separation that cannot be carried out in a reasonable time. This value is derived here from edge a_1, a_7 using the equation $A_7 = A_1 + l(a_1, a_7)$. Town a_7, however, can also be reached from town a_2. The value of A_7 is then given by $A_2 + l(a_2, a_7)$ and is equal to 13 ; this replaces the original value 900. In this way, all of the nodes are checked until one is satisfied that each town is reached in the cheapest possible way. The optimal path is then found by retracing the steps that led to the final value for A_{11}. In the graph in Fig. 23.1 this is the path a_1, a_5, a_8, a_{11} with a total value of 6.

The graph in Fig. 23.1 is, in fact, the graph obtained for the separation of Ca, Co and Th on a cation-exchange column. One arrives at this graph by replacing nodes a_1 - a_{11} by CaThCo//, Ca//Th/Co, Th//Ca/Co, Co//Ca/Th, Ca Th//Co, Ca Co//Th, Th Co//Ca, Ca//Th/Co, Co//Ca/Th, Th//Ca/Co and //Ca/Th/Co, respectively. The weights of the edges are distribution coefficients obtained from the literature as explained above. The conclusion in this particular instance is that one must first reach situation a_5, i.e., Ca Th//Co, meaning that one must first elute Co. As the weight of the edge is 0, this means that at least one solvent has been described in the literature that permits one to elute Co with a distribution coefficient of 0 without eluting Ca and Th. The following steps are the elution of Th (distribution coefficient = 2) and Ca (distribution coefficient = 4).

23.2. GRAPH THEORETICAL CLASSIFICATION METHODS

In section 18.4.3 it was seen how the minimal spanning tree of a graph can be used for classification purposes. This is also the case for the branch and bound procedures in section 22.2. In this section, a third OR procedure, which can be used to carry out a classification, is given. It was introduced into analytical chemistry by Massart and Kaufman (1975) ; it was applied to a TLC problem, but it should be useful in all instances where an identification of a substance is desired. Suppose that one has to develop an identification scheme for a large group of substances. Instead of the approach in Chapter 18, where one investigates such a group as a whole, one can also reason that one should concentrate on those groups of substances that are hardest to separate. The methods developed for separating those substances probably permit the separation of the other substances also and, if not, it should be relatively easy to find methods that do permit this. The problem is therefore reduced to deciding which are the groups that are hardest to separate. This can be done by using communications networks. Such networks are used by sociologists to study communication patterns and, for example, in a complex organization to identify the sets of people between whom communications exists. In Fig. 23.2 the existence of direct communication between individuals from a population of eight people, A-H, is denoted by an edge. G is linked directly to C and indirectly to F, but not, either directly or indirectly, to A. In this way, one can distinguish two sets

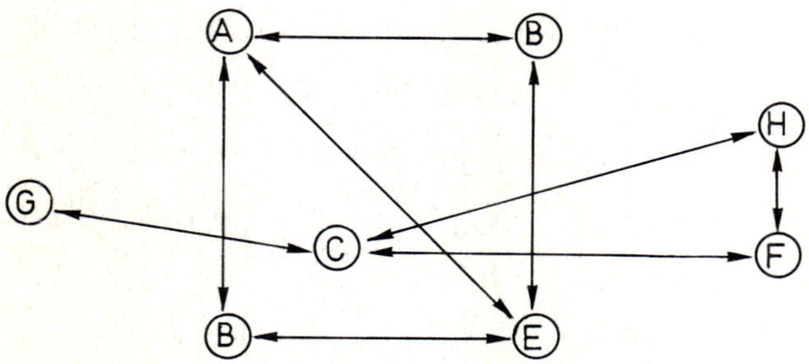

Fig. 23.2 A communications network (adapted from Massart and Kaufman, 1975)

of nodes that are connected in some way to each other, namely the sets ABDE and CFGH. In graph theoretical terms {ABDE} is a connected graph, while {ABCDEFGH} is not, and the OR problem is therefore to find the connected components of the latter graph, which is a trivial problem.

Returning now to the TLC problem, one observes that there is an analogy between distinguishing sets of people close enough to communicate and between sets of substances that are so alike that they are hard to separate by TLC. By considering the substances as nodes and by joining by an edge two substances that are hard to separate, a communication network can be constructed. It remains to define the term "hard to separate". This can be decided only from already existing TLC data. If there are n TLC systems in which the hR_F values are known, a distance between two substances A and B can be computed, in the same way as in Chapter 18, by using the equation

$$D_{AB} = \sqrt{\sum_{k=1}^{n} \left[(hR_F)_{A,k} - (hR_F)_{B,k}\right]^2} \tag{23.1}$$

If D_{AB} is smaller than some arbitrary pre-determined value, A and B are termed "hard to separate".

This was applied (Massart and Kaufman, 1975) to a set of 33 antibiotics using the data from 11 TLC systems and led to the isolation of six "hard to separate" groups. The potential value of this procedure for classification purposes can be deduced from the composition of these groups. The tetracyclines, the penicillins and the rifamycins are found to constitute three of the groups and the oligosaccharides, dihydrostreptomycin, neomycin, paromomycin and streptomycin are also found in one group. One observes that this classification makes chemical sense.

23.3. DYNAMIC PROGRAMMING

The problem in section 23.1 can also be solved using dynamic programming. This is a method of sequential optimization based on Bellman's (1957) principle of optimality. Graphs permit a clear representation of the method, but they are by no means necessary. The second example in this section does not use graphs (although this would have been possible).

Bellman's principle of optimality can be stated in the following way : "a policy is optimal when, at a given stage and whatever the preceding decisions, the decisions which remain to be taken constitute an optimal policy taking into account the state of the system". This can best be explained by using the following example. Suppose that three towns A, B and C must be joined by a highway going from A to C through B. From B to C there are two possible courses. The first, through D, is cheaper than the second, through E. It is then clear that for the total course (between A and C) to be optimal it is necessary that the decision to be taken at stage B (between D and E) should also be optimal. In other words, if road ABDC is optimal from A to C, then so must be the road BDC from B to C. Hence the optimal policy for AC is composed of optimal sub-policies for AB and BC. This can now be applied to the ion-exchange problem. To achieve this the nodes of the graph in Fig. 23.1 must be organized in stages. This is shown in Fig. 23.3.

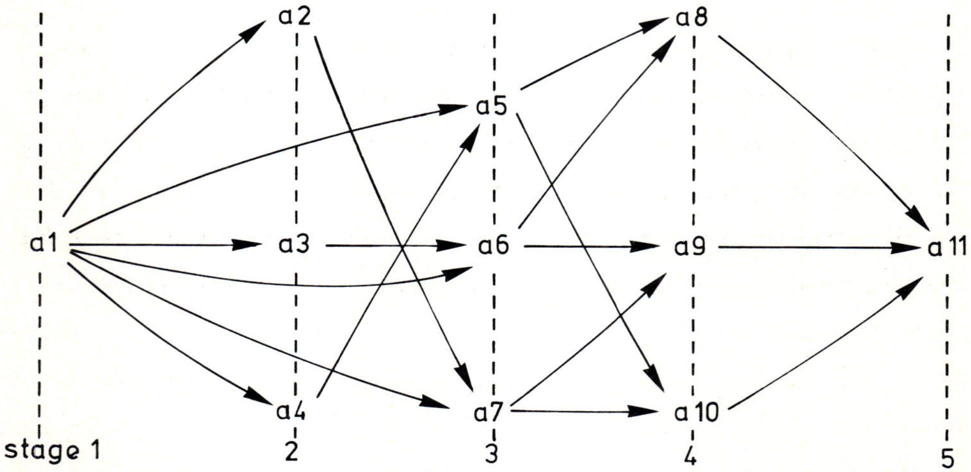

Fig. 23.3. Graph of Fig. 23.1 rearranged in such a way that dynamic programming can be applied.

Points 1 - 11 are towns and a highway must be built from town 1 to town 11. Several pathways are possible, as shown in the figure. Bellman's principle implies, for example, that if town 8 is reached best by way of town 5 and if the optimal pathway from 1 to 11 includes 8, then it also necessarily includes 5. The optimal policy consists of optimal sub-policies such as reaching 8 through 5.

In the present instance, one determines the optimal sub-policy for each stage. For stage 2, the solution is trivial as only one pathway is possible from 1 to 2, 3 and 4 and the best sub-policy of going from 1 to 2 is by using the only possible path with a value of 3. For stage 3, there are several alternatives. Town 5, for example, can be reached directly from 1 or indirectly by way of 4. The cumulative values of the edges constituting these pathways are 0 and 910, respectively. Clearly the optimal sub-policy for 5 is to go directly from 1 to 5. One proceeds in the same way, first for each of the towns on subsequent levels, until one arrives at the final town (town 11).

For example, the best sub-policy for town 8 is by way of 5. As the optimal sub-policy for 5 is to go from 1 to 5, this necessarily means that 8 is best reached by going from 1 through 5 to 8. When one arrives at 11 the best total policy is obtained. In this example, this means going from 1 by way of 5 and 8 or 10 to 11.

To apply this to the ion-exchange problem, one only has to name the nodes according to the separation situation (such as AB//C, C//A/B, etc.). Details can be found in the paper by Massart et al. (1973).

An interesting application concerning the optimization of the analysis of nuclear materials in safeguard systems was described by Bouchey et al. (1971). They used dynamic programming to minimize variance on the measurement of "material unaccounted for" (MUF), a material balance of special nuclear materials. MUF is the difference between the material introduced into a system and the amount of material removed, and is a criterion for determining diversion or loss of these materials. The determination of MUF requires the analysis of the materials present at different stages of the fuel cycling process, of wastes, etc.

On each of these stages i, an error, characterized by its variance s_{1i}^2, is made. By carrying out n_i repeat analyses, the variance on the mean of the result obtained for stage i decreases by a factor $\sqrt{n_i}$, but the costs increase. The total variance s_t^2 on the estimation of MUF is equal to the sum of the variances obtained at each stage and the total cost, C, is equal to the sum of the costs incurred for the determinations carried out at each stage. If there is no cost constraint, one will simply carry out as many replicate determinations at each stage as is felt to be necessary in order to obtain an adequate precision on MUF. If there is a cost constraint, one has to decide, however, how much money (or effort) to allocate to each stage in order to obtain the minimal value for s_t^2, i.e., one has to choose how many determinations should be carried out at each stage. In other words, an optimal combination of n_i values has to be selected.

Let us consider this optimization in greater detail by using the example given in Bouchey et al.'s paper (see Table 23.I).

Table 23.I

Values of constants in the optimization problem given by Bouchey et al. (1971). For symbols, see text.

Measurement stage or point (i)	N_i	s_{1i}^2	s_{2i}^2	C_i
1	50	0.21	1.0	10
2	80	0.50	3.5	5
3	100	0.10	0.06	5
4	200	0.84	7.00	3
5	500	0.20	0.44	8

It was shown that the total variance is

$$s_t^2 = \sum_{i=1}^{M} s_i^2 = \sum_{i=1}^{M} \frac{N_i^2}{n_i} \left[s_{1i}^2 + s_{2i}^2 \left(1 - \frac{n_i - 1}{N_i - 1}\right) \right] \tag{23.2}$$

where

M = total number of measurement points (5 in the example) ;

s_i^2 = the variance obtained by carrying out n_i replicate determinations at measurement stage i ;

s_{1i}^2 = the variance due to the analytical imprecision ;

s_{2i}^2 = the variance due to varying amounts in the N_i items available at measurement point i. One subjects n_i of these N_i items to analysis.

The cost constraint can be written as

$$\sum_{i=1}^{M} C_i\, n_i \leqslant C \tag{23.3}$$

where C_i is the cost of carrying out one determination at stage i and C is the total allowed cost. In the example, C = 300 (dollars).

One can now apply Bellman's principle and determine optimal sub-policies. To do this, one first determines the optimal sub-policy for measurement stage 1. This is a trivial task : clearly, if one allocates 100 dollars to this measurement point, the optimal strategy will consist in measuring 10 items, i.e., n_1 = 10. This is done for every possible amount of dollars allocated (per 10 dollars, for example).

In a second step, one determines the optimal sub-policies for each amount of dollars that can be assigned to measurement points 1 + 2. As an example, consider the calculations for 60 dollars. The possible combinations are (n_1 = 1, n_2 = 10, s_t^2 = 5330), (n_1 = 2, n_2 = 8, s_t^2 = 4439), (n_1 = 3, n_2 = 6, s_t^2 = 5004), (n_1 = 4, n_2 = 4, s_t^2 = 6905), (n_1 = 5, n_2 = 2, s_t^2 = 13222). The best sub-policy is n_1 = 2, n_2 = 8. This means that, if the final optimal strategy consists of using 240 dollars for the last three measurement points and 60 for the first two, it will necessarily consist of a solution where n_1 = 2 and n_2 = 8. It also means that the strategies containing n_1 = 2, n_2 = 8 are always better than those with n_1 = 1, n_2 = 10 ; n_1 = 3, n_2 = 6, etc. Only the former should be taken into account in further calculations and the latter possibilities can be eliminated.

Consider now the third step of the calculation, in which the best sub-policy for the three first measurement points together is determined. Again, one computes the optimal sub-policy for each amount of dollars per 10 dollars.

For 70 dollars, for example, one of the possibilities is to allocate 10 dollars to the third measurement point and 60 to the first two. These 60 dollars will necessarily be distributed between the two points, so that $n_1 = 2$ and $n_2 = 8$.

The complete optimal solution is $n_1 = 2$, $n_2 = 8$, $n_3 = 2$, $n_4 = 34$, $n_5 = 16$. The calculation can be speeded up by using a computer, but in this instance a more formal treatment of the problem is needed, as given in Bouchey et al.'s paper. Their paper gives a good introduction to dynamic programming and its terminology and can be read before turning to the more specialized literature, such as the books by Jacobs (1967) and Hadley (1964).

23.4. SEQUENCING AND COORDINATION PROBLEMS

Two types of problems are discussed here. The first consists of the coordination of several "jobs", some of which can be carried out simultaneously, while others are carried out according to a specified sequence, in such a way that the whole project is achieved in the shortest possible time.

The second problem arises when several "jobs" have to be carried out and one needs to determine the optimal sequence of these jobs. Sequencing problems are important OR problems and their solution is discussed in textbooks, such as that by Ackoff and Sasieni (1968). In this section, only one example (the toxicological laboratory example) will be discussed, because it permits the illustration of a class of OR methods called heuristic methods.

23.4.1. The PERT technique

Many projects, including the development of analytical methods and related problems, necessitate a number of specific decisions and activities. If these activities are not planned and coordinated, the project will take longer than necessary. Methods such as PERT (Program Evaluation and Review Technique) can serve as aids in scheduling these activities. Consider the following problem, which occurred in the laboratory of one of the authors (D.L.M.).

The laboratory was presented with the task of analysing plant samples for fluoride and stating how grave it thought the extent of pollution was. It was found that there was no officially accepted method in the author's country for the determination of fluoride in plants and that there were no generally accepted normal levels of fluoride. Therefore, the laboratory was faced with a double task, before coming to a conclusion, namely the development of a method, the results of which could be proved to be sufficiently accurate, and the determination of normal values. The following tasks were undertaken :

(1) The most promising method (an oxygen-flask destruction method followed by potentiometric determination of fluoride) was selected.

(2) This method was subjected to the usual preliminary checks on accuracy, precision, limit of detection, etc.

(3) After carrying out step 2, it was recognized that in order to produce data that would be generally accepted, sufficient proof of accuracy would be necessary. Therefore, a standard material with known fluoride content was obtained. This took 2 months to locate and obtain.

(4) It was also decided that the method would have to be calibrated with another method used by a government agency. To contact this agency, arrange the details of the intercomparison, obtain the samples and compare the results took several months.

(5) At about the time step 3 was finished, it was decided to collect a large number of samples of grass at random from all over the country in order to arrive at a normal value for fluoride in grass. The collection of these samples took at least 1 month.

(6) Preparation and analysis of the samples.

(7) Interpretation of the results and writing of the report.

In the planning of this project, stages 3 and 4 were initiated too late so that the project took many months longer to complete than was strictly necessary.

A breakdown of the project in stages using the PERT programming technique would have shown that stages 3 and 4 would probably be the most time consuming and should have been started earlier. This conclusion could also have been

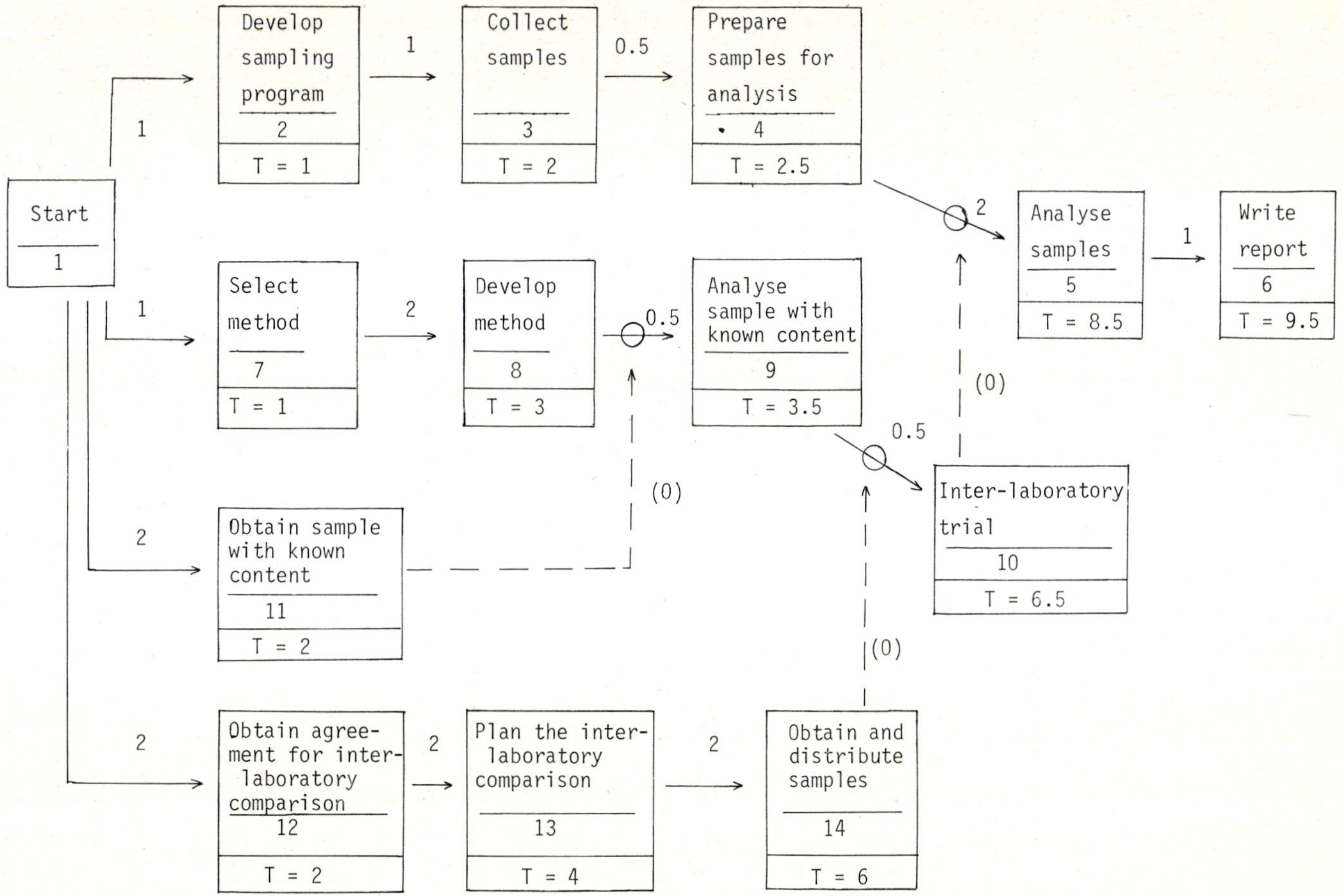

Fig. 23.4. A PERT network. The times are given in months. T is the time that is required in order to accomplish each task under optimal planning conditions.

obtained, of course, without using the PERT technique. Nevertheless, it constitutes a very useful formalization of project planning and it cannot be denied that the kind of time-consuming error described here is common in laboratories engaged in more or less complex projects. In Fig. 23.4 a very rudimentary PERT network is given that describes the necessary steps in a logical sequence. The facts that certain activities must be completed, before some others can be undertaken and that certain steps can be carried out in parallel are taken into account.

The nodes in this graph represent points in time (events) and the edges represent constraints indicating that an event must precede another event. The values of the edges are given by the times needed to carry out the tasks necessary to reach the nodes. The cumulative time to reach a stage is given under the corresponding node. These are the sums of all expected times leading to the event under consideration. If the event is reached by two paths, the cumulative time for the longest one is taken. From the example given, it is clear that the total time necessary is determined by the sequence of tasks 1-12-13-14-10-5-6. This is the critical path in the network. The PERT network also makes it possible to examine the consequence of a delay on any of the other tasks of the project on the total duration.

Examples relating exclusively to the development of analytical methods are not found in the literature. Applications in which the carrying out of analytical determinations is included as a task in a larger project can be found in papers by Goulden (1974) and Kahan and Karas (1976). They concern PERT networks for a residue analysis programme and the development of a new food product, respectively.

As indicated above, the PERT network given here is rudimentary. More complete information can be obtained from Levin and Kirkpatrick (1966) and Moder and Phillips (1964). In particular, the time estimation is carried out in a more complex manner than is shown in Fig. 23.4. Usually, one makes three estimates of time, one optimistic, a, one pessimistic, b, and one that is considered as the most likely, m. From these three estimates, one obtains a mean or expected time

$$t = \frac{a + 4m + b}{6}$$

These mean times are then used to calculate the cumulative times.

23.4.2. An heuristic sequencing method

The placing of parts of a procedure in the right or optimal order is one of the important tasks of the analytical chemist. Let us consider, for example, the classical dichotomous approach to qualitative analysis, in which one tries to identify an element or a substance by splitting up the group of possible substances into two smaller groups and determining to which of the two groups the unknown belongs. The process is then repeated in this group until only one possibility is left. There are many variations on this theme and one of these will now be illustrated by the example of a toxicological laboratory.

In some instances, the toxicologist will have n possibilities and he will examine them one at a time until the substance is identified. This is a special case of the dichotomous approach, as the group is split up into groups consisting of n-1 and 1 members, n-2 and 1, and so on until the unknown is in the 1 member group. The time needed for the n qualitative determinations is not the same and it is assumed that the probability of occurrence of each of the substances is known. What is the optimal sequence, i.e., the sequence that will lead on the average to an identification in the shortest time ? Should one start with a very fast procedure for an infrequently found drug, so that probably one will have to carry out a second determination for another compound, or should one begin with a lengthy procedure for a substance that is encountered frequently, so that the chances are high that one will be able to stop after this step ?

Suppose that the probability p_i, $i = 1, \ldots, n$ of the occurrence of each poison and the execution time t_i, $i = 1, \ldots, n$ for the methods that permit one to identify them are known. Only one poison occurs in the sample, each method allows the identification of only one poison and the symptoms do not yield any prior information. In this simple case, an equally simple technique allows one to determine the sequence of methods that minimizes the mathematical expectation of the sums of the times necessary for the execution of the methods that lead to the identification of the unknown.

The method requires two steps : (a) the methods are ranged first in the order of decreasing probability p_i, and (b) the methods i and i+1 are inverted when $p_i\ t_{i+1} - p_{i+1}\ t_i < 0$. Step (b) is repeated until no more inversions are possible. It can be shown that the resulting sequence is optimal. This technique has the structure typical of heuristic methods. However, heuristic methods do not necessarily yield completely optimal results. In this particular case, however, the optimal solution is obtained.

It should not be necessary to say that the picture given here of the toxicological laboratory is very much simplified and that, in practice, many other factors (such as varying amounts of prior information) must be taken into consideration. Nevertheless, in many instances qualitative schemes are based on the type of considerations given above and therefore they should be amenable to models (although these will be usually more complex) such as that described here.

REFERENCES

R.L. Ackoff and M.W. Sasieni, Fundamentals of Operations Research, Wiley, New York, 1968.

R. Bellman, Dynamic Programming, Princeton Univ. Press, Princeton, N.J., 1957.

G.D. Bouchey, B.V. Koen and C.S. Beightler, Nuclear Technology, 12 (september) (1971) 18.

R. Goulden, Analyst, 99 (1974) 929.

G. Hadley, Non-linear and Dynamic Programming, Addison Wesley, Reading, Massachusetts, 1964.

O.L.R. Jacobs, An Introduction to Dynamic Programming, Chapman and Hall, London, 1967.

G. Kahan and A.J. Karas, Food Technol., (5) (1976) 86.

A. Kaufmann, Introduction à la Combinatorique en vue des applications, Dunod, Paris, 1968.

R.I. Levin and C.A. Kirkpatrick, Planning and control with PERT/CPM, McGraw-Hill, New York, 1966.

D.L. Massart, C. Janssens, L. Kaufman and R. Smits, Anal. Chem., 44 (1972) 2390.

D.L. Massart, C. Janssens, L. Kaufman and R. Smits, Z. anal. Chem., 264 (1973) 273.

D.L. Massart and L. Kaufman, Anal. Chem., 47 (1975) 1244A.

J.J. Moder and C.R. Philips, Project Management with CPM and PERT, Reinhold Publishing, New York, 1964.

Chapter 24

MULTICRITERIA ANALYSIS

24.1. INTRODUCTION

In Chapter 1, it was observed that in many instances there is more than one optimization criterion. These criteria are often conflicting or interrelated, as shown in section 9.4. To make an optimal decision, one is forced to make a compromise between several criteria. A simple example is the choice of an instrument, such as a spectrophotometer. One has to make a compromise between cost and quality of the apparatus. A more costly apparatus usually has a larger resolving power, which has a direct bearing on characteristics such as precision, accuracy and information content of the spectra.

This is a very common situation whenever decisions have to be made (politics, economics, engineering, etc., and analytical chemistry). In all of the preceding chapters optimization problems have been discussed with one criterion (unicriterion analysis). A recent trend in OR is the study of multicriteria analysis.

There are several possible approaches, of which only three will be presented here. This chapter follows to a large extent the presentation of multicriteria analysis given by Brans (1976).

24.2. UTILITY FUNCTIONS

Consider p criteria, $f_1, f_2, \ldots, f_p$. For a particular decision x, these take the values $f_1(x), f_2(x), \ldots, f_p(x)$. Suppose that it is possible to express numerically the importance of the criteria by weights with the coefficients $\lambda_1, \lambda_2, \ldots, \lambda_p$, then one obtains a function

$$N(x) = \sum_{h=1}^{p} \lambda_h f_h(x) \qquad (24.1)$$

which is called a utility function. The multicriteria problem is then reduced to the unicriterion problem of optimizing N(x), which can be solved using more classical techniques. Unfortunately, this approach is subject to three very important disadvantages :

(a) Let us consider an arbitrary decision $\tilde{x}$. One can prove that there are an infinite number of utility functions that have this solution as the optimal one, so that any decision can be justified a posteriori with a certain utility function.

(b) Very often the optimal solution will be identical with the one obtained for one of the associated unicriterion problems. This means, in fact, that one of the criteria is given such a weight that the other criteria are neglected.

(c) It is extremely difficult to give a priori weights for all of the criteria valid over the whole range of values that these criteria can take.

The first two difficulties can be partially eliminated by using a related technique called "goal programming" (see for example Ijiri, 1965). Suppose that for each criterion a certain value is given as the goal value. Let these be called f'_1, f'_2, ..., f'_p and let the vector combining these criteria be called F'. This is, in fact, what is done implicitly by an analytical chemist when he selects a procedure. He will consider what the ideal precision, accuracy, cost, etc., are and he will try to find the procedure that approaches this ideal most closely.

In the same way, in the goal programming method, one investigates whether the ideal solution F' satisfies the constraints that are imposed. If not, one determines the solution that satisfies the constraints which is closest to the ideal. The value of the criteria for a solution x is called F(x). The problem is then reduced to the unicriterion problem of minimizing $d\left[F', F(x)\right]$ for feasible solutions x, where d represents a distance (for the mathematical significance of the distance concept, see section 18.7).

24.3. OUTRANKING RELATIONS

Methods using outranking relations are used only when the number of possible solutions is finite. The object is no longer to find the optimal solution but rather a set of solutions that are better than other solutions. These solutions are said to outrank the others.

Here we shall discuss the ELECTRE I method proposed by Roy (1972). To explain the method, we shall consider a simple example. Suppose one has to choose between seven procedures. In Table 24.I are given the criteria, the weights assigned to these criteria and the values the criteria can take.

Table 24.I

Table of criteria

h criterion	weights	possible values
1 time	$\lambda_1 = 5$	120, 60, 30, 15, 5 min
2 precision (relative)	$\lambda_2 = 4$	$\pm$ 1, $\pm$ 3, $\pm$ 10%
3 are toxic reagents used	$\lambda_3 = 3$	yes, no
4 free from interference	$\lambda_4 = 3$	yes, no

From the values taken by the weights, one observes that the person carrying out the selection thinks time the most and toxicity and interferences the least important criteria.

In Table 24.II the values taken by the criteria are given for the possible procedures.

Table 24.II

Evaluation of procedures

Criteria	1	2	3	4	5	6	7
1	120	60	60	30	30	30	5
2	1	1	3	3	3	10	10
3	n	y	n	n	y	n	y
4	y	y	y	n	y	y	n

One now compares each procedure with each other procedure. This is done in two steps. In the first step, one notes in what respect the procedures differ. Suppose we compare procedures 1 and 2, then 1 is better according to criterion 3, which we denote by writing N^+ : 3, and worse according to criterion 1 (N^- : 1). The results are given in Table 24.III.

Table 24.III

Comparison of the procedures according to the criteria

	1	2	3	4	5	6	7
1	-	N^+: 3 N^-: 1	N^+: 2 N^-: 1	N^+: 2,4 N^-: 1	N^+: 2,3 N^-: 1	N^+: 2 N^-: 1	N^+: 2,3,4 N^-: 1
2	-	-	N^+: 2 N^-: 3	N^+: 2,4 N^-: 1,3	N^+: 2 N^-: 1	N^+: 2 N^-: 1,3	N^+: 2,4 N^-: 1
3	-	-	-	N^+: 4 N^-: 1	N^+: 3 N^-: 1	N^+: 2 N^-: 1	N^+: 2,3,4 N^-: 1
4	-	-	-	-	N^+: 3 N^-: 4	N^+: 2 N^-: 4	N^+: 2,3 N^-: 1
5	-	-	-	-	-	N^+: 2 N^-: 3	N^+: 2,4 N^-: 1
6	-	-	-	-	-	-	N^+: 3,4 N^-: 1
7	-	-	-	-	-	-	-

In the second step, one takes into account the weights in order to arrive at a numerical expression. The preference ratio is given by

$$P = \frac{\sum_{h \in N^+} \lambda_h}{\sum_{h \in N^-} \lambda_h} \tag{24.2}$$

For example, for the comparison of 1 and 2, this becomes

$$P = \frac{3}{5} = 0.6$$

We can now construct Table 24.IV, where only the values that exceed 1 are given. To return to the example, as P = 0.6 for the comparison of 1 and 2, then P = 1/0.6 = 1.67 for the comparison of 2 and 1.

Table 24.IV

Numerical comparison of the procedures

	1	2	3	4	5	6	7
1	-	-	-	1.40	1.40	-	2.00
2	1.67	-	1.33	-	-	-	1.40
3	1.25	-	-	-	-	-	2.00
4	-	1.14	1.67	-	-	1.33	1.40
5	-	1.25	1.67	-	-	1.33	1.40
6	1.25	2.00	1.25	-	-	-	1.20
7	-	-	-	-	-	-	-

Until now, we have taken into account only the fact that one procedure has a better value for some criterion or not. It is possible that one procedure is so much worse according to one criterion that, even when it is better in all other respects, one does not wish to conclude that it is better. To do this, one adds discrepancy conditions. In the present example, these are 120 min compared with 5 or 15 min and 60 min compared with 5 min for criterion 1 and 10% compared with 1% for criterion 2. Consider, for example, the comparison 1/7 : P = 2.00, so that 1 is considered to be better than 7. However, one of the discrepancy conditions (120 versus 5 min) is fulfilled, so that one reserves a conclusion. All of the comparisons for which there is a discrepancy condition are deleted from Table 24.IV, yielding Table 24.V.

Table 24.V

Comparison of the procedures, taking discrepancy conditions into account

	1	2	3	4	5	6	7
1	-	-	-	1.40	1.40	-	-
2	1.67	-	1.33	-	-	-	-
3	1.25	-	-	-	-	-	-
4	-	1.14	1.67	-	-	1.33	1.40
5	-	1.25	1.67	-	-	1.33	1.40
6	-	-	1.25	-	-	-	1.20
7	-	-	-	-	-	-	-

At this stage, one introduces a dominance threshold, T, which must be at least 1 and is usually higher. The philosophy is that it is preferable not to judge one procedure to be better than the other when only a slight difference between both is obtained. In this way, one takes into account the uncertainty involved in choosing the λ_h values and the fact that some other criteria may have been overlooked. In the present example, T is considered to be 1.33 and all values that do not exceed this threshold are eliminated. This yields Table 24.VI, in which one now has a summary of those instances where one procedure is clearly better than another. These procedures dominate the others (symbol D). For example, in Table 24.VI one observes that procedure 1 dominates procedures 4 and 5.

Table 24.VI

Dominance Table

	1	2	3	4	5	6	7
1	-	-	-	D	D	-	-
2	D	-	D	-	-	-	-
3	-	-	-	-	-	-	-
4	-	-	D	-	-	D	D
5	-	-	D	-	-	D	D
6	-	-	-	-	-	-	-
7	-	-	-	-	-	-	-

From this table, one constructs a dominance graph, where 1 → 2 means that 1 dominates 2. This graph is shown in Fig. 24.1.

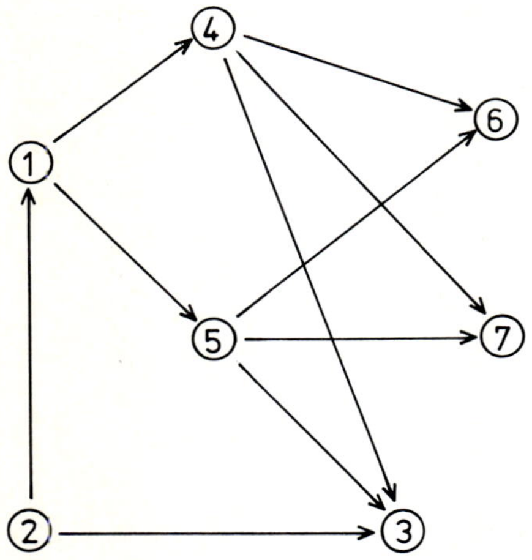

Fig. 24.1. Dominance graph.

In the graph, one then determines the kernel, which is defined as a set of nodes such that :

(a) no node of the kernel is dominated by another node of the nucleus ;

(b) all nodes outside the kernel are dominated by at least one node of the kernel.

In the present instance, this set is composed of the nodes 2, 4 and 5, which means that one of the procedures 2, 4 and 5 should be selected while procedures 1, 3, 6 and 7 should be eliminated from consideration.

The ELECTRE I method is not subject to the disadvantages (a) and (b) of the utility function method and it is much less subject to disadvantage (c). There is still the difficulty of choosing weight coefficients, but these are used in a much less absolute way than in the utility function approach. Furthermore, it is possible to take all weights equal to 1.

24.4. INTERACTIVE METHODS

A consequence of the complexity of the relationship between several conflicting objectives or criteria is that it is very difficult to have a global view of their relative values. A trial carried out in order to improve this view has recently led to the development of interactive methods that can be considered to be the most modern type of methods in multicriteria analysis. They allow a better understanding of possible compromises between objectives.

In the more classical methods based upon goal programming, utility functions or outranking relations, all information concerning variables, objectives, constraints and eventual weights must be known before a solution is obtained. On the other hand, interactive methods are based upon a dialogue between the decider and the researcher. In this dialogue, a sequential process is set up, during which a solution is proposed to the decider who in turn gives information about his view on the values of this solution for the different objectives. A qualification of his opinion then allows the researcher to compute a new solution, which is in turn examined by the decider. This process is then repeated until either the decider accepts one of the solutions or he concludes that no compromise is possible.

Several interactive algorithms have been proposed for the multicriteria linear programme and several other multicriteria problems. A discussion of these algorithms can be found in the books by Wallenius (1975) and Zeleny (1976).

REFERENCES

J.P. Brans, Internal Report University of Brussels, CSOOTW/079 (in flemish), 1976.

Y. Ijiri, Management Goals and Accounting for Control, North-Holland, Amsterdam, 1965.

B. Roy, Revue METRA, 11 (1) (1972) 121.

J. Wallenius, Interactive Multiple Criteria Decision Methods. An Investigation and an Approach, Helsinki School of Economics, Helsinki, 1975.

M. Zeleny (Editor), Multiple Criteria Decision Making : Kyoto 1975, Springer, Berlin, 1976.

PART IV

REQUIREMENTS FOR ANALYTICAL PROCEDURES

Chapter 25

THE DIAGNOSTIC VALUE OF A TEST

25.1. INTRODUCTION

In section 2.1.3, it was concluded that in some instances an increase in the precision of the analytical determination is not effective because there are other and more important causes of variability. In the agricultural analytical laboratory, the main cause of variability is the heterogeneous composition of the soil, resulting in a sampling error. Similarly, the sampling of lots (wagon loads, shipments, etc.) of many bulk products such as ores, coal and fertilizers leads to errors that are directly related to the fluctuations in the composition within the lot. An estimate of these errors can be given only if the magnitude of these fluctuations is known. Such a knowledge can usually be obtained from past experience, i.e., through the analysis of many samples. The variance of the results for many samples taken from a lot is an estimate of the real variance, σ_t^2, which is the sum of the variance σ_x^2 as a measure for the inhomogeneity within the lot and the variance σ^2 as a measure of the precision of the analytical procedure. Drawing one sample from the lot and analysing it leads to a composition of the entire lot with a precision σ_t. Reduction of this precision by $\sqrt{n}$ can be obtained by drawing and analysing n samples from the lot. It is also possible to draw n samples and to combine them to give a gross sample before the analysis. Then only σ_x is reduced by a factor $\sqrt{n}$. For a discussion of sampling of inhomogeneous (bulk) products and sampling strategies, we refer to Baule and Benedetti-Pichler (1928), Duncan (1962), Visman (1969) and Ingamells and Switzer (1973).

It is important to stress that for the design of an optimal sampling strategy an exact formulation of the problem in terms of the inhomogeneities, σ_x, and the required precision, σ_t, is necessary. Some aspects of sampling will be treated in Chapter 26.

In this section we shall consider the clinical laboratory, where the cause of the variability is usually the biological variability (the variability that a population displays in the values of physiological parameters). Whereas the sampling and analysis of heterogeneous (bulk) products must lead to a representative value of the average composition of the lot, determinations in the clinical laboratory are aimed at reliable figures for individuals. Drawing more samples from one individual does not eliminate the difference of the parameters between individuals. The value of a test with a certain precision for diagnostic purposes depends to a large extent on the biological variability.

This type of problem is not confined to the clinical laboratory. When the control laboratory of a government institution renders a judgment on whether or not a certain food sample contains a dangerous amount of mycotoxins, this can also be viewed as a diagnostic problem. The same is also true for polluted or unpolluted water. In this section, the word "diagnostic" will be used in the context of a clinical laboratory, but one should bear in mind that a large part of what is discussed here should also be valid for other diagnostic problems.

When physiological parameters are estimated by chemical tests, the distribution of the observed values is determined by both the physiological variability and the analytical error distributions. The purpose of carrying out such tests is, of course, to decide whether or not the sample being analysed belongs to a healthy or an ill person. In general, routine clinical laboratories proceed by assuming that values within certain limits (usually the mean of the observed or, when it is available, of the physiological distribution plus and minus twice the standard deviation) are probably normal (the normal range), while those outside these limits probably indicate illness. Since $\sigma_t^2 = \sigma_x^2 + \sigma^2$ it is intuitively clear that tests with a very low precision will contribute considerably to the variability. They will therefore lead to a high uncertainty in the diagnosis and the precision will be an optimization criterion. On the other hand, it is also evident that when a very precise method is used, a further increase in precision will hardly contribute to the reduction of the uncertainty in the diagnosis. In many instances, it is easy to decide whether or not a test with higher precision is really wanted. The question which arises then is how to investigate the effect of the precision in the also frequent instance that an easy decision is not possible, and how to do this in a formal way. Several workers have proposed general rules, e.g., the error should be smaller than one quarter of the physiological range (Tonks, 1963). Other workers, such as Westgard et al. (1974), have developed statistical criteria for judging the precision and accuracy of new methods. These can then be related to medical requirements. Still other workers have presented calculations concerning the magnitude of the number of diagnostic errors. One example is a study by Acland and Lipton (1971). They asked the following question : if the test is performed on the healthy members of a population (the true analytical values lie within the normal range), what proportion of the actual results will lie outside the limits of the normal range as a result of analytical error ? If σ is the precision of the method, σ_x the standard deviation of the population and $r = \sigma/\sigma_x$, then Table 25.I is obtained.

Table 25.I

Probability that a "normal" sample (normal range : mean $\pm$ 2 σ_x) should yield an abnormal result due to analytical error (from Acland and Lipton, 1971)

r	probability
0.1	0.01
0.2	0.01
0.3	0.02
0.4	0.03
0.5	0.05
0.6	0.06
0.8	0.10
1.0	0.14

Studies such as these give some idea of the effect of analytical error. They do not take into account, however, several factors that influence the value of analytical tests.

A more complete formal analysis of the effect of precision on the diagnostic value of a test can be carried out using Bayes' theorem.

In clinical chemistry, the simplest possible objective is to separate two classes, namely people in the physiological and non-physiological states (not ill and ill, or normal and abnormal). The values of a physiological parameter are part of different (often Gaussian or near-Gaussian) distributions for persons in the physiological state and the non-physiological state. This is illustrated by Fig. 25.1 adapted from Martin et al. (1975). When values higher than C are found, one assumes that the person is probably ill, and when it is lower, that the person is probably healthy. Even without analytical errors, classification errors are made in doing so. In Fig. 25.1, about 5% of the healthy persons are classified as ill, and these 5% therefore constitute false-positive results.

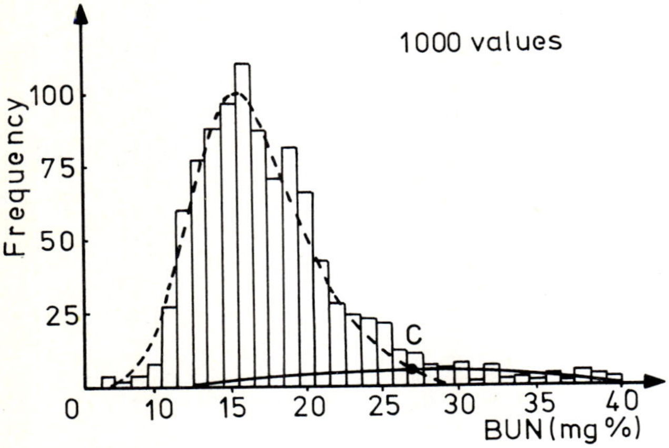

Fig. 25.1. A "normal" and an "abnormal" population (adapted from Martin et al., 1975).

In statistical terminology, this is called the α-error (see section 3.2). On the other hand, 50% of ill persons are not detected. There are, therefore, 50% false negatives or the β-error is 0.5. As the analytical error influences the overlap between both distributions, it also influences the number of misclassifications.

In the example in Fig. 25.1, the number and type of misclassifications also depend on the *a priori* probabilities of both classes (ill, not ill). In the example, one knows from prior experience that 10% of the people are ill and 90% are not. This knowledge must also be taken into account in calculating the effect of the precision of the method, which can be done using Bayes' theorem.

Bayesian theory is concerned with calculating the probabilities of various mutually exclusive hypotheses (or events), H_1, H_2, ..., H_n, connected with an event A. This event occurs itself with a certain probability depending on which of the hypotheses is true. The probability for A, given that event H_1 has occurred, is symbolized by $P(A/H_1)$ and is called the conditional probability of A given that H_1 has occurred.

Under the condition of mutual exclusivity and the condition that either H_1, H_2, ... or H_n occurs, i.e., $P(H_1) + \ldots + P(H_n) = 1$, it can be shown that

$$P(H_1/A) = \frac{P(H_1)\ P(A/H_1)}{P(H_1)\ P(A/H_1) + P(H_2)\ P(A/H_2) + \ldots + P(H_n)\ P(A/H_n)} \qquad (25.1)$$

where $P(H_1/A)$ is the probability that event H_1 will occur when A occurs. This is called Bayes' rule or theorem. As an example, let

H_1 = the fact that a person is healthy ;
H_2 = the fact that a person is ill ;
A = the finding of a negative value, i.e., a value less than C (Fig. 25.1)

Eqn. 25.1 then becomes

$$P(H_1/A) = \frac{P(H_1)\ P(A/H_1)}{P(H_1)\ P(A/H_1) + P(H_2)\ P(A/H_2)} \qquad (25.2)$$

The *a priori* probability that a person is healthy, $P(H_1)$, is 0.90 while $P(H_2) = 0.10$. As 5% of all healthy persons have a false-positive value, the probability $P(A/H_1)$ that a negative result will be obtained for a healthy person is 0.95. The probability $P(A/H_2)$ that a negative result will be obtained for an ill person is then 0.5.

Eqn. 25.2 yields

$$P(H_1/A) = \frac{(0.90)\ (0.95)}{(0.90)\ (0.95) + (0.10)\ (0.5)} = 0.945$$

Therefore, finding a value less than C indicates that there is 94.5% chance of a person being healthy.

By appropiate assignation of the symbols of eqn. 25.2, one can also ask other questions, such as (Martin et al., 1975 and Hall, 1969) :

What is the probability that the patient is healthy when a value higher than C is obtained ? The result is 0.474 and gives the fraction of false-positive values.

What is the probability that the patient is ill with a value below C ? The result is 0.055 and gives the fraction of false-negative values.

What is the probability that the patient is ill with a value higher than C ? The result is 0.526, meaning that only about half of the ill population is detected with this test.

These numbers characterize the performance of a test as a diagnostic tool. This is discussed further in section 25.3. It should be noted that combinations of tests can (and usually are) employed for diagnostic purposes. In this chapter, the discussion is confined, however, to the diagnostic value of individual tests.

Let us now investigate the effect of the precision. The variances of the observed distributions are the sums of the variances of the physiological and non-physiological distributions and the analytical error distribution. Fig. 25.2 shows the effect of two different precisions on the BUN test. When

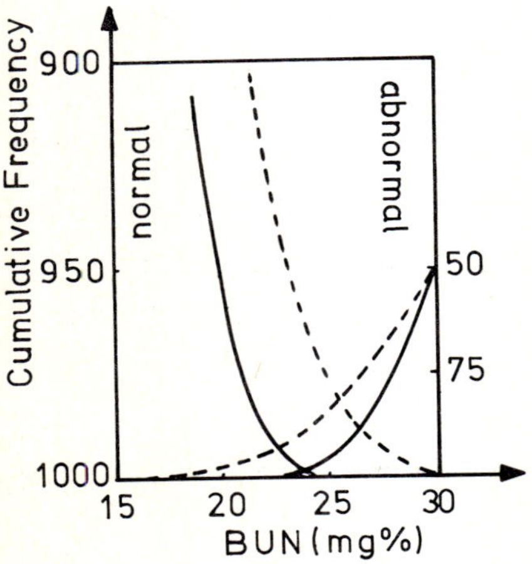

Fig. 25.2. Effect of the precision on the overlap between distributions. Solid curve, standard deviation 3 mg-% ; dashed curve, 5 mg-% (from Martin et al., 1975).

these distributions are known, one can calculate the number of false-positive and false-negative results. Of course, their proportion increases when the precision becomes worse. The extent to which this can be tolerated is usually a matter of subjective judgement. Formal methods to decide on the value of a particular test with a particular precision have, however, also been studied and are discussed in section 25.3 and following sections.

25.2. THE RULE OF BAYES

Let us consider n possible and equally probable outcomes of an experiment and an event E consisting of a of the outcomes. The probability of the event E is then defined as

$$p = \text{Probability (E)} = P(E) = \frac{a}{n} \quad (25.3)$$

The conditional probability of an event E given another event F is defined as the probability that E occurs, given that F occurs and therefore that the only ways E may occur are those included in F. This conditional probability is denoted by $P(E|F)$ and can be calculated by means of the following equation

$$P(E/F) = \frac{P(E.F)}{P(F)} \quad (25.4)$$

where E.F is the event E and F occur simultaneously. Consider, for example, a population of n people containing n_F healthy people and n_E people with glucose values in serum less than x mg/l ("low glucose value"). The events "to be healthy" and "to have a low glucose value" are denoted by F and E, respectively.

The probabilities of these events are given by

$$P(F) = \frac{n_F}{n} \quad (25.5)$$

and

$$P(E) = \frac{n_E}{n} \tag{25.6}$$

We may now restrict our attention to the healthy people and inquire how many have a low glucose value.

If the number of healthy people with a low glucose value is n_{EF}, then the probability of having a low glucose value for healthy people is given by the equation

$$P\text{ (for a healthy person to have a low glucose value)} = \frac{n_{EF}}{n_F} \tag{25.7}$$

This probability is denoted by $P(E|F)$. We then have

$$P(E/F) = \frac{n_{EF}}{n_F} = \frac{n_{EF}/n}{n_F/n} = \frac{P(E.F)}{P(F)} \tag{25.8}$$

where E.F is the event to be healthy and to have a low glucose value.

Two events are called mutually exclusive if they cannot occur simultaneously. If A and B are mutually exclusive, then

$$P(A.B) = 0 \tag{25.9}$$

A set of events $A_1, A_2, \ldots, A_k$ are called mutually exclusive if each pair of events from the set form a mutually exclusive pair, i.e., if for all i and j

$$P(A_i.A_j) = 0 \tag{25.10}$$

A set of events $A_1, A_2, \ldots, A_k$ are called exhaustive if every sample point belongs to one of the events.

Consider an event E and a set $A_1, A_2, \ldots, A_k$ of mutually exclusive and exhaustive events.

Consider one of the events A_i. We are interested in the probability $P(A_i|E)$. We know that

$$P(A_i/E) = \frac{P(A_i.E)}{P(E)} \tag{25.11}$$

The following results are also immediate

$$P(E/A_i) = \frac{P(E.A_i)}{P(A_i)} \tag{25.12}$$

and

$$P(A_i.E) = P(A_i).P(E/A_i) \tag{25.13}$$

Furthermore, the event E must occur together with one of the events A_i, as these events are mutually exclusive and exhaustive. Therefore,

$$E = E.A_1 \cup E.A_2 \cup \ldots \cup E.A_k \tag{25.14}$$

where U denotes the union of sets.

We can also immediately see that

$$P(E) = P(E.A_1) + P(E.A_2) + \ldots + P(E.A_k) = \sum_{i=1}^{k} P(E.A_i) \tag{25.15}$$

Introducing results 25.13 and 25.15 into eqn. 25.11, we obtain

$$P(A_i/E) = \frac{P(A_i).P(E/A_i)}{\sum_{i=1}^{k} P(E.A_i)} \tag{25.16}$$

Introducing eqn. 25.13 once again, we obtain

$$P(A_i/E) = \frac{P(A_i)\ P(E/A_i)}{\sum_{i=1}^{k} P(A_i)\ P(E/A_i)} \tag{25.17}$$

This last equation is called Bayes's rule.

25.3. OPTIMAL DICHOTOMOUS DECISIONS

The result of a test or a combination of tests often leads to a dichotomous decision. For example, according to a positive or a negative result, one may decide that a person is ill or healthy and should therefore undergo a particular therapy or not. It is clear that in some instances this will be the wrong decision.

In medical decision theory, one defines the following quantities (see, for example, among many others, Vecchio, 1966)

$$\text{Sensitivity} = \frac{\text{diseased persons positive to the test}}{\text{all diseased persons tested}} \times 100 \qquad (25.18)$$

$$\text{Specificity} = \frac{\text{non-diseased persons negative to the test}}{\text{all diseased persons tested}} \times 100 \qquad (25.19)$$

It should be noted here :

- that the terms specificity and sensitivity used here are defined in another way than when used in the analytical sense (Chapters 6 and 7) ;

- that the terminology "positive to the test" means here that the result of the test leads to the conclusion that the person is diseased.

Until now, we have supposed more or less implicitly that the decision limit is situated at the crossing point of both distributions (point A in Fig. 25.3).

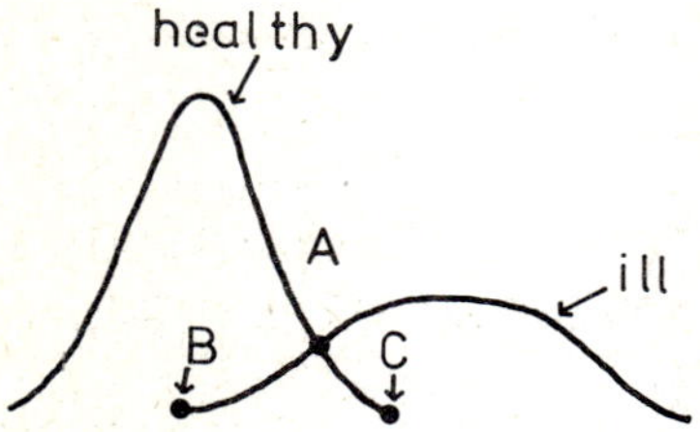

Fig. 25.3. The selection of a cut-off point.

However, this is not necessarily so. One might decide that all persons that could be ill should undergo a therapy and that therefore the cut-off point should be B. On the other hand, with a very dangerous (or costly) therapy, one might prefer to treat only those people who definitely have the disease. The cut-off point should then be C.

Clearly, the sensitivity and the specificity of the test depend on this decision point. For instance, if point C is chosen, the sensitivity is smaller but the specificity is higher. In fact, sensitivity and specificity are related. This is discussed further in section 25.4.

When one wants to evaluate the effect of the analytical precision, one should know the relative importance of sensitivity and selectivity. This question has been considered by Lindberg and Watson (1974). They stated it as : "how many normals may be falsely classified (at least temporarily) in order to reduce by one the number of diseased persons misclassified as normal ?" If one agrees that, for a certain kind of problem, the importance of correctly diagnosing the illness is A times higher than falsely classifying a healthy person as ill, then one can calculate the effect of the selection of a particular cut-off point. By multiplying the number of falsely healthies by A and adding this to the number of falsely positives, one arrives at a criterion which Lindberg and Watson call the "Loss to Society". When one plots the "Loss to Society" as a function of the cut-off point for a particular value of A, one obtains a figure such as Fig. 25.4. The minimum obtained in this curve is, of course, the preferred decision value.

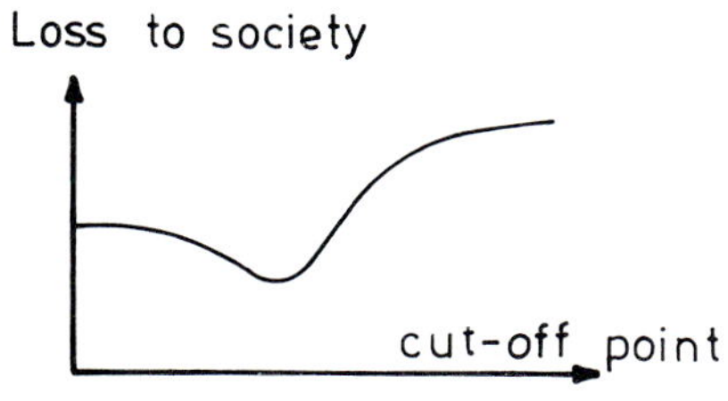

Fig. 25.4. The "Loss to Society" as a function of the cut-off point (from Lindberg and Watson, 1974).

The effect of the precision around the decision point can now be calculated. An example, again taken from Lindberg and Watson's paper, is given in Fig. 25.5.

One observes that there is only a small increase in the loss up to a certain point. From then on, the increase is more serious because at relatively small values of the analytical errors the Loss is determined by the overlap of the pure (error-less) "ill" and "not ill" distributions.

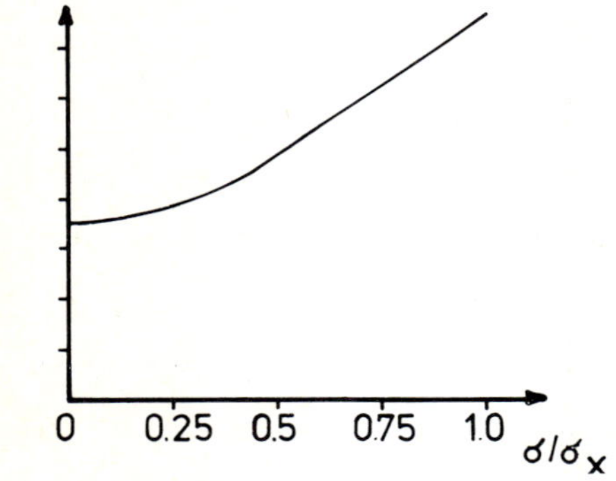

Fig. 25.5. The "Loss to Society" as a function of the precision of the test for A = 10 (from Lindberg and Watson, 1974).

This agrees in a sense with the rule proposed by Tonks (see section 25.1). It is indeed possible to state that a certain amount of error can be tolerated for a particular test. How large this amount may be depends, however, on the overlap of the ill and not ill distributions, the relative prevalence of both states and the importance attached to both kinds (α and β) of error. This problem is essentially the same as that of the decision limit discussed in Chapter 7. Bayes' rule was not applied in that chapter but this could have been done. In the same way Bayes' rule can be applied when a combination of tests is used instead of a single one (pattern recognition).

25.4. THE ROC CURVE

25.4.1. Use in the selection of a cut-off point

As seen in the preceding section, sensitivity and specificity are related and therefore so are true and false positive ratios. The former is the sensitivity/100 and the latter is equal to $\left(1 - \frac{\text{specificity}}{100}\right)$. In general, the more sensitive a test is, the less specific it becomes. To analytical chemists this conclusion will certainly not come as a surprise. On the other hand, it appears that analytical chemists have not tried to describe the phenomenon in a formal way.

This has been done in signal detection theory, which has been developed in the psychophysical sciences (Green and Swets, 1966) where the relationship between sensitivity and specificity is of central interest. One of the more interesting applications of signal detection theory is the interpretation of roentgenograms. The physician is presented with a complex image and has to decide, for example, whether this image indicates tuberculosis or not (see, for example, Lusted, 1971). It has been found that a physician disagrees with a colleague once out of three times and that being presented (without knowing it) with the same image, he disagrees with his first conclusion once out of five times (Yerushalmy, 1969) ! Clearly, there are large observer errors and this phenomenon has been studied by several authors. One observes, for example, that some physicians rarely miss an abnormal image, but that their conclusions are frequently falsely positive. On the other hand, physicians who conclude only infrequently that tuberculosis is present when this is not true, are apt to miss more often a case of the disease.

It is not difficult to think of immediate analogies in analytical chemistry. These are evident, for example, in practical courses in analytical chemistry, where the first excercises are often of a qualitative nature and students are confronted with the difficult decision of whether a certain colour, indicating a particular metal ion, is present or not. Every one who has taught such a course knows that there are students who record the presence of the colour only when it is undeniable. Such a student has a very small score of false positives and a large score of false negatives. His overscrupulous colleague, who always detects a hint of the colour to be observed, rarely misses one of the ions present but detects many ions that are, in fact, absent. The same argument can now be repeated for tests where a decision limit has been established. A decision limit such as C in Fig. 25.3 causes many false negatives but less true positives.

In signal detection theory the relationship between false positives and true positives is given by the so-called receiver operating characteristic curve (ROC curves). Three hypothetical ROC curves are given in Fig. 25.6. One observes that at one end of the curve, one finds zero true positives and zero false positives, i.e., a situation of absolute specificity and zero sensitivity, while at the other end there is absolute sensitivity and zero specificity. Clearly, neither of these extreme situations is of practical interest and the decision point must be situated somewhere in between. As discussed higher this cut-off point must depend on :

- the relative prevalence of the two populations to be discriminated ;
- the relative importance of the mistakes (false positives versus false negatives).

It has been shown (see McNeil et al., 1975a) that the optimal position on the ROC curve occurs where the slope of the curve is equal to

$$\frac{P(H_1)}{P(H_2)} \cdot \frac{C_{fp}}{C_{fn}} \qquad (25.20)$$

where

$P(H_1)$ = probability that a patient does not suffer from a particular illness ;
$P(H_2)$ = probability that a patient suffers from a particular illness ;
C_{fp} = cost of a false-positive diagnosis ;
C_{fn} = cost of a false-negative diagnosis.

If $P(H_1)/P(H_2) = 9$ as in eqn. 25.1 and $C_{fp}/C_{fn} = 0.1$ (a false negative is considered to be 10 times more costly than a false positive), then the optimal slope on the ROC curve is given by 0.9. This yields a point on the ROC curve corresponding with a particular true positive/false positive ratio. If the underlying probability distributions of the test results for ill and non-ill persons are known, then this allows one to determine the optimal cut-off point or decision limit.

25.4.2. Use in rating a test

The shape of a ROC curve also gives an indication of the value of a test. If the test is to have any value, the fraction of true positives must always be higher than the fraction of false positives (except in the two extreme situations, where both are zero or unity). In general, the higher the true positives/false positives ratio, the better is the discriminatory power of a test.

The "discriminatory power" of a test is determined by the extent of overlap between the distributions obtained for the test values for the two populations ("ill" and "not ill"). The extent of overlap itself is determined by the difference between the maxima of the two distributions and by the variance of the distributions. In signal detection theory one defines the parameter d' as the ratio of the difference between maxima and the common standard deviation. This parameter can be considered to be a quality criterion for a test. Fig. 25.6 gives the ROC curves for d' = 0, 1, 2 and 3 (and equal variance).

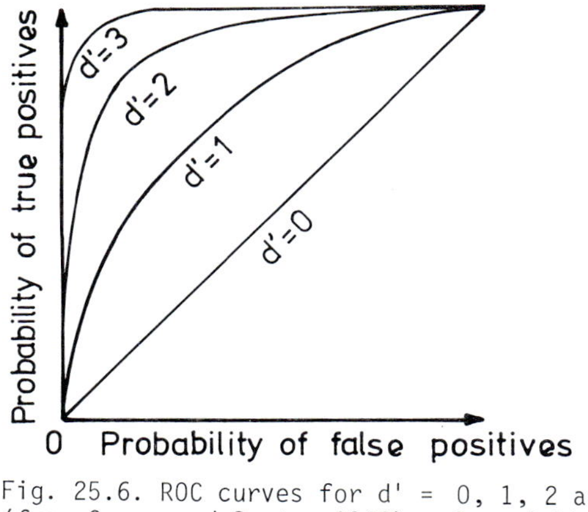

Fig. 25.6. ROC curves for d' = 0, 1, 2 and 3 (and equal variance) (from Green and Swets, 1966). Copyright John Wiley and Sons. Reprinted by permission of John Wiley and Sons.

In Fig. 25.6, the test yielding the ROC curve with d' = 3 is clearly a better test than that with the ROC curve with d' = 2, because at every point the true positive/false positive ratio is higher for the first than for the second test.

Very often the underlying distributions are not known, so that a direct determination of d' is not possible. The question then arises of how to determine d' from experimental ROC curves. It is easier to do this by plotting the results on double probability paper.

In section 2.2.6, it was explained that a normal curve plotted on probability paper becomes a straight line. If two overlapping normal curves are plotted on double probability paper, a straight line also results for the cumulative relative frequency as a function of the upper class boundaries.

In the present case this means that in such a graph a straight line is obtained when plotting the true positives as a function of the false positives. In Fig. 25.7 this has been done for the three ROC curves of Fig. 25.6. It should be noted that the slope of these lines is unity and that this is true only when the two distributions have identical variances. Indeed, the difference between

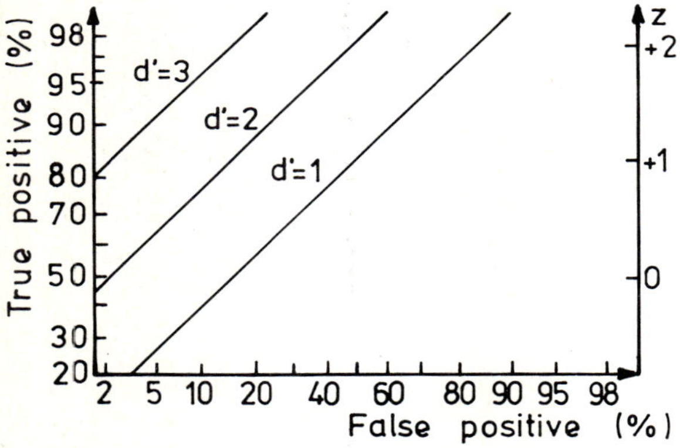

Fig. 25.7. Plot of the ROC curves of Fig. 25.7 on probability-probability paper (from Green and Swets, 1966). Copyright John Wiley and Sons. Reprinted by permission of John Wiley and Sons.

16% and 50% in a single probability plot gives the standard deviation or, in other words, the slope of such a plot is proportional to the magnitude of the standard deviation. In a double probability plot, the slope is then the ratio of the standard deviations of the two underlying distributions (see also Lusted, 1971). The test with d' = 3 yields a "higher" line than that with d' = 2 in the double probability plot. The question remains of how to express this difference. To obtain comparable values, one expresses d' in standard deviation units (i.e., one uses the z-scale of Chapter 2) ; d' is then obtained in the probability-probability graph by reading the z-value on the false-positive axis where P (true positive) = 0.5. This point is the mean of the "ill" or "positive" distribution and reading the z-score on the false-positive axis amounts to placing the origin of the z-axis at the point where the mean of the "not ill" distribution is situated (Green and Swets, 1966) (see Fig. 25.8). One can verify that this convention is important only when the slope is not unity. When it is unity, one reads the same value at the P = 0.5 points on both the false- and true-positive coordinates. When it is not, this indicates that the variances of the two distributions are not equal. As the result is expressed in standard deviation units, it depends on which distribution is chosen to obtain the normalized result, so that a convention is necessary. In Fig. 25.8, the d' values are 1.6 and 2.05. Another convention consists in determining the intersection point of the ROC line with the negative diagonal, i.e., the line drawn from the upper left corner to the 0.5, 0.5 point. The (absolute value of the) normal deviate of the intersection point multiplied by 2 is called the sensitivity index, d'_e. The intersection of the ROC line with the diagonal is used because on the diagonal the absolute values of the normal deviates of both axes are identical. This procedure therefore constitutes a kind of normalization by averaging the variances of both distributions (Lusted, 1971).

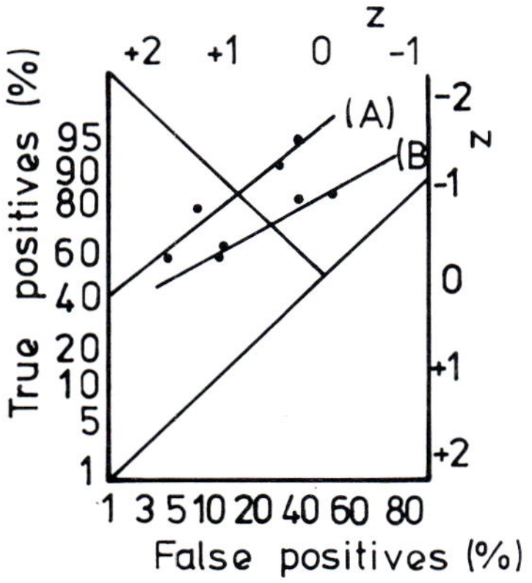

Fig. 25.8. An application of ROC curves to the estimation of the effect of training. A represents the interpretation of mammograms by radiologists, B by paramedical personnel (from Lusted, 1971).

Both d' and d'_e characterize the performance of tests as diagnostic tools. There seem to be no applications in analytical chemistry. Lusted (1971) summarized some very interesting applications of the ROC curve to signal detectability in the interpretations of roentgenograms, or similar images. The ROC lines in Fig. 25.8 depict the interpretation of mammograms. Curve A represents the performance of four radiologists and curve B the performance of six paramedical personnel. The d'_e of the radiologist group is about 1.9 while for the paramedical group it is only about 1.1, thereby proving the superiority of training of the former group. The same principles are used to compare error rates for detecting nodules on film. Direct viewing (d'_e = 1.9) proved superior to television viewing (d'_e = 0.98). However, contrast enhancement made a better tool of television viewing (d'_e = 1.46).

Although ROC curves have never been used in analytical chemistry, the same principles can probably be applied. The interpretation of the colouring of a wool thread by wine, to decide whether or not prohibited synthetic colouring material is present in the wine sample, is of the same order of difficulty as the interpretation of the roentgenograms and there is no doubt that trained personnel perform better than untrained personnel. It therefore seems that ROC curves should permit

(a) the evaluation of the reliability of the conclusions (analytical or not) obtained by qualitative tests ;

(b) the evaluation of the performance of an analyst in carrying out these tests ;

(c) the evaluation of the effect of training schemes, the experience and the professional competence on the performance of the analytical chemists.

ROC curves are graphs of probabilities of true-positive answers compared with probabilities of false-positive answers. For example, a ROC curve allows one to estimate the probability of occurrence of the "ill state" if a certain test result is obtained. To express this in a more direct analytical-chemical analogy, a ROC curve permits one to estimate the probability that one of two possible drugs is really present when a certain R_F value is obtained. We have investigated a similar situation in Chapter 8 (information) and therefore one would expect there to be a link between ROC curves and information theory. This link has been discussed by Metz et al. (1973), who evaluated ROC curve data in terms of information theory and considered applications in radiography. For a detailed discussion of the relationship between both, readers are referred to their paper.

Until now, ROC curves have not been investigated in analytical chemistry. From the discussion in this section it appears, however, that research on ROC curves might prove useful.

25.5. COST - BENEFIT CONSIDERATIONS

As discussed in Chapter 9, the criteria according to which one evaluates analytical procedures are interrelated. Very often, a better performance according to one criterion results in a worsening of another criterion. The problem of optimizing methods with more than one criterion was discussed in Chapter 24, where multicriteria analysis was introduced.

In this section, we discuss the relationship between the cost and value of a test. In general, the more sophisticated or elaborated a procedure, the more information one expects and the more it costs. A common question in economics is the following : "given a certain procedure (for example, a polymer synthesis procedure), if more dollars are spent a better (or more of a) product will be obtained. How are additional costs and benefits related ?". This kind of question is answered by cost-benefit calculations. In the same way, an analytical chemist might ask, "given a certain procedure for determining lead in drinking water, spending X dollars can yield a faster method with a better precision. Is it worthwhile to spend the X dollars ?". In the context of analytical chemistry, one or two applications of cost-benefit analysis can be cited.

An interesting study concerns the so-called six sequential guiac protocol, which consists of six sequential tests on stool for occult blood, which indicates colonic cancer. Neuhauser and Lewicki (1975) asked the question : "what do we gain from the sixth stool guiac ?". The answer is contained in Table 25.II.

Table 25.II

Marginal cost of guiac stool tests

Number of tests	Marginal cost (US $)
1	1175
2	5492
3	49150
4	469534
5	4724695
6	47107214

Neuhauser and Lewicki (1975). Reprinted by permission.

The marginal cost can be defined here as the additional cost per additional case detected. In other words, the sixth test permits one to detect cancers not detected by the five first tests at a price of nearly 50 million dollars per case detected ! Other more or less relevant articles have been written by Durbridge et al. (1976) (on the evaluation of benefits from screening tests carried out immediately on admission to hospital) and by McNeil et al. (1975b) (on cost-effectiveness calculations in the diagnosis and treatment of hypertensive renovascular disease).

A very different, but equally interesting paper by Brown (1969) investigated the economic benefit to Australia of research on atomic-absorption spectroscopy. The costs considered were those encountered by research institutes in supporting the research ; the benefits included items such as productivity gains within assaying laboratories and licence fees. It is not possible to discuss here the methodology of these rather elaborate economic calculations, but it is interesting to note that the final conclusion was positive : research on atomic-absorption spectroscopy has yielded substantial economic benefits and should continue to do so during the next 10 years. It is also noteworthy that Brown concluded that cost-benefit analysis of research programmes is subject to the following difficulties :

(a) the results may take many years before making an economic impact ;

(b) the benefits are usually diffused throughout the economy ;

(c) the outcome of a research programme is very uncertain.

REFERENCES

J.D. Acland and S. Lipton, J. Clin. Pathol., 24 (1971) 369.
B. Baule and A. Benedetti-Pichler, Z. anal. Chem., 74 (1928) 442.
A.W. Brown, Econ. Record, 45 (1969) 158.
A.J. Duncan, Techmometrics, 4 (1962) 319.
T.C. Durbridge, F. Edwards, R.G. Edwards and M. Atkinson, Clin. Chem., 22 (1976) 968.
D.M. Green and J.A. Swets, Signal Detection Theory and Psychophysics, Wiley, New York, 1966.
G.H. Hall (1969), Lancet, 1 (1969) 531.
C.O. Ingamells and P. Switzer, Talanta, 20 (1973) 547.
D.A.B. Lindberg and F.R. Watson, Meth. Inf. Med., 13 (1974) 151.
L.B. Lusted, Science, 171 (1971) 1217.
H.F. Martin, B.J. Gudzinowicz and H. Fanger, Normal Values in Clinical Chemistry, Marcel Dekker, New York, 1975.
B.J. McNeil, E. Keeler and S.J. Adelstein, N. Engl. J. Med., 293 (1975a) 211.
B.J. McNeil, P.D. Varady, B.A. Burrows and S.J. Adelstein, N. Engl. J. Med., 293 (1975b) 216.
C.E. Metz, D.J. Goodenough and K. Rossmann, Radiology, 109 (1973) 297.
D. Neuhauser and M. Lewicki, N. Engl. J. Med., 293 (1975) 226.
D.B. Tonks, Clin. Chem., 9 (1963) 217.
T.J. Vecchio, N. Engl. J. Med., 274 (1966) 1171.
J. Visman, Materials Research and Standards 9, no 11 (1969) 9, 51, 62.
J.O. Westgard, R.N. Carey and S. Wold, Clin. Chem., 20 (1974) 825.
J. Yerushalmy,Radiological Clinics of North America, 7 (1969) 381.

Chapter 26

REQUIREMENTS FOR PROCESS MONITORING

26.1. INTRODUCTION

Many analytical problems are monitoring problems, for instance the sampling of a product stream and subsequently analysing the samples in order to obtain a picture of the variability of the composition of the product with time. It may also be necessary to estimate from the composition of the samples the average composition of the product during a certain time interval, i.e., the composition of a lot produced during that time interval.

In this chapter, the term monitoring is used in the sense of observing, registering or displaying the process by either continuously or intermittently measuring a process variable, for instance a concentration in a product stream. The term process is used in a general sense and is not restricted to industrial processes. The problem of sampling and analysing a product stream leaving a chemical plant is not essentially different from measuring one or more pollutants in a river. The monitoring of a chromatographic process by a detector is also a problem that belongs to this category.

In section 25.1 it was observed that the average composition of a lot can be determined with a precision σ_t depending on the inhomogeneities within the lot as expressed by the variance σ_x^2, and the precision σ of the analytical procedure. If only one sample is drawn from the lot, the following relationship holds

$$\sigma_t^2 = \sigma_x^2 + \sigma^2 \qquad (26.1)$$

If n samples are drawn and analysed individually, the precision reduces to

$$\sigma_t^2 = (\sigma_x^2 + \sigma^2)/n \qquad (26.2)$$

and for the analysis of a gross sample obtained upon combining and mixing these n samples, σ_t is given by

$$\sigma_t^2 = \sigma_x^2/n + \sigma^2 \tag{26.3}$$

In order to decide which sampling strategy should be used, information about the lot in terms of the variance σ_x^2 has to be available (for instance from past experience).

However, eqns. 26.1, 26.2 and 26.3 cannot be used when the samples are correlated. Common sense dictates the use of a sampling scheme in which the samples are drawn from parts that are well spread over the entire lot. Samples taken from nearly the same location are expected to be more alike, and thus correlated, than samples taken from locations far apart. The influence of correlations upon the requirements for sampling and analysing, in order to describe adequately the variations in the composition, is discussed in this chapter. Thereby the general, three-dimensional problem is reduced to a one-dimensional problem. A lot can be considered as a finite part of a process and inhomogeneous lots arise from processes that produce product streams of varying composition. Sampling such streams, for instance by taking samples from a conveyor belt, is often easier than sampling wagon loads (Van der Mooren, 1967).

For sampling and analysing streams, a wide variety of analysers (detectors, monitors) are available. These analysers either operate continuously or intermittently and thus also sample the product stream continuously or intermittently (Chapter 10). Although such process analysers are often used on line with the process, often adequate monitoring of the process is possible by taking samples and transporting them to the laboratory for the analysis.

In general, the object of the analysis will be to reconstruct the variations in the composition. However, in a number of instances the analyst is interested in the average composition during a certain time interval (the composition of the lot) and the position of maxima or peak areas (gas chromatography).

The selection of an analyser for process monitoring, whether operating continuously or not, obviously requires a knowledge of the characteristics of the analyser. It also requires a knowledge of the process to be monitored. In section 26.2 we consider the type of knowledge that has to be gathered in order to make a sensible selection.

Although the analytical problem will be discussed in terms of measuring variations as a function of time, the treatment applies equally when the time variable is replaced by the position (sampling a river at several locations). Sampling in more dimensions (surfaces, soils, etc.) is more complex but not fundamentally different.

For general reading and mathematical details, the reader is referred to the literature on process control, electronics and systems theory (see Chapter 10).

26.2. DESCRIPTION OF THE FLUCTUATIONS

In general, a stream must be sampled continuously or frequently and the samples analysed precisely in order to obtain the information required for a description of that stream in terms of variations in the composition. "Slow" monitors tend to obscure "fast" fluctuations, and "noisy" monitors add fluctuations to those arising from the process to be monitored. Whereas the time can be, but not necessarily is, an important criterion for the selection of a procedure for the analysis of discrete samples, it is imperative to include the time parameters when considering the problem of process monitoring.

In Part I the characteristics of analytical procedures, or of analysers, were discussed, in Chapter 10 with special emphasis on the characterization of continuous analysers. In this section, ways of presenting process fluctuations are described. Thereby, a distinction has to be made between "deterministic" and "stochastic" fluctuations. An example of the first category is the chromatographic process, where the elution profiles can be described (approximately) by the gaussian function

$$x(t) = x(t_R) \exp \left[-(t - t_R)^2/2\sigma_g^2\right] \qquad (26.4)$$

where x(t) is the concentration of the eluted component at time t, $x(t_R)$ the (maximal) concentration at the retention time t_R and σ_g^2 the variance of the gaussian peak and thus a measure of the width of that peak.

Stochastic fluctuations cannot be described analytically (in a mathematical sense) and are more conveniently represented by the autocovariance or autocorrelation functions that were introduced in Chapter 10 for the purpose of describing noise. An example of such an autocorrelation function (autocorrelogram) is presented in Fig. 26.1, taken from a paper by Müskens and Hensgens (1977).

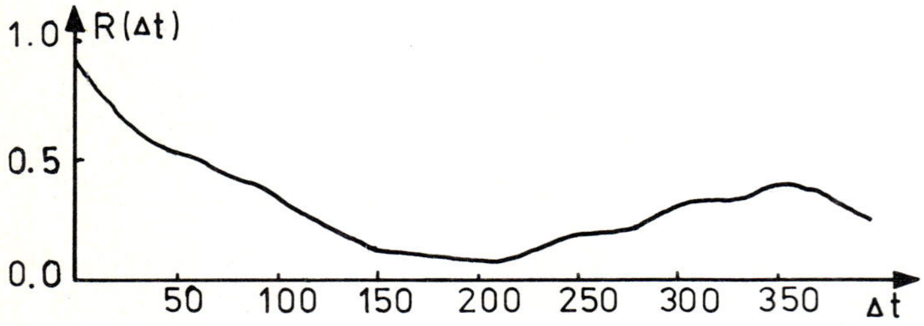

Fig. 26.1. Autocorrelogram of the NH_4^+ concentration in the River Rhine at Bimmen during the period 1971-75 (Müskens and Hensgens, 1977).

An analysis of this autocorrelation function reveals an annual periodicity and an exponential component with a time constant τ_x of about 120 days. This value cannot be considered to be very reliable as the period during which the measurements were gathered was relatively short with respect to the correlation time. Inspecting the correlogram by eye leads to a lower value, probably as low as 50 days. This, however, hardly affects the conclusions drawn in this chapter. The function also shows a rapid decrease for values of $\Delta t < 1$ day. This rapid decrease can be ascribed to the experimental errors. In fact, the autocorrelation function in Fig. 26.1 is not the true function for the river but includes the autocorrelation of the noise of the analytical procedure. For a complete analysis of the data, the reader is referred to the original literature. It is important to observe that Fig. 26.1 represents an *a posteriori* model of the river (with respect to the NH_4^+ fluctuations). Examples of correlograms of several processes were given by Vandeginste et al. (1976) and Van der Grinten and Lenoir (1973).

As a first approximation, the (symmetrical) autocovariance function of the NH_4^+ variations can be considered to be exponential

$$R(\Delta t) = s_x^2 \exp\left(- |\Delta t|/\tau_x\right) \qquad (26.5)$$

It appears that exponential autocovariance functions give a satisfactory description of many processes (Van der Grinten and Lenoir, 1973).

Both the gaussian function for describing the elution profile (eqn. 26.4) and the autocovariance function of the stochastic fluctuations are representations of x(t) in the time domain. As has been indicated in Chapter 10, the autocovariance function can be converted into the power spectrum through a Fourier transform. Although x(t) cannot be reconstructed from the power spectrum, the spectrum defines the (average) contributions of the periodic sine and cosine functions of different frequencies to x(t). For an exponential covariance function this contribution is the same for angular frequencies ω below $\omega_x = 1/\tau_x$. From this frequency onwards, the contribution rapidly decreases.

Whereas the Fourier transform of the autocovariance function yields the power spectrum, the Fourier transform of the fluctuations x(t) yields the frequency spectrum, i.e., the amplitudes of the periodic sine and cosine waves from which the function x(t) can be reconstructed. For a (symmetrical) gaussian profile with the origin at t_R ($t_R = 0$), the cosine transform

$$x(\omega) = F\left[x(t)\right] = \int_0^{\infty} x(t) \cos(\omega t)\, dt \qquad (26.6)$$

yields the frequency spectrum

$$x(\omega) = F\left[x(t)\right] = \sigma_g \sqrt{\pi/2}\, \exp\left(- \omega^2 \sigma_g^2 / 2\right) \qquad (26.7)$$

which is also gaussian, with a standard deviation equal to the reciprocal of the width of the gaussian elution profile σ_g (Westerberg, 1969). x(t) can be recovered from $F\left[x(t)\right]$ through the inverse Fourier transform

$$x(t) = \int_0^{\infty} F\left[x(t)\right] \cos \omega t \; d\omega \qquad (26.8)$$

It is important to observe that $F\left[x(t)\right]$ has a value of virtually zero for frequencies ω of about three or four times $1/\sigma_g$. To put it differently, the elution profile virtually does not contain frequencies higher than about $3/\sigma_g$ to $4/\sigma_g$ radians per second (rps) or $3/2\pi\sigma_g$ to $4/2\pi\sigma_g$ cycles per second (cps). For recovering the signal from the continuous frequency spectrum $x(\omega)$, the integration limit in eqn. 26.8 can be replaced by this upper frequency.

Whereas the Fourier integral of eqn. 26.8 covers the gaussian elution profile between $-\infty$ and $+\infty$, the Fourier expansion as used by McWilliam and Bolton (1969a) and by Goedert and Guiochon (1973) represents the gaussian peak $x(t)$ between $t = -\frac{1}{2} T = -\pi\sigma_g$ and $t = +\frac{1}{2} T = +\pi\sigma_g$ by

$$x(t) = 0.3989 + 0.4839 \cos 2\pi \frac{t}{\tau} + 0.1080 \cos 4\pi \frac{t}{\tau} + 0.0089 \cos 6\pi \frac{t}{\tau} + 0.0002 \cos 8\pi \frac{t}{\tau} \qquad (26.9)$$

It can be seen from this equation that the contribution (amplitude) of the cosine terms to $x(t)$ decreases rapidly with increasing frequency. The amplitude of the fourth harmonic, with a frequency of $4/T$ cps or $8\pi/T$ rps, is only a few ‰ compared with the amplitude of the basic frequency. Recovering $x(t)$ with eqn. 26.9 and thus neglecting the fifth and higher harmonics yields a virtually perfect gaussian profile. In agreement with the results of Westerberg, the Fourier expansion of eqn. 26.9 yields a highest frequency of $4/T = 4/2\pi\sigma_g$ cps. For a 6σ peak width of 1 sec this corresponds to roughly 4 cps.

The Fourier analysis as shown for the gaussian profile can be applied to any curve. In practice, there always will be an upper frequency that contributes to the fluctuations. However, in many instances the transforms and expansions have to be obtained numerically.

26.3. MONITORING WITH CONTINUOUS ANALYSERS

26.3.1. The distortion of a gaussian peak

The shape of a gas chromatographic elution profile as measured by the detector and shown by the recorder depends on many factors. One of these factors is the finite time constant of the detector (and the recorder) used for monitoring the effluent of the column. The influence of the detector response upon a symmetrical, gaussian profile emerging from the column has been the subject of many studies (for instance, Gladney et al., 1969, McWilliam and Bolton, 1969a,b, 1971, Anderson et al., 1970, Grushka, 1972, Goedert and Guiochon, 1973, Chesler and Cram, 1973, Pauls and Rogers, 1977). These studies were mainly initiated by the difficulties that were met in the data processing of chromatograms with skewed peaks. Most of these studies relate to the distortion of a gaussian peak by a detector with a first-order response and thus characterized by a single time constant τ_a (see Chapter 10).

The distortion can most easily be expressed by the so-called convolution integral, i.e.

$$y(t) = S \int_{-\infty}^{+\infty} x(t-t')\, h(t')\, dt' \qquad (26.10)$$

where S, x(t), y(t) and h(t) are the sensitivity, the input, the output and the pulse response of the detector, respectively, and t' is a dummy variable. For a gaussian input and an exponential (first-order) response, the convolution integral becomes

$$y(t) = S \int_{0}^{\infty} x(t_R) \exp\left[-(t-t'-t_R)^2/2\sigma_g^2\right] \cdot \frac{1}{\tau_a} \exp(-t'/\tau_a)\, dt' \qquad (26.11)$$

where $x(t_R)$ is the concentration in the maximum of the undistorted peak, σ_g is the width (standard deviation) of the gaussian peak and τ_a the time constant of the detector. It can be seen that y(t) is determined by the entire profile eluted before the time t to an extent that is given by the exponential response of the detector. The integral in eqn. 26.11 is taken between 0 and ∞, corresponding with the physically significant part of the exponential function. Eqn. 26.11 cannot be solved analytically. A numerical solution for several values of τ_a/σ_g is presented graphically in Fig. 26.2, taken from a paper by McWilliam and Bolton (1969a).

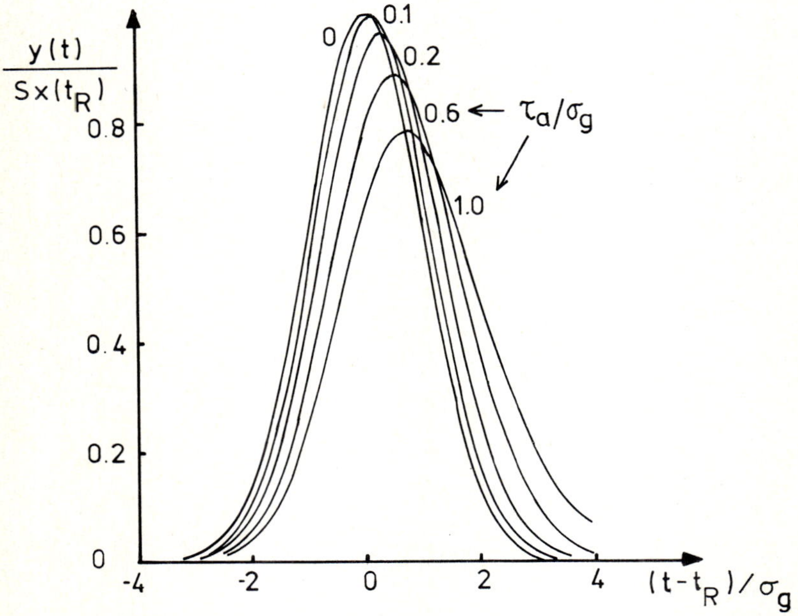

Fig. 26.2. Distortion of a gaussian peak for several values of τ_a/σ_g (reprinted with permission from McWilliam and Bolton, 1969a).

The distorted peak is skewed and broader than the original peak. The maximum of the distorted peak for all values of τ_a/σ_g lies on the trailing track of the undistorted peak. Although this distortion does not affect the peak area, peak distortion should be avoided as much as possible for the accurate determination of retention times and automatic processing of chromatograms with (partially) overlapping peaks. Avoidance is better than a cure, even if the mathematical process of deconvolution (the reverse of convolution, see for instance Den Harder and De Galan, 1974) in a number of instances can be applied to restore the true peak shape or simply to sharpen the peaks (Kirmse and Westerberg, 1971). Peak distortion is visible for τ_a/σ_g ratios exceeding 0.1. Hence the time constant of the detector, in order to avoid skewing, should certainly be less than 1/60th of the width of the peak (about 6 σ_g).

Peak distortion can also be studied by making use of the presentation of the peak in the frequency domain. A time constant τ_a corresponds to a bandwidth or break frequency of $\omega = 1/\tau_a$ rps or $1/2\ \pi\tau_a$ cps. It is clear that the components of the gaussian curve with frequencies highe than $1/\tau_a$ rps become smaller in amplitude and shift in phase. As a result, the shape of the peak changes. As has been shown by McWilliam and Bolton (1969a), calculation of the reduction of amplitudes and phase shifts leads to the same results as the convolution method. At a frequency $\omega = 1/\tau_a$ the amplitude is reduced by a factor $\sqrt{2}$ and to be safe $1/\tau_a$ should be at least 10 times the highest frequency contained in the gaussian peak. This leads to roughly the same rule to be set for the selection of the detector. However, in critical situations, an accurate analysis of the problem has to replace this simple rule.

The principles described above can be applied to other analytical methods, provided that a first-order response of the detector can be assumed. Higher order responses were discussed by McWilliam and Bolton (1969b).

26.3.2. Measuring stochastic fluctuations

The measurement and distortion of stochastic fluctuations can be described in essentially the same way as that of the gaussian elution profile. However, because of the stochasticity, a description in terms of the convolution integral of eqn. 26.10 in order to arrive at the requirements for a monitor to be used for measuring these fluctuations hardly makes sense. However, a description in the frequency domain yields useful results.

As has been shown, the fluctuating process is described adequately by its power spectrum. If we consider a process characterized by an exponential autocorrelation function with a correlation time τ_x, the power spectrum is given by (Van der Grinten and Lenoir, 1973)

$$\sigma_x^2\ (\omega) = \frac{2\sigma_x^2\ \tau_x}{1-\omega^2\tau_x^2} \tag{26.12}$$

The integral taken over all values of ω yields the variance σ_x^2.
For a first-order response of the monitor, the variance as seen by the monitor is given by the integral (for a sensitivity S = 1)

$$\sigma_y^2 = \int_0^\infty \frac{2\sigma_x^2\ \tau_x}{1-\omega^2\tau_x^2} \cdot \frac{1}{1+\omega^2\tau_a^2} \cdot d\omega \tag{26.13}$$

where the power of the (high-frequency) components is reduced to an extent that depends on the time constant τ_a of the monitor. Solving the integral of eqn. 26.13 leads to the simple result (Van der Grinten and Lenoir, 1973)

$$\sigma_y^2 = \sigma_x^2\ \frac{\tau_x}{\tau_x + \tau_a} \tag{26.14}$$

It is clear that for fast analysers ($\tau_a \ll \tau_x$) the variations observed are equal to the true variations. For slow analysers the observed variations will be reduced and even vanish for very large values of τ_a. For $\tau_a = 0.1\ \tau_x$, $\sigma_y^2 \approx 0.9\ \sigma_x^2$, and thus $\sigma_y \approx 0.95\ \sigma_x$. For $\tau_a = 0.01\ \tau_x$, $\sigma_y^2 \approx 0.99\ \sigma_x^2$, and thus $\sigma_y \approx 0.995\ \sigma_x$. Obviously, as a rule one might state that, if one is interested in a reliable estimate of the standard deviation, the time constant of the monitor should be smaller by a factor of 10 - 100 than the correlation time or time constant of the process. Using this argument, one might arrive at the conclusion that the River Rhine (Fig. 26.1) needs to be sampled and analysed with a monitor having a time constant of about 1 day to 1 week in order to obtain a reliable picture of the standard deviation of the NH_4^+ concentration (assuming that the *a posteriori* calculated correlogram still holds). In this case, any continuous analyser will be satisfactory as far as the time constants are concerned.

Making sure that the fast fluctuations are recorded faithfully poses somewhat stricter requirements. The integrand of eqn. 26.13 can be used to estimate the effects for any frequency.

26.4. SHANNON'S SAMPLING METHOD - DISCRETE SAMPLING

In the preceding section, an account was given of the requirements for continuous (flow) analysers for the precise display of the fluctuations of the process through the measurement of one of its properties. In many instances the process is sampled discontinuously, either because the analyser operates discontinuously (for instance, a process gas chromatograph) or because the samples have to be analysed in the laboratory. However, it is possible to reconstruct the true variations from the values obtained from samples that are taken at (regularly spaced) intervals. To this end, the sampling or analysing time t_a or its reciprocal, the sampling frequency, has to be adjusted to the properties (time constant) of the process to be monitored.

Rewriting eqn. 26.6 as the complex Fourier integral leads to the complex frequency spectrum

$$\mathcal{X}(\omega) = \int_{-\infty}^{+\infty} x(t) \exp(-j\omega t)\, dt \qquad (26.15)$$

where $j = \sqrt{-1}$. The function $x(t)$ can be obtained from the inverse Fourier transform

$$\mathbf{x}(t) = \int_{-\omega_m}^{+\omega_m} \mathcal{X}(\omega) \exp(j\omega t)\, d\omega \qquad (26.16)$$

where the integration boundaries can be set equal to ω_m, the highest frequency component of $x(t)$. The value of x for $t = n/2\nu_m = n\pi/\omega_m$ (n integer) is given by ($\omega = 2\pi\nu$)

$$x\left(\frac{n}{2\nu_m}\right) = x\left(\frac{n\pi}{\omega_m}\right) = \int_{-\omega_m}^{+\omega_m} \mathcal{X}(\omega) \exp\left(\frac{j\pi n\omega}{\omega_m'}\right) d\omega \qquad (26.17)$$

Development of the frequency spectrum in a Fourier series with a basic frequency $\nu_o = 2\pi\omega_o = 1/T$ cps leads to

$$x(\omega) = \sum_{n=1}^{\omega_m/\omega_o} c_n \exp\left(-\frac{j\pi n\omega}{\omega_m}\right) \tag{26.18}$$

with

$$c_n = \frac{\pi}{\omega_m} \int_{-\omega_m}^{+\omega_m} x(\omega) \exp\left(\frac{j\pi n\omega}{\omega_m}\right) d\omega \tag{26.19}$$

Combination of eqns. 26.19 and 26.17 leads to the equality

$$x\left(\frac{n}{2\nu_m}\right) = 2\nu_m c_n \tag{26.20}$$

Thus the n sampled values of x(t) determine the coefficients c_n, which in turn determine the frequency spectrum and therefore also the complete function x(t) between the values t = 0 and t = T. The sampling interval should be equal to $\Delta t = 1/2\ \nu_m$ and the sampling frequency 2 ν_m, twice the highest frequency contained in the function x(t). This rule governing the sampling interval in order to restore from the sampled values of x(t) the original function is called Shannon's sampling theorem. The corresponding sampling interval $\Delta t = 1/2\ \nu_m$ is called the Nyquist interval. Application of Shannon's sampling theorem to the gaussian elution profile leads to approximately eight samples per 6 σ-width (Westerberg, 1969). Thus a peak of 1 sec requires a sampling frequency of 8 cps.

Some remarks should be made about regaining x(t) from the sampled values. It is clear that such a reconstruction should be made via the frequency spectrum, but it is also evident that this would involve lengthy calculations. It can be shown that it is easier to obtain x(t) by making use of the so-called cardinal function $(\frac{\sin x}{x})$ (see, for instance, Gore, 1960). We shall not discuss the use of this function as it is seldom applied in analytical chemistry.

In the data processing of chromatograms, spectra, etc., it is common practice to make use of curve-fitting techniques. The interval chosen for digitizing the detector response (for storage in a digital memory) is usually made smaller than the Nyquist interval, permitting the recovery of x(t) through a linear, quadratic or cubic interpolation between the sampled values. At the same time, the digitized values can be smoothed with the result that the influence of the detector noise is reduced to some extent (digital filtering).

The sampling theorem of Shannon and the corresponding Nyquist interval define the minimal number of samples required for reconstruction of the continuous fluctuations, whatever the method that is chosen for this reconstruction. If the sampling rate is too small, information will be lost and the reconstruction is in error. A discussion of these errors for some deterministic signals has been given by Kelly and Horlick (1973) and by Horlick and Yuwen (1976). Table 26.I gives an indication of the errors that are made in estimating peak heights for a set of profiles frequently met in analytical chemistry.

Table 26.I

Minimal number of samples required for a given accuracy (Kelly and Horlick, 1973)

Maximum error, % (% of peak height)	Triangle	Exponential	Lorentz	Gaussian
10	6	20	6	3
1	40	330	36	6
0.1	360	4500	150	9
0.01	3200	51000	630	11
0.001	...	...	2600	14

26.5. INFLUENCE OF ERRORS

In the previous sections it has been assumed that the fluctuating property can be recorded or reconstructed properly when the time constant of the monitor or the sampling frequency meets certain requirements. However, if the measurements are in error, the recorded or reconstructed function x(t) is also in error. It is common practice in the processing of digitized spectra to choose a higher digitizing (sampling) frequency than the frequency corresponding to Shannon's sampling theorem. An extensive treatment of (analogue and digital) filtering procedures cannot be given here. However, it is sufficient to note that usually the power spectrum of the noise of monitors extends to much higher frequencies than the power spectrum or frequency spectrum of the process to be monitored. If the time constant τ_a or the sampling frequency T_a is very much smaller than the correlation time of the process, a filter can be used to remove the noise and still retain the variations of the process variable. Instead of the time constant of the monitor, the time constant of the filter determines whether these variations can be monitored adequately. The same applies to the filtering of sampled values. If the power spectra of noise and variations x(t) overlap, means other than filtering have to be used to recover the true variations (see, for instance, Hieftje, 1972a,b).

26.6 DETERMINING AVERAGE COMPOSITIONS

A category of problems frequently met in analytical chemistry is the sampling and analysis of a process stream in order to determine the average composition of that stream over a limited period of time T. The average composition is often a measure of the quality of the batch or lot produced during that period. The sampling requirements for the case of uncorrelated samples have already been indicated in section 26.1. If a certain precision is required, the number of samples depends simply on the variance found within the lot and on the precision of the analytical procedure (eqns. 26.1, 26.2 and 26.3). An entirely different situation arises for samples that are correlated. This problem has been discussed by Müskens and Kateman (1978). However, the equations to be used for calculating the precision for different sampling rates, correlations and lot sizes are rather awkward. Therefore, the results of Müskens and Kateman (1978) will be presented graphically.

The problem can be formulated as follows. A lot is a part (from $t = 0$ to T) of a process (stream). The process is characterized by a correlation time or time constant τ_x and a variance σ_x^2. Then, the real mean of the lot is

$$\mu = \frac{1}{T} \int_0^T x(t)\, dt \qquad (26.21)$$

where x(t) is the real composition at time t. The lot is sampled every t_a seconds (or hours, etc.). Every sample is taken during a time interval G and thus corresponds to the average composition during that time interval. The number of samples combined into a gross sample is n ; consequently $n = T/t_a$. Müskens and Kateman calculated σ_{est}^2 as a measure for the uncertainty of estimating the real average composition of the lot through the analysis of the gross sample ; σ_{est} is the sampling error that arises from the variations within the lot and it depends on σ_x^2, τ_x, T, n (or t_a) and G. The relationship between the relative lot size, T/τ_x, and the relative sample size, G/T, for different values of n in order to obtain a precision $\sigma_{est} = 0.1\ \sigma_x$ is represented in Fig. 26.3. The precision of the analytical procedure has not been taken into account. However, the total precision can easily be calculated as the sampling errors will usually be independent of the precision of the analytical procedure.

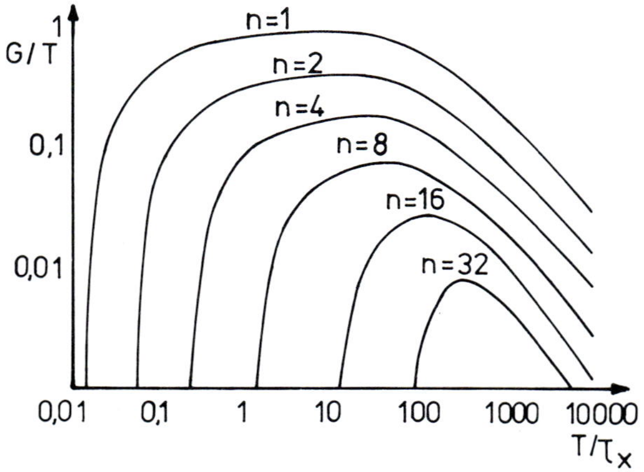

Fig. 26.3. Number of samples (n) required for obtaining a precision equal to one tenth of the process variations (σ_x) as a function of the relative sample size (G/T) and relative lot size (τ/τ_x) (Müskens and Kateman, 1978).

From Fig. 26.3, which applies to processes with exponential correlation functions, some important conclusions can be drawn.

(a) For small lots (small with respect to the correlation time) the composition within the lot hardly changes due to the correlation and only one or a few small samples are required in order to obtain a precise value of the composition of the lot. However, different lots will have different compositions.

(b) In contrast to the small lots, the concentration within large lots will fluctuate appreciably. In order to obtain a certain precision, one can take many small (uncorrelated) samples (compare with eqn. 26.2 or 26.3). It is also possible to analyse one large sample with the effect that within this large sample many fluctuations are included and thus an average value is obtained.

(c) For medium-sized lots a certain precision can be obtained by either taking many small samples or less and larger samples. The resulting gross sample will be smaller when it is composed of many small samples.

(d) For a given sample size the number of samples required will initially increase with the size of the lot [see also (a)] and subsequently decrease. For large lots, or uncorrelated processes, the mean composition of a sample is equal to the mean of the lot.

The total precision, σ_t, of determining the average NH_4^+ concentration (annual mean) in the River Rhine at Bimmen (see Fig. 26.1) as a function of the number of samples, n, was calculated by Müskens and Kateman (1978) with the equations

$$\sigma_t^2 = \sigma_{est}^2 + \sigma^2/n \qquad (26.22)$$

and

$$\Delta_{est} = t_n \sigma_t \qquad (26.23)$$

for a confidence level of 0.05. t_n is the value of Student's t for this confidence level and n determinations. The results are plotted in Fig. 26.4 for small samples (G = 0) and for integrated samples between the sampling actions ($G = t_a$). For the last situation, σ_{est}^2 tends to be zero as $G = t_a$ is equivalent to continuous sampling and taking an integrated sample every t_a (seconds, hours).

Then the value of Δ_{est} will be equal to $t_n\ \sigma/\sqrt{n}$.

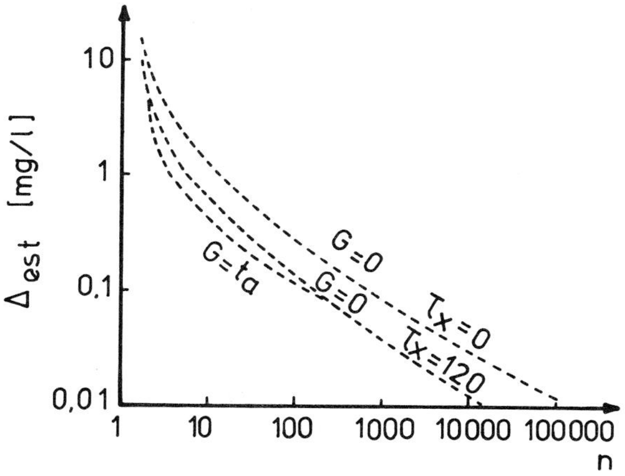

Fig. 26.4. Estimated precision (Δ_{est}) as a function of the number of samples for processes with correlation times (τ_x) of 0 and 120 days (Müskens and Kateman, 1978).

From Fig. 26.4, conclusions can be drawn about the sampling strategy. Müskens and Kateman concluded that, of course, the best strategy (Δ_{est} = min.) would involve the collection of a sample over the whole year and a precise analysis of this sample. If no precise method is available or if the samples cannot be preserved, it follows that in order to obtain a precision of 0.1 mg/l (average about 3 mg/l) would involve taking an (integrated) sample every 3 days. If no correlation data were known and one consequently had to assume a value of zero for τ_x, the sampling frequency should be increased by roughly a factor 10, clearly illustrating the benefit of an *a priori* knowledge concerning the process.

One final remark should be made. There should be no doubt about the *a priori* knowledge of the process being the same as the *a posteriori* description of the process as derived from (a large number of) measurements made in the past. Of course, this remark not only applies to this section but is also valid for all applications mentioned in this (and the next) chapter.

REFERENCES

A.H. Anderson, T.C. Gibbs and A.B. Littlewood, J. Chromatogr. Sci., 8 (1970) 640.
B. Baule and A. Benedetti-Pickler, Z. anal. Chem., 74 (1928) 442.
S.N. Chesler and S.P. Cram, Anal. Chem., 45 (1973) 1355.
H.M. Gladney, B.F. Dowden and J.D. Swalen, Anal. Chem., 41 (1969) 883.
M. Goedert and G. Guiochon, Chromatographia, 6 (1973) 76.
W.C. Gore, Operations Research and Systems Engineering, in C.D. Flagk, W.H. Huggins and R.H. Roy (Editors), John Hopkins, Baltimore, Md., 1960, p. 604.
E. Grushka, Anal. Chem., 44 (1972) 1733.
A. den Harder and L. de Galan, Anal. Chem., 46 (1974) 1464.
G.M. Hieftje, Anal. Chem., 44 (1972a) 81A.
G.M. Hieftje, Anal. Chem., 44 (1972b) 69A.
G. Horlick and W.K. Yuen, Anal. Chem., 48 (1976) 1643.
P.C. Kelly and G. Horlick, Anal. Chem., 45 (1973) 518.
D.W. Kirmse and A.W. Westerberg, Anal. Chem., 43 (1971) 1035.
I.G. McWilliam and H.C. Bolton, Anal. Chem., 41 (1969a) 1755.
I.G. McWilliam and H.C. Bolton, Anal. Chem., 41 (1969b) 1762.
I.G. McWilliam and H.C. Bolton, Anal. Chem., 43 (1971) 883.
P.J.W.M. Müskens and N.G.J. Hensgens, Water Res., 11 (1977) 509.
P.J.W.M. Müskens and G. Kateman, Anal. Chim. Acta, Computer Techniques and Optimisation, 1 (1978) 1, 11.
R.E. Pauls and L.B. Rogers, Anal. Chem., 49 (1977) 625.
B.G.M. Vandeginste, P.J.M. Salemink and J.C. Duinker, Neth. J. Sea Res., 10 (1976) 59.
P.M.E.M. van der Grinten and J.M.H. Lenoir, Statistische Procesbeheersing, Spectrum, Utrecht/Antwerp, 1973.
A.L. van der Mooren, Thesis, Delft, 1967.
A.W. Westerberg, Anal. Chem., 41 (1969) 1770.

Chapter 27

REQUIREMENTS FOR PROCESS CONTROL

27.1. INTRODUCTION

The requirements for the use of analysers for process monitoring were considered in Chapter 26. It appeared that process variations in many instances can be reconstructed from the measurements, even if the analyser does not respond infinitely rapidly or when the process is sampled intermittently. However, often in industrial practice process variables have to be kept constant, or within certain limits, for instance in order to manufacture a product of constant quality. To this end, the process has to be controlled and the control action often has to be preceded by a measurement. The analytical chemist is often faced with the development of procedures or analysers to be used, either on-line or off-line, for this purpose. The requirements for such procedures or analysers are set by the nature of the process, the extent to which the disturbances are to be reduced by the control action and by the quality of the controller. Van der Grinten (1963a, b, 1965, 1966, 1973) has developed criteria that can serve as a guide for the analytical chemist who is faced with the problems of how precisely, how rapidly and how frequently to analyse in order to make successful process control possible or at least to select the best procedure or analyser for the process control.

If a control action is to be based on the result of an analytical measurement, the quality of the control will depend on the quality of the result of the measurement. Imprecise results will lead to imprecise actions and a result that comes too late for the control action is useless.

Van der Grinten has defined the quality of the control, the control efficiency or the controllability (factor) by the relationship

$$r^2 = \frac{\overline{x(t)^2} - \overline{x_c(t)^2}}{\overline{x(t)^2}} = \frac{\sigma_x^2 - \sigma_{xc}^2}{\sigma_x^2} \tag{27.1}$$

where x(t) is the quantity, for instance a concentration in a product stream, upon which the control action will be based. x(t) is taken with respect to the average value which is equal to the desired value, the so-called set point. Thus $\overline{x(t)^2}$ is the average squared deviation from that set point. Whereas x(t) represents the value for the uncontrolled process, $x_c(t)$ refers to the optimal controlled process. The controllability r reaches a value of unity in perfect control (no disturbances left after control). When r = 0 the control action has no effect, so $\sigma_x = \sigma_{xc}$.

In practice, the controllability is determined by the pattern of the fluctuations, here again assumed to be first order with an exponential correlation function (see Chapters 10 and 26), and by the characteristics of the controller, of the analyser and of the (chemical or physical) process. The contribution of the analyser to the control efficiency will be indicated by the measurability (factor) m. For a perfect controller and process, the quality of the control loop is limited only by the "imperfections" of the analyser (precision, dead time, etc.). In this case

$$r = m \tag{27.2}$$

27.2. MEASURABILITY

27.2.1. Dead time

The (exponential) autocovariance function (eqn. 26.5) can be considered as a function that permits the forecasting of future disturbances. If the deviation at time t = 0 is given by x(0), this forecast of the disturbance x(t) can be made by

$$x(t) = x(0)e^{-t/\tau_x} \tag{27.3}$$

In fact, x(t) is the average or most probable disturbance that can be expected at time t when the disturbance at t = 0 is known to be x(0). This behaviour is illustrated in Fig. 27.1.

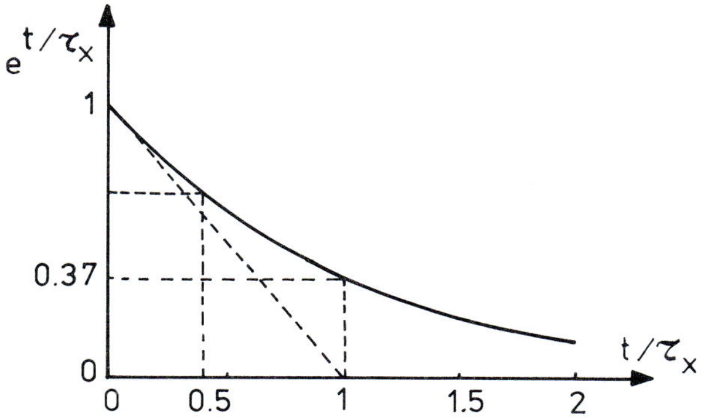

Fig. 27.1. Average change of disturbance x(t) (van der Grinten, 1965).

The disturbance at time t will be known at time $t + t_d$ if at time t the process is sampled and the dead time of the analyser is t_d. Obviously the controller should therefore not act upon the disturbance at time t, but upon the most probable disturbance at time $t + t_d$. The efficiency of the control action is given by the following equation, expressing the fact that only the deviations from $t + t_d$ onwards can be accounted for

$$m_d = \exp(-t_d/\tau_x) \qquad (27.4)$$

The measurability factor m_d resulting from a finite dead time decreases exponentially with that dead time. This equation can be applied to analysers sampling either continuously or intermittently.

27.2.2. Sampling time

An analyser with a sampling time t_a has a similar effect upon the control efficiency as the dead time t_d. If at time t a control action should be taken, it has to be based upon the measurement of a sample taken at time t - t'. When the analyser is ready for sampling, t' will be zero. It also can happen that for the control action the sample composition at time $t - t_a$ has to be used, i.e., when the analyser is almost ready for the next sampling. According to van der Grinten (1963a,b) the best control action consequently requires a forecast over an average length of time $\frac{1}{2} t_a$. This leads to a measurability factor of

$$m_a \approx \exp\left(-\frac{1}{2} t_a/\tau_x\right) \qquad (27.5)$$

For continuous sampling, t_a reduces to zero and thus $m_a = 1$.

27.2.3. Precision

Owing to imprecise measurements with continuous analysers, the control action may be in error. The measurability factor due to such imperfections, according to van der Grinten (1963a,b) is given approximately by

$$m_p \approx 1 - \frac{\sigma}{\sigma_x} \sqrt{\frac{\tau_a}{\tau_x}} \qquad (27.6)$$

Apparently m_p increases with increasing precision (decreasing σ) and decreasing time constant τ_a. For (continuous) analysers with a bandwidth ($1/\tau_a$) much larger than the bandwidth of the process variations ($1/\tau_x$), it is possible to decrease σ by filtering the continuous signal (true variations and analytical noise) and still retain the true variations. However, the filtering process for $\tau_a < \tau_x$ can never lead to a noise reduction smaller than $\sigma \sqrt{\tau_a/\tau_x}$ (eqn. 26.14). Thus, even a perfect controller cannot reduce the disturbances to less than $\overline{x_c(t)^2} = \sigma^2 \tau_a/\tau_x$, which explains qualitatively eqn. 27.6.

Eqn. 27.6 applies to continuous analysers. For practical purposes this equation can also be used for analysers with intermittent sampling ; then τ_a in this equation has to be replaced by the sampling time t_a.

27.2.4. The total measurability

The total measurability, apart from some factors of minor importance (samples gathered during a certain interval of time or integrated samples), is approximately given by

$$m \approx m_d \, m_a \, m_p \qquad (27.7)$$

The measurability as a function of the characteristics of the analyser is represented in Figs. 27.2 and 27.3.

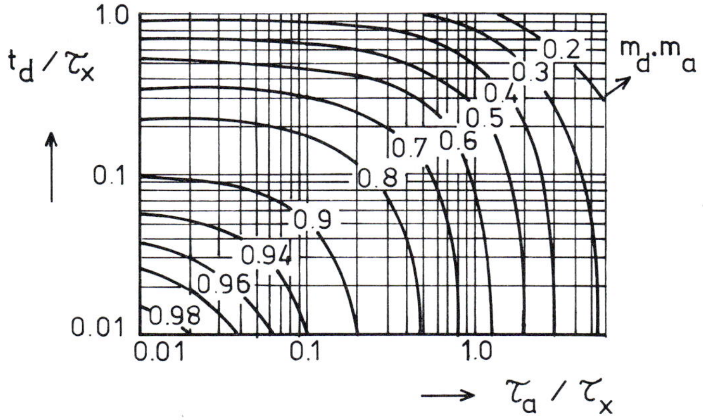

Fig. 27.2. Measurability as a function of dead time and time constant for first-order processes. (van der Grinten and Lenoir, 1973).

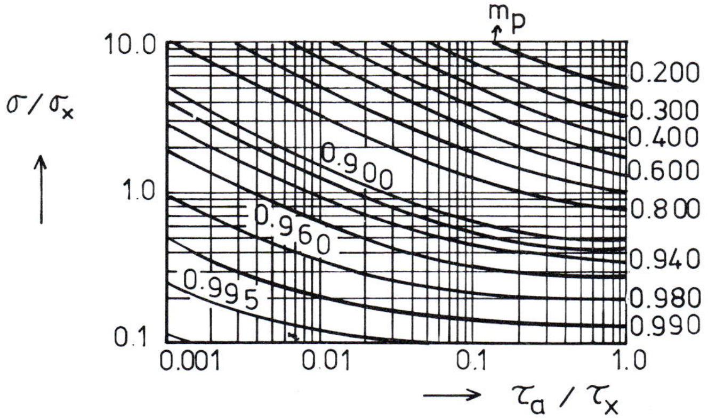

Fig. 27.3. Measurability as a function of precision and time constant for first-order processes. (van der Grinten and Lenoir, 1973).

These figures can be used for estimating the measurability graphically. As expected, the control efficiency decreases rapidly with increasing τ_a, t_a, t_d

and σ. It is clear that these constants have to be considered in relation to the pattern of fluctuations (see equations for m and Figs. 27.2 and 27.3). Even large time constants, etc., will not prevent a satisfactory control action of a "slow" process and ultimate precision usually is not required for the control of "noisy" processes.

27.3. SOME APPLICATIONS

Van der Grinten (1963a, b, 1965, 1966), van der Grinten and Lenoir (1973) and also Leemans (1971) described a number of applications illustrating clearly the importance of the above to the analytical chemist. We feel that the analytical chemist should cooperate with the control engineer in selecting the optimal analyser. Although the equations that can be used for comparing analytical procedures for process control are fairly simple, the underlying mathematics are complex and cannot be dealt with in this chapter. Because of this and also because the process variables are required for making the right choice, the problem can hardly be solved by the analytical chemist alone. The applications given here merely serve as an illustration that the analytical chemist should be aware of his possible contributions to the solution of the kind of problems described in this chapter.

The applications also illustrate clearly the interaction between the time parameters and the precision. Although a rigorous solution of the control problem is not always possible because of the lack of a satisfactory model of the process to be controlled, the principles are generally valid. Almost every analysis is part of a control loop and there always has to be a balance between time and precision of the analysis. If the optimal situation cannot be calculated exactly, it has to be approached intuitively. Probably very helpful for developing such an intuitive approach will be the simulation game described by van den Akker and Kateman (1976). This game is based upon the principles described in this chapter. In this context, we also refer to a paper by Vandeginste and Janse (1977).

Van der Grinten and Lenoir (1973) described the use of a process chromatograph for the control of a process with τ_x = 250 min and σ_x = 5%. The component to be measured has as retention time t_R = 5.9 min, whereas the total chromatographic run required 20.7 min. The concentration of the component can be estimated from the peak height with a precision of 3% (σ). As the sampling time is 20.7 min, a measurability m_p = 0.91 can be calculated or determined graphically from Fig. 27.2. With a dead time t_d = 5.9 and a sampling time t_a = 20.7, measurabilities m_d = 0.98 and m_a = 0.96 are obtained. Hence the total measurability m = 0.98 . 0.96 . 0.91 = 0.86, corresponding in a reduction of disturbances from 5% to 2.6%.

The precision can be increased to virtually σ = 0 and thus m_p = 1 when the peak area of the component is normalized with respect to the area of all peaks in the chromatogram. However, then t_d is increased from 5.9 to 20.7 min and consequently m_d decreases from 0.98 to 0.93. The resulting m is then 0.89 with a corresponding reduction of the disturbances from 5% to 2.3%, slightly better than for the analysis making use of the peak height. Clearly there is a trade-off between precision and speed.

In Table 9.I and Fig. 9.1 a number of analytical procedures for the determination of nitrogen were compared in considering the cost of the analysis (Leemans, 1971). The same procedures can be used for a comparison when used for controlling a process for the manufacture of a nitrogenous fertilizer with τ_x = 66 min and σ_x = 1.2% N. The cost of the analysis in order to reach a certain measurability is plotted in Fig. 27.4. Whereas m_d and m_p cannot be influenced, m_a can be increased by decreasing t_a (taking samples more frequently). The measurability factors for a sampling time t_a = 30 min (m_a = 0.80) are given in Table 27.I. Some conclusions of Leemans can be summarized as follows

(1) The classical distillation yields the smallest measurability factor in spite of its high precision. This is caused by the large dead time of the analysis.

(2) With the sacrifice of (part of) the precision, the measurability is increased dramatically by using the faster automatic distillation procedure.

(3) The best performance is obtained with a non-specific imprecise procedure. It is an ideal method for on-line control, not only yielding the highest measurability but also being cheaper than all other analytical procedures.

Table 27.I

Measurability of some Analytical Techniques for Analysis of Nitrogen (adapted from Leemans, 1971)

A sampling frequency of 2 samples/h is assumed, which means that m = 0.80.

Criterion and analytical technique	Dead time of analysis, min	Standard deviation of analysis, % N	m_d	m_p	m
Total N, classical distillation	75	0.17	0.32	0.99	0.24
Total N, DSM automated analyser	12	0.25	0.84	0.97	0.65
NO_3-N, Technicon Autoanalyser	15.5	0.51	0.79	0.92	0.58
NO_3-N, ion-selective electrode	10	0.76	0.86	0.85	0.58
NH_4NO_3 : $CaCO_3$ ratio, X-ray diffraction	8	0.8	0.89	0.83	0.59
Total N, fast neutron--activation analysis	5	0.17	0.93	0.99	0.74
Specific gravity, γ-ray absorption	1	0.64	0.98	0.88	0.69

Undoubtedly the cost of analysis has to be related to the benefit of the control action. This action leads to a more constant product. In this case, the nitrogen content of the fertilizer should be 22%. If the standard deviation of the disturbances is σ_{xc}, the set point of (22 + 2 σ_{xc})% has to be chosen in order to guarantee with reasonable certainty a product containing at least 22% N. An increase in m therefore reduces the process costs, as is illustrated in Fig. 27.4. The difference between the process costs and analysis costs defines the optimal procedure, unless other selection criteria have to be taken into account. It should be observed that the best procedure for the process described in this section is not necessarily the best procedure for other processes.

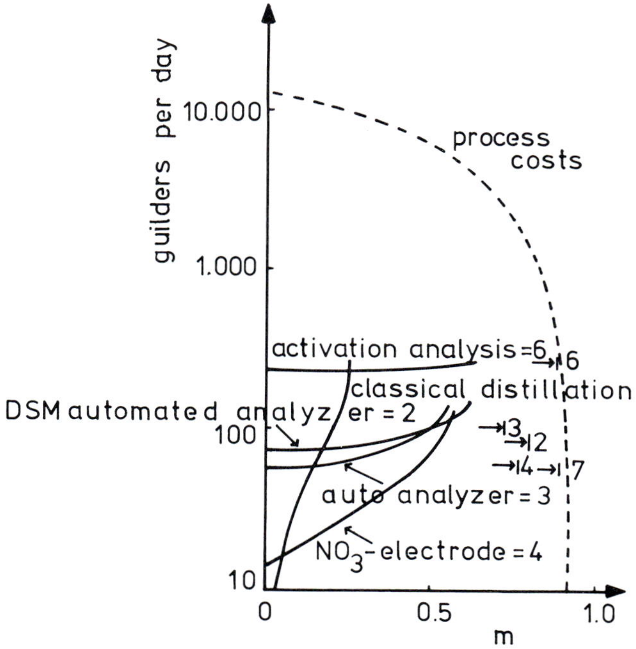

Fig. 27.4. Process costs due to process fluctuations (dotted line) and analysis costs (full line) as a function of the measurability factor. The horizontal lines (with arrows) refer to the fully automated techniques (reprinted with permission from Leemans, 1971. Copyright American Chemical Society).

REFERENCES

F.A. Leemans, Anal. Chem., 43 (11) (1971) 36A.
B.G.M. Vandeginste and T.A.H.M. Janse, Z. anal. Chem., 206 (1977) 321.
M.J.A. van den Akker and G. Kateman, Z. anal. Chem., 202 (1976) 97.
P.M.E.M. van der Grinten, Contr. Eng., 10 (No 10) (1963a) 87.
P.M.E.M. van der Grinten, Contr. Eng., 10 (No 12) (1963b) 51.
P.M.E.M. van der Grinten, ISA J., 12 (No 12) (1965) 48.
P.M.E.M. van der Grinten, ISA J., 13 (No 2) (1966) 58.
P.M.E.M. van der Grinten and J.M.H. Lenoir, Statistische Procesbeheersing, Spectrum, Utrecht/Antwerp, 1973.

PART V

SYSTEMS APPROACH IN ANALYTICAL CHEMISTRY

Chapter 28

ANALYTICAL CHEMISTRY AND SYSTEMS THEORY

28.1. THE SCOPE OF ANALYTICAL CHEMISTRY

Analytical chemistry can be regarded as a scientific discipline, unique in character. It can also be considered as being the sum of a set of sub-disciplines such as spectroscopy and chromatography. Yet another way of looking at analytical chemistry leads to the opinion that this branch of chemistry is nothing but an application of physics, physical chemistry, mathematics, etc., in order to arrive at methods suitable for tackling analytical problems. Apparently the definition of analytical chemistry depends on the angle from which the field is observed. Well over twenty definitions of analytical chemistry have been reported, reflecting the different opinions about this discipline. With these different points of view in mind, one is tempted to agree with the simple statement that analytical chemistry is what the analytical chemist does. However, it is of increasing importance to question what the analytical chemist is expected to do. In our opinion, the answer is given by a combination of the definitions of Gottschalk (1972) and Kaiser (1974), stating that analytical chemists have to produce qualified, relevant information about products and processes in an optimal way. These definitions have been stimuli in writing this book.

The formal methods for optimization, selection and classification as described in the preceding chapters are, at least in principle, generally applicable. However, in order to make use of this general applicability, it is necessary to stress the agreements rather than the differences between the several methods that are in use in analytical chemistry. Certainly, there are many differences. For instance, chromatography has not much in common with spectroscopy when one looks at the fundamentals underlying these methods. However, these methods also show a number of striking agreements, probably even more than might be seen at a first glance. A discussion of the common features is the aim of Part V of this book.

The most important factor to be borne in mind is the reason for applying analytical methods, which is the solution of analytical problems. All analytical methods, or rather all well described analytical procedures, serve essentially the same purpose. They are in use for determining the identities and/or amounts of compounds, elements or ions.

Procedures differ in the way the determination is effected. The way of describing the performance of procedures is, or rather should be, the same for each procedure. A judgement about the applicability is possible only when the performance is given in standardized terms such as accuracy, precision and information. Classification, comparison, selection, improvement and optimization require the use of well defined and generally accepted criteria. Performance or characterization parameters as described in Part I can be and are used as such.

Apart from the common purpose of the development and application of all analytical procedures, i.e., the solution of analytical problems, it can be observed that the general structure of all analytical procedures is essentially the same. Four steps can be distinguished, viz., the sampling, the sample preparation or clean-up required prior to the next step, the measurement(s), and finally the conversion of the results of the measurement(s) into the analytical results. The actual nature of each of these four steps may vary widely from procedure to procedure, but the function of each step is essentially the same for every procedure.

The essential aspects are clearly recognized if the analysis is described as follows : a sufficiently representative sample (1) is to be treated (2) in such a way that the measurement (3) can yield meaningful analytical results (4). The structure is given schematically in Fig. 28.1. Each of the parts of the procedure, or all four parts together, should be subjected to a control action in order to provide analytical results of a given quality. One of the controls in the analytical laboratory is a (repeated) calibration of the procedure. This particular control refers to only one of the quality parameters, i.e., the reliability (Chapter 5). Procedures have been developed for the control of other characteristics such as the precision. Other control actions are, for instance, the maintenance of constant temperature and pressure.

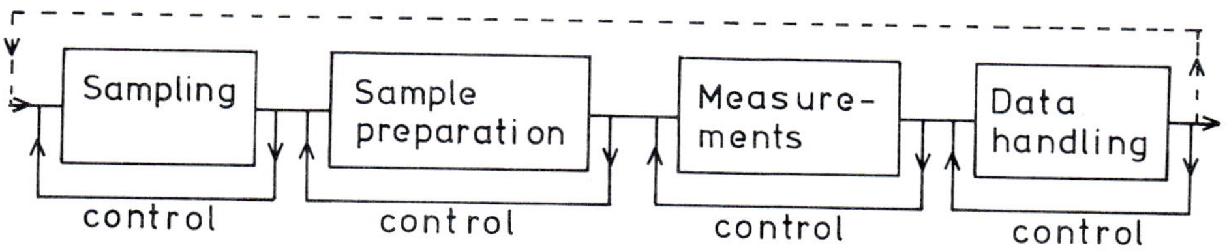

Fig. 28.1. Structure of an analytical procedure.

The measurement(s) as part of an analytical procedure can be considered as the heart of the procedure. It is therefore not surprising that the bulk of the analytical literature deals with the measurement, which to a large extent defines the possibilities and limitations for solving analytical problems. The study of the analytical measurement, whether based upon empirical facts or theoretical considerations, is usually carried out by specialists, each using the language associated with the sub-discipline. It is therefore not surprising that similarities have often been obscured and differences have been augmented.

However, the outputs of widely different instruments in use in the analytical laboratory are to a large extent identical in principle. Also, the transformation of the measurements into analytical results shows a parallelism between different procedures.

Basically, with only a few exceptions, the output appears as a spectrum (image), a two-dimensional picture showing peaks and valleys. (In some instances, e.g., titrations and polarography, the peaks emerge when plotting the first derivative of the output signal). Usually the positions of the peaks in the spectrum mark the identities of the compounds, elements or ions present in the (pre-treated) sample, whereas the peak heights or areas are correlated to the amounts of these components. Whether one considers infrared spectra, polarograms or gas chromatograms, the information about the identities is drawn from the location of the peaks. Information about amounts is obtained from peak areas or heights (with some exceptions, for instance titration curves).

The reduction of the two-dimensional picture to meaningful analytical results follows some general lines. These similarities clearly emerge when dealing with automated data processing techniques such as smoothing, curve fitting and pattern recognition, and are generally applicable to procedures producing two-dimensional pictures (two-dimensional methods) (Kienitz and Kaiser, 1968). Being aware of these similarities probably can save much effort in research and development as the techniques developed for optimizing one method can often easily be adapted for use with another analytical method.

A two-dimensional analytical procedure can often be extended to a more-dimensional or reduced to a one-dimensional analytical procedure. Whereas in the two-dimensional analytical procedure the output is measured as a function of one variable (time, wavelength, voltage, etc.), measurement of the output as a function of two (or more) variables leads to a similar but somewhat more complex picture. Examples can be found in mass spectrometry, where the ion current can be measured as a function of the magnetic field strength and of the ionization energy, and in neutron-activation analysis when the intensity of the radiation is not only measured as a function of the energy of the radiation but also the decay is taken into account.

Reducing a two-dimensional to a one-dimensional method makes sense when either information with respect to the identities or information with respect to the amounts (concentrations) is available. A light absorption measured at one wavelength can be used for a quantitative analysis if the identity of the component is known. A refractive index can be used for identification if only one component is present (100%) in the sample or for a quantitative analysis if the two components in the sample are known. It is therefore not surprising that in the routine laboratory one-dimensional procedures play an important role.

Looking for and stressing the similarities between different analytical procedures, together with the notion that all procedures are to be used for solving problems, leads to a generalized picture of analytical chemistry, analytical procedures and (probably) analytical problems. The function of analytical procedures and (probably) the formulation of the analytical problem can be given in general (mathematical) terms, although physically and chemically large differences exist. It will undoubtedly facilitate the use of the techniques described in this book.

28.2. SYSTEMS THEORY

The use of terms beginning with the prefix "systems", just like terms such as information and communication, looks like being a new fashion in analytical chemistry. The question arises of whether the use of systems theory, systems engineering, systems analysis and a systems approach adds a new dimension to analytical chemistry and whether it facilitates the solution of analytical problems. To begin with, it might be concluded that many of the thoughts and methods put forward by the advocates of systems theory are not really new and can be regarded as new wrappings for old ways of thinking and solving problems. However, such a conclusion would probably be an underestimation of the value of systems theory in modern science and technology. At present, its real value in analytical chemistry is difficult to estimate. Reading texts on systems theory, of which we shall mention only two by Von Bertalanffy (1950, 1956), who proposed the general systems theory, is encouraging. A few aspects that are of importance will be treated in this section ; it would require too many pages to give a detailed picture.

In a way, modern science can be characterized by an ever increasing specialization, unavoidable because of the increasing amount and diversity of skills and knowledge required for the solution of problems. Both theory and practice are becoming more complex. A huge amount of scientific literature is being published and has to be digested in order for workers to become familiar with even relatively small areas of scientific progress. Communication between scientists is easy when they are active in the same (sub-)discipline and confusion easily arises when workers from different disciplines meet. Different "languages" and different ways of thinking often inhibit the progress of the interdisciplinary research that is required for the solution of many problems in modern science. Systems theory aims at providing a common language and offering approaches that can be used in the whole scientific world.

In addition to the facilitation of interdisciplinary approaches through the introduction of a common language, methods developed in one branch of science might easily be adapted for use in other branches. As has been stressed by Von Bertalanffy, problems of widely different natures, and thus the solutions, are often very similar when one looks at them more closely. For instance, models of physical systems in a number of instances can be employed in sociology. Control and other aspects are met in the living organism as well as in industrial systems, etc. Duplication of research efforts can be avoided, provided that the problems and solutions are presented in a language familiar to all scientists.

An important aspect met in systems theory, systems engineering, etc., is the notion that the whole is more than the sum of the parts. In fact, it adds a new dimension to many ways of thinking. The behaviour of a system cannot be derived from the behaviour of the component elements unless the relationships between the elements are known.

A few remarks should be made about some other theories, methods and approaches with the prefix "systems". Owing to the confusion in the literature, exact definitions cannot be given. Systems theory appears to include systems engineering and operational research. Systems theoretical considerations can be largely verbal or highly mathematical and abstract (see, for instance, Zadeh and Desoer, 1963). The development of (mathematical) models for general use is characteristic of systems theory. Systems engineering can be regarded as a means of attacking real problems and designing real systems by making use of such models. It is characterized by an integral (systems) approach. The term operations research is usually reserved for the generally applicable techniques used in the process of systems engineering. Statistics, information theory, cybernetics, etc., are usually not considered as operations research techniques, although these theories and associated techniques clearly play an important role in systems engineering.

From these very concise remarks about systems theory and the remarks made about analytical chemistry in the preceding section, it is not surprising that several workers in analytical chemistry have started to explore the applicability of systems theory and related disciplines to analytical problems. The terms and definitions used in systems theory have been summarized and made available to the analytical chemist by the Arbeitskreis "Automation in der Analyse", convener G. Gottschalk (1971) (English translation by I.L. Marr, 1973). These terms and definitions are presented in Table 28.I. Unfortunately, the

Table 28.I.

Basic terms of systems theory - Definitions (from Arbeitskreis, 1973)

System	demarcated arrangements of a set of elements and a set of relationships between these elements
Element	given or chosen relevant components of a specific system
Relationship	given or chosen coupling of the elements of a specific system
Function	behaviour patterns and effects of a system
Structure	known relationships between the elements of a system which lead to specific functions
Organization	breakdown of a system into subsystems with relevant relationships between them. Subsystems can also have the appearance of elements
Feedback	function by means of a closed sequence of relationships
Black box	system with structure unknown at the time, but with given magnitudes of input and output
Model	system which represents in part, functions and/or structures of a real or an abstract original system
Input-output analysis	method of elucidation of functions of a system based on investigation of the relationships between the input and output
Trial-and-error method	method of stepwise elucidation of functional relationships in a system making use of established facts
Simulation	copying of a specific function of a system by means of a functional model

Arbeitskreis "Automation in der Analyse" has illustrated the terms and definitions by taking examples that are not really relevant to the analytical chemist. However, it should be borne in mind that a presentation of analytical chemistry in terms of systems theory requires a considerable effort. Some results of such efforts have appeared in the analytical literature, to a large extent in papers by Gottschalk (1972), Malissa (1966, 1974), Malissa and Jellinek (1969) and the Arbeitskreis (1972). In the work of other analytical chemists, systems theoretical thoughts are more implicit (see for instance, Kaiser, 1973, and many papers dealing with aspects of automation).

Looking at the problems in analytical chemistry, and more in particular at the problems described in this book, we have to describe two systems, viz., the analytical procedure and the analytical laboratory. In the following chapters we shall discuss these systems.

REFERENCES

Arbeitskreis "Automation in der Analyse", Z. anal. Chem., 256 (1971) 257.
Arbeitskreis "Automation in der Analyse", Z. anal. Chem., 261 (1972) 1.
Arbeitskreis "Automation in der Analyse", Talanta, 20 (1973) 811.
G. Gottschalk, Z. anal. Chem., 258 (1972) 1.
H. Kaiser, in Methodicum Chimicum, Band I : Analytik, Teil 1, G. Thieme, Stuttgart and Academic Press, New York, 1973, p. 1-20.
R. Kaiser, Z. anal. Chem., 272 (1974) 186.
H. Kienitz and R. Kaiser, Z. anal. Chem., 237 (1968) 241.
H. Malissa, Z. anal. Chem., 222 (1966) 100.
H. Malissa, Z. anal. Chem., 271 (1974) 97.
H. Malissa and G. Jellinek, Z. anal. Chem., 247 (1969) 1.
L. von Bertalanffy, Brit. J. Phil. Sci., 1 (1950) 134.
L. von Bertalanffy, General Systems, 1 (1956) 1.
L.A. Zadeh and C.A. Desoer, Linear System Theory, McGraw-Hill, New York, 1963.

Chapter 29

THE ANALYTICAL PROCEDURE

29.1. THE BLACK BOX

In the preceding chapter, the black box was defined as a system with an unknown (internal) structure, but with given magnitudes of input and output. Referring to an analytical procedure or an analytical instrument as a black box means that nothing is known about the physical, chemical, mechanical or electronic components or processes that convert the sample with an unknown composition into a sample with a known composition. A substantial part of the research effort in analytical chemistry has been and still is devoted to the elucidation of the unknown structure of black boxes or, to put it differently, to turn black boxes into white, or at least grey boxes. Such an elucidation satisfies human curiosity and often leads to a procedure with a superior performance. However, from an analytical point of view, procedures can be, and often actually are, equally useful when the internal structure is not (fully) known to the user. Moreover, for an analytical chemist faced with widely different problems and procedures, it is virtually impossible to be (entirely) familiar with the physical and chemical principles that underlie the procedures and with the details of the design of the instruments. Even if procedures and equipment are white boxes to some scientists they may well appear to be black boxes to others.

A black box is useful for the analyst if, and only if, its output can be used to arrive at the (approximate) composition of the unknown sample. To put it differently, the input-output relation or the calibration function has to be known. However, every analyst is aware of the influence of parameters (also called descriptors ; Kaiser, 1973) such as temperature, volume of reagent and wavelength on the measurement. Clearly the black box is not adequately described by the input-output relation between the composition of the sample and the measurement alone. Parameters that influence this relation should be specified, and often kept constant, in order for useful analytical results to be obtained.

The analytical procedure as a black box can be a description in words of what has to be done in order to determine the composition of the sample (the analytical recipe). Such a description can be supplemented or replaced with a more schematic model as presented in Fig. 29.1.

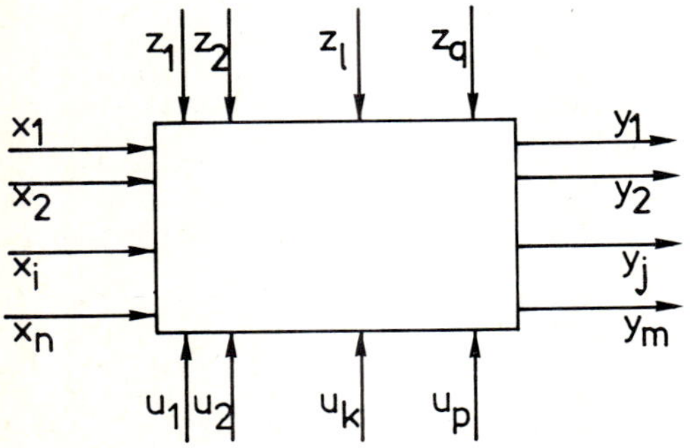

Fig. 29.1. The analytical procedure as a black box.

Essential for this model are the nature (units) of the input variables $x_1, \ldots, x_i, \ldots, x_n$ representing the composition (concentrations, amounts, identities) and of the output variables $y_1, \ldots, y_j, \ldots, y_m$ representing the measurements (voltages, readings). Of major importance are the input-output (x - y) relations that are required in order to arrive at the analytical results from the measurements. The u variables that have to be specified are those which influence the measurements and consequently the x - y relations. Because of this influence they have to be controlled and are called the controllable variables. Input variables that cannot be kept constant are indicated by z_l. The origin of these non-controllable variables is often unknown. They lead to (stochastic) fluctuations in the output variables.

The general picture of Fig. 29.1 is reduced to a model with one x and one y variable in a one-dimensional analysis. Usually such one-dimensional analyses can be used for only one type of sample. The type of sample influences the calibration function. It can be considered as a controllable variable.

A closer inspection of the analytical procedure as a system, i.e., a systems analysis, reveals several types of input and output. We arrived at sets of input and output variables that are relevant when looking at the calibration function. However, other inputs and outputs exist. For instance, materials flow in and out of the apparatus and energy and skills are required to produce results. These aspects will not be dealt with here, as they are less relevant in the context of this book, although they are essential in the design of instruments, the organization of the laboratory, etc.

A few remarks must be made about a particular input and output, namely that connected with information. Some of the principles of information theory have been introduced in Chapter 8. Information obtained from an analytical procedure has been defined as the difference between the uncertainty pertaining to the composition before and after analysis. The uncertainty before analysis (pre-information) is an input parameter and the uncertainty remaining after the analysis is an output parameter. These parameters do not refer to the composition of the sample, but to the (number of) possible compositions. The corresponding input-output relation, i.e., the difference between the uncertainties, cannot be used for arriving at the composition of the sample. It merely refers, depending on the analytical problem, to the number of different compositions that can be discriminated by the application of the analytical procedure. It runs parallel to the application of information theory in communication theory, i.e., the distinction between several possible messages when these are transferred through a noisy channel (telephone, etc.). Representing the analytical procedure as a noisy channel, the process of analysis can be represented by Fig. 29.2 (for a comparison with a communication diagram, see Shannon and Weaver, 1949). The composition is coded as a (physical) property. This property is measured and noise is added. Decoding is possible when the relationship between x and y is known and when the relevant signal is not obscured by the noise.

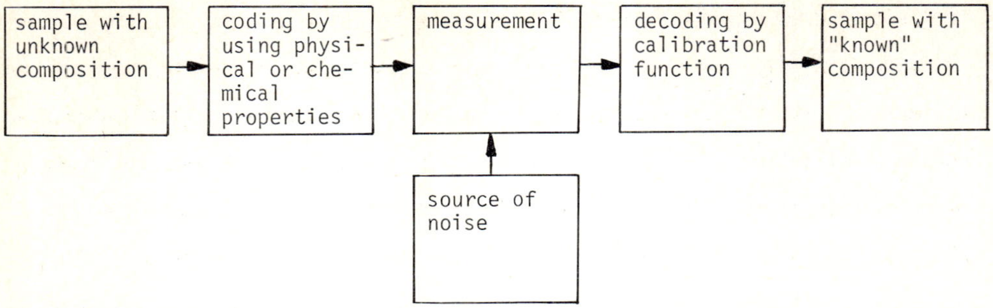

Fig. 29.2. The analytical procedure as a communication process.

29.2. SOME INPUT AND OUTPUT VARIABLES AND THEIR RELATIONS

Let us return to the x and y variables that are relevant for describing the relations between the measurements and the compositions. The x variables can, in quantitative analysis, be expressed as either concentrations or amounts. In a number of instances we have represented the variables $x_1, \ldots, x_n$ by the vector $\vec{x}$, defining the composition in the space of compositions (Chapter 17). The presentation of x variables in qualitative analysis is more complicated. In contrast to the quantitative composition, the identity cannot be represented by a set of continuous variables. The n-dimensional space of quantitative compositions (with the n concentrations as coordinates) as used in this book has the property that closely related samples will correspond with adjacent points in that space, whereas widely different samples are represented by points separated by large distances. This property of the space of compositions is desirable when considering the calibration function.

For qualitative analysis it is desirable to define a space of compositions (identities) with the same property (mixtures will not be considered). Depending on the analytical problem, each composition should be represented by a distinct point (or vector) or should cluster with points representing similar compositions (for instance, all alcohols should be represented by a cluster of points). For several reasons such a space is difficult to define, but other means of achieving the same goal are available. These means are the several "codes" that have been invented to represent the identity of a chemical compound.

The most widely used codes are the name or formula of the chemical compound. However, not all formulae and names identify chemical compounds unambiguously. The same molecular (elemental) formula, for instance, can represent different chemical compounds. Use of the systematic IUPAC nomenclature or the complete structural formula can prevent ambiguities, but these names and formulae are not easily handled by computers. Other codes that are more suitable for computer handling, i.e., for retrieval of, in particular, organic compounds, have been designed (for reviews, see Lynch et al., 1971 ; Ash and Hyde, 1975). Three main categories of structural representation can be distinguished.

The codes belonging to the first category are the so-called fragmentation codes. These codes do not describe the entire structure of the molecule, but rather indicate the presence of certain portions of the structure, for instance functional groups. Numbers or letters, or their combinations, are used to encode the several possible structural elements. The code of a chemical compound consists of a combination of such elements. However, the relative position of the several structural elements is not encoded. Consequently, it is impossible to obtain the entire structure from the code, but it is easy to select from a set of coded structures those which are similar.

The second category of codes consists of connection (connectivity) tables. Every atom, apart from hydrogen, in the chemical structure is given an arbitrary number. In its simplest form, the structure is represented by a table with the nature of each atom, the numbers of neighbouring atoms and the nature of the bonds (single, double, etc.). For computer use the connection tables can be linearized (sequence of symbols). In this code all atoms are considered to be equally important. The code does not lead to a unique representation of chemical structures, although the introduction of a set of rules for numbering the atoms can convert the connection table into a unique coding system. Usually, structural elements cannot easily be recognized when inspecting a connection table. However, computer programs have been developed for recognizing certain structural elements (typically not functional groups, but rather parts of the skeleton).

Frequently used are the so-called linear notations, especially the Wiswesser line notation and the IUPAC notation developed by Dyson. Through the use of a detailed set of rules a compact code, that is economical to store, is achieved. In this code some important chemical features are highlighted, e.g., ring structures. Every structure is represented by only one code and consequently every compound can be retrieved by its code. Some structural elements, e.g., those related to the encoding rules, are easily recognized.

It can be concluded that codes other than the common structural formulae and names are available for uniquely representing molecular structures. Some of them can be converted into each other. The codes that have been developed for computer handling of information systems also enable one to search for structures with common structural features. The coding systems have some properties that can be compared with the space of compositions as introduced for quantitative analysis, although they are not strict mathematical formulations of a "space of identities".

The output variables that are of primary interest are the y variables representing the results of the measurements and that can be used to arrive at the composition of the sample. These variables can be represented by vectors or points in a space of measurements, in both quantitative analysis (Chapter 17) and qualitative analysis.

For quantitative analysis the principal input-output (x-y) relation is the calibration function. To a large extent it defines the applicability of the analytical procedure. For a one-component quantitative analysis this input-output relation usually is given as $S = y/x$, the sensitivity of the analytical procedure (Chapter 6). If the sensitivity is zero, the analytical procedure is useless, although a value differing from zero does not always correspond to a useful procedure. This is particularly so if there is a large influence of the z variables upon the x variable (noise).

The sensitivity of a continuous procedure is equally defined by y/x. However, the x-y relation is time dependent. For a first-order response, the influence of x upon y is governed by the sensitivity S and first-order time constant τ (Chapter 10).

The concept of the sensitivity representing the x-y relations can also be applied to multi-component analyses and leads to the general formula in matrix notation (for a more elaborate description, see Chapter 17, eqn. 17.20)

$$|S| = \left| \left(\begin{bmatrix} S_{11} & \cdots & S_{1n} \\ \cdot & \cdots & \cdot \\ \cdot & \cdots S_{ji} & \cdot \\ \cdot & \cdots & \cdot \\ S_{m1} & \cdots & S_{mn} \end{bmatrix}' \begin{bmatrix} S_{11} & \cdots & S_{1n} \\ \cdot & \cdots & \cdot \\ \cdot & \cdots S_{ji} & \cdot \\ \cdot & \cdots & \cdot \\ S_{m1} & \cdots & S_{mn} \end{bmatrix} \right) \right|^{\frac{1}{2}} \qquad (29.1)$$

where S_{ji} are partial sensitivities and $|S|$ is the sensitivity of the multi-component procedure. Again, a sensitivity of zero corresponds to a useless method. The "quality" of the procedure increases with increasing sensitivity, provided that the number of dimensions (measurements) and the errors remain the same. As has been shown in Chapter 17, the sensitivity can be used as a criterion for selecting the best set of wavelengths for a multi-component spectrophotometric procedure. In principle, this method can be considered as a method of feature selection.

An equivalent of the sensitivity as defined by eqn. 29.1 for qualitative analysis does not exist, owing to the lack of a strict mathematical formulation of the space of identities. Consequently, the input-output relation for a qualitative analysis is more complicated. Usually it consists of a table of chemical compounds represented by their names, formulae or codes together with the corresponding spectra or physical or chemical properties. The quality of the procedure is determined essentially by the extent to which spectra (or physical properties) can be used to identify chemical compounds. If each compound has a unique spectrum, an identification is possible. In those instances where similar compounds have similar spectra, certain structural features can be derived from the spectrum. If such similarities exist, interpretation rules or structure correlation tables can be used to arrive at the identity of chemical compounds from certain spectral features. The design of such rules can be based upon either theoretical considerations or experience. Rules expressing the relationship between spectral features and structural elements can also be established by formal methods based upon the study of clusters [pattern (re)cognition, cluster analysis ; see Chapters 16, 18 and 20] . Although not essential, the use of linear codes for representing the chemical structure will facilitate these studies, especially when large numbers of spectra are used.

The u variables were introduced because of their influence upon the measurements (for constant x, y varies with u). In fact, the u variables correspond to the knobs on the apparatus. The apparatus itself also can be a u variable, just like the volume of a pipette, the amount and strengths of reagents, etc. It is obvious that each procedure has its own set of u variables. All of these controllable variables specify the conditions under which the procedure has to be carried out. In fact, it is a description or part of a description of the procedure. If the box is completely black, a large number of possible u variables exist. Some knowledge about the black box can be of help in selecting the variables that might influence the measurements. Even then the exact influence may be unknown and consequently the optimal setting of the knobs (optimal conditions) is difficult to predict. In Part II, on experimental design, methods were discussed that can be used to establish the optimal conditions.

The exact variations of the z variables is usually not known and consequently the z-y relations remain undetermined. Only the variations in y are studied by using statistical methods.

29.3. THE COMBINATION OF BLACK BOXES

In a number of instances it is advantageous to divide the black box into a set of black boxes (subsystems). In the schematic representation of the analytical procedure as shown in Fig. 28.1, four black boxes can be distinguished, viz., the sampling, the sample preparation, the measurement(s) and the data handling. Each of these subsystems has (an) input(s) and (an) output(s) and consequently (a set of) input - output relation(s). The output of a certain black box is the input of the next. Some input-output relations of the whole system can be calculated from the input-output relations of the subsystems. This is especially helpful for the design of analytical procedures from parts with known properties or when looking at bottle-necks in the chain of subsystems.

The total sensitivity of the procedure is equal to the product of the sensitivities of the parts, i.e., $S_t = S_1.S_2.S_3...$ Each sensitivity is expressed in units corresponding to the units used for the relevant input and output. Thus the sensitivity of a dilution is simply a constant (less than unity).

Whereas the time lag or dead time, t_d, clearly has additive properties [thus t_d (total) = t_d (sampling) + t_d (sample preparation) + ...], the frequency of analysis or its reciprocal, t_a, is not uniquely related to, for instance, the sampling frequency or, the measuring frequency. However, the design of a procedure (or organization of a laboratory) should obviously always lead to the same frequencies of sampling, sample preparation, etc. At least the average frequencies should be the same. It obviously makes no sense to gather samples with a larger frequency than the measuring frequency.

Time constants of continuous subprocedures can be used to predict the time constant(s) of the whole procedure. The mathematics are complicated and cannot be dealt with here. However, in many instances it is safe to state that the time constant of the whole procedure is equal to the largest time constant found in the chain of subprocedures.

Although the exact z-y relations are in principle unknown, and usually need not be known, it often is desirable to detect the sources of the z fluctuations. For this detection, use can be made of the additive property of the variance, viz., σ^2 (total) = σ^2 (sampling) + σ^2 (sample introduction) + ... Thus the variance of the output y can be divided into parts that can be ascribed to the several sources. ANOVA (Chapter 4) can be used to estimate these contributions as it is possible to determine from a set of experiments the variance contributed by the measurement, the sum of the variances of the measurement and the sample preparation and the total variance. It is clearly impossible, and not necessary, to estimate the variances of the subsystems directly.

In the preceding parts the total system has been regarded as a set of subsystems that are connected in series. It is also possible to connect black boxes in parallel. For quantitative analysis this represents a multi-component analysis : a procedure for the determination of a certain element is combined with a procedure suitable for estimating another. Such combined systems, in fact, behave independently (however, in a laboratory organisation they cannot be considered as independent, see Chapter 30). For qualitative analysis a different situation arises. The combination of, for instance, two procedures, each yielding a partial identification, may be required for a full identification. For such combinations the correlation between the physical properties or spectra obtained from the individual procedures has to be taken into account. Information theoretical studies permit the evaluation of the usefulness of the combined system for identification purposes. As has been shown in Chapters 8 and 17, the total amount of information is not equal to the information obtained from the individual procedures.

29.4. AN EXAMPLE

Although the use of block diagrams to represent schematically structures that are not easily described in words is widespread in science and technology, analytical recipes are usually described in words rather than in diagrams. The description of analytical recipes using, or supplementing them with, such diagrams has certain advantages. Let us consider the recipe for the complexometric determination of iron (III) as used by Malissa and Jellinek (1969) to illustrate the use of a symbolic language (see the next section). The procedure (of Vorliček and Vydra) is described concisely as follows :

Remove metallic iron by treating the powdered sample (Renn slag) (1 g) for 20 h with $FeCl_3$ solution (6%) (50 ml). Filter the mixture, wash the residue with hot water (100 ml) and heat it in a platinum crucible with HCl-HF (1:1) for 1-2 h on a sand-bath ; repeat the procedure, if necessary. Dilute the resulting solution with doubly distilled water (oxygen free) to 150 ml, add H_3BO_3 (2-3 g) and adjust the pH to 1.5-2 by adding NaOH solution (10%). Heat the solution at 60-70° and titrate with 0.05 M EDTA in a nitrogen atmosphere, potentiometrically or by the dead-stop method.

This description has a large number of u variables, viz., variables that (apparently) influence the calibration function. These u variables are the nature and amount of sample, the pre-treatment of the sample (powdering), the amount and strength of the $FeCl_3$ solution, the time required for removal of metallic iron, the material of the crucible, the amount and temperature of the water to be used for washing the residue, etc. Altogether there are about 30 controllable variables.

Replacing the description in words by a list of u variables provides a check list of variables that influence the performance characteristics. Information with respect to the course of analysis is lost unless the analytical procedure is split into parts and for each part the corresponding u variables are given (see the next section).

Essentially this description is a black box, providing information about the controllable factors. Although it might easily be included, it does not provide information on the performance characteristics of the procedure (sensitivity, precision, detection limit, time parameters) and as such the description is not complete. Although this black box can be considered to be adequately described when aiming at the application of the procedure, it is to be considered as incomplete for the purpose of comparison with other procedures. For the analyst familiar with related procedures, the box probably is not completely black and he may be able to estimate the performance characteristics from his experience with related procedures. Communicating an analytical procedure to those who are not familiar with the principles of the procedure is possible with a black box, provided that a careful description is given.

29.5. SOME OTHER WAYS OF DESCRIBING ANALYTICAL PROCEDURES

The course of the analysis is obscure when the procedure is represented by the model in Fig. 29.1. More details about the course of the procedure can be included when the system is divided into subsystems. A very rigorous division into subsystems has been proposed by Malissa and Jellinek (1969) and by Malissa and Simeonov (1978). Their symbolic language is aimed at retaining all information required for performing the procedure. The symbolic representation of the procedure described in the previous section is shown in Fig. 29.3. It resembles the representation of a procedure by Fig. 28.1, although there are some important differences. The symbols in Fig. 29.3 are black boxes that yield information on the several u parameters and are in fact representations of the unit operations that are required for performing the analysis (heating, filtration, etc.). Input and output of materials are clearly indicated by arrows. The whole scheme is designed to yield the same information as the written text.

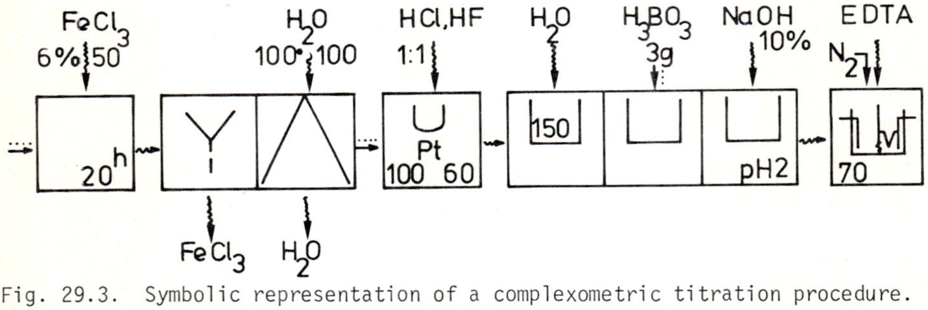

Fig. 29.3. Symbolic representation of a complexometric titration procedure.

In order to be able to describe a wide variety of procedures, a large number of symbols unambiguously representing the various unit operations are required (the semantics of the symbolic language). In addition, a set of rules has to be designed in order to connect the various unit operations (the grammar). In our opinion, for a full description of analytical procedures a rather complicated language is required and for that reason we doubt whether eventually the goal of more simply and clearly representing analytical procedures will be reached.

Own experience indicates that such a symbolic language is useful for wet chemical analysis, but fails when instrumental and more dimensional procedures are considered. Examples of the comparison of procedures using the symbolic language has been given by Gottschalk (1972) and by Ortner and Scherer (1977).

Malissa and Jellinek (1969) discussed another (computer) language, aiming at a facilitation of the automation (computerization) of analytical procedures. The same recipe for determining iron (III) iron in that language is shown in Table 29.I. It is easy, even without an explanation, to read Table 29.I. Again, the representation of the procedure is a black box.

Table 29.I.

Analytical procedure in computer language (Malissa and Jellinek, 1969)

Step		
0	START	
1	SAMPL S1	L1 : $FeCl_3$ solution (6%) ; 50 ml
2	ADD L1	L2 : Water ; 100°C ; 100 ml
3	SOLV (1200)	L3 : HCl/HF solution (1:1)
4	FILT	L4 : Water ; doubly distilled
5	WASH L2	L5 : NaOH (10%)
6	ADD L3	L6 : EDTA (0.05 M)
7	HEAT (100;60)	S1 : Sample ; 1 g
8	DILUT L4 (150)	S2 : H_3BO_3 ; 3 g
9	ADD S2	
10	ADD L5	
11	IF (PH.LT.2.0) GO TO 10	
12	TITR L6 (70)	
13	END	

Another computer language aimed at laboratory automation has been described by Toren et al. (1972). However, the future application of these languages is uncertain. Laboratory automation may well develop along other lines as a result of the introduction of microprocessors.

29.6. QUALITY CONTROL

Analysing is a process, either continuous or discontinuous, that has to be kept under control. The "quality" of the analytical results produced by that process has to be guaranteed. Therefore, in the scheme in Fig. 28.1 control actions have been included. Usually the u variables are kept constant or varied in a specified way in order to prevent the production of incorrect results. For many analytical procedures this type of control is not sufficient to avoid (systematic) errors. Apparently in such cases not all u variables have been specified. In such instances a more or less frequent calibration is required. Considering the problem from that angle, the process of calibration is a control action. Several other methods have been developed to keep the analysing process under control, for instance, the use of control charts (Chapter 5).

29.7. THE ANALYTICAL PROCEDURE AS A SUBSYSTEM

Analytical procedures are used for solving analytical problems. Some of these (generalized) problems have been discussed in Part IV. In general, analytical results are required in order to be able to make decisions. Analytical chemistry helps one to decide whether actions should be taken. Analytical results from the clinical laboratory will or will not be followed by therapy ; in the research laboratory, results will be of help in guiding the research, etc.

Although we have considered the analytical procedure as a separate system, it clearly shows interactions with the environment. A general model for these interactions is not available, although there are strong indications that every analytical procedure is part of a control loop and thus helps to regulate processes, whether these processes be the therapy of patients or research activities. The general goal of the analysis will be to optimize these processes.

Defining such optimization problems requires communication with scientists of other disciplines. In that context, analytical procedures should be represented by black boxes. For that purpose, the function and characteristics as described in this chapter are certainly more important than the internal elements of and relationships within the black box. Although the picture of the analytical procedure presented is far from complete, it is worth stimulating the development of generalized pictures.

REFERENCES

J.E. Ash and E. Hyde, Chemical Information Systems, Wiley, New York, 1975.
G. Gottschalk, Z. anal. Chem., 258 (1972) 1.
H. Kaiser, in Methodicum Chimicum, Band I : Analytik, Teil I, p. 1, G. Thieme, Stuttgart and Academic Press, New York, 1973.
M.F. Lynch, J.M. Harrison, W.G. Town and J.E. Ash, Computer Handling of Chemical Structure Information, Macdonald, London and Elsevier, New York, 1971.
H. Malissa and G. Jellinek, Z. anal. Chem., 247 (1969) 1.
H. Malissa and V. Simeonov, Z. anal. Chem., 289 (1978) 257.
H.M. Ortner and V. Scherer, Talanta, 24 (1977) 215.
C.E. Shannon and W. Weaver, The Mathematical Theory of Information, Univ. Illinois Press, Urbana, Ill., 1949.
E.C. Toren, R.N. Carcy, A.E. Sherry and J.E. Davies, Anal. Chem., 44 (1972) 339.

Chapter 30

THE ANALYTICAL LABORATORY

30.1. INTRODUCTION

The analytical laboratory is a complex system. Although this complex system in a way resembles the analytical procedure - samples enter into and analytical results emerge from the system - it is impossible to improve or optimize the analytical laboratory by merely looking at the inputs (x variables) and outputs (y variables), changing the controllable u variables and seeing whether the performance of the laboratory improves. The laboratory cannot be considered as a black box when aiming at optimization. It is inevitable that one must consider the internal structure or the organization of the analytical laboratory.

The performance characteristics that are to be used for laboratory optimization are usually the same as those used for optimizing the analytical procedure. However, there can be differences. For instance, the time that elapses between the acceptance of a sample by the laboratory and the termination of the analysis is usually not equal to the time lag of the analytical procedure (Chapter 21). The rate of arrival of the samples, especially with irregular rates of arrival, has a pronounced influence on the length of the queue and consequently on the waiting times. It also is clear that such effects will influence the cost per analysis. In some instances the relationships between the performance characteristics and the corresponding characteristics of the laboratory are simple, but in many other instances they are complicated or even obscure.

The performance characteristics of a laboratory are strongly influenced by the structure of the system. Usually the complex laboratory system consists of many elements or sub-systems : analytical procedures and instruments of different kinds, personnel with different tasks and skills and nowadays also laboratory computers. Between the procedures and people, several interactions (relationships) exist, for instance a manual procedure does not produce information without a technician. The whole set of elements and relationships, i.e. the organization of men and machines, largely influences the performance of the laboratory.

Studies of this performance or attempts to optimize it require the construction of a laboratory model, in particular a model that can predict the effect of a change in the organization. Experimentation with the real laboratory often is too costly and can lead to disappointments. Models comprising procedures and instruments with their characteristics, human behaviour and interactions between procedures, between people and between procedures and people have not been described in the analytical literature, although some aspects have been studied, such as communications between the individuals in a research laboratory (Allen, 1971 ; Frost and Whitley, 1971). The increasing need for analytical information in many sectors of science and society can only be met if it is produced more economically (and if the need is thoroughly questioned). It therefore is not surprising that attempts have been made (but probably not always published) to construct simplified models that represent some aspects of the analytical laboratory and that can yield a valuable contribution to the problem of laboratory optimization. As has been remarked in Chapter 21, the very construction of such simplified models can contribute to a better understanding of what is happening in the laboratory and consequently to a better organization.

30.2. SOME NOTES ON THE ORGANIZATION OF THE LABORATORY

In a paper on analytical laboratory organizational design, Cook (1976) uses the following definition of an analytical laboratory organization : "An analytical chemistry laboratory organization is the rational coordination of the activities of a number of people for the achievement of some common explicit analyses or analytical goals, through division of labour and function and through a hierarchy of authority and responsibility". This definition is a modification of one given by Schein (1970), who comments : "The organization is a complex social system which must be studied as a total system".

Although we can observe that the definition is clearly in line with the systems approach, it obviously refers, as do many other definitions of organization, predominantly to the personal aspects, i.e., the functions, tasks and responsibilities to be assigned in relation to the objectives of the laboratory. For laboratory optimization, a definition comprising all elements, equipment and personnel is required, but obviously is more difficult to handle. Nevertheless, a discussion of a laboratory organization in a restricted sense is possible and fruitful.

For an organization change (or design) to be succesful, according to Cook (1976), the objectives of the proposed change must be clearly defined. The objectives can be grouped into three areas, corporate, personnel and government :

corporate :

- management attitude towards reorganization
- changing corporate goals and objectives
- short- and long-term economic forecasts of major business
- computing capacity available for automation
- recognition of the nature of the corporate business one supports - manufacturing, pilot plant and/or research

personnel :

- staff requirements - age distribution
- new breed of specialists
- job rotation
- job progression
- psychological contract

government :

- government regulations.

It is evident that the optimal laboratory organization strongly depends on the environment of the laboratory (industry, government, university). Probably there are as many laboratory organizations as there are analytical laboratories, although it is possible to distinguish between some main types. Cook discusses seven options, ranging from small to large laboratories and from routine (control) to research laboratories.

Laboratory organizations are often represented by schematic block diagrams representing functions, tasks and responsibilities of the several departments and/or people. Considering the number and nature of factors that should be taken into account, laboratory optimization from the point of view of organization (even in this restricted sense) is rather complicated and by and large cannot be completely tackled with formal optimization techniques.

However, Goulden (1974) discusses several management studies and techniques that may be helpful in analytical research, development and service. He considers - more or less in parallel with Cook - three essential components : the work or tasks to be undertaken, the organization necessary to effect that work and the people by whom the work will be done. He draws attention to some more or less formal methods that are of use in planning (project selection, evaluation and control), to the several views on organization and to factors that are related to people (their skills, motivation, etc.).

Because of the difficulty of applying formal techniques to organizational models of the analytical laboratory, we shall refrain from an extensive discussion of this topic. This, of course, does not imply that we consider the laboratory organization to be of minor importance. It is beyond doubt that a laboratory with reliable analytical equipment and highly skilled technicians but poorly organized will produce analytical information inefficiently.

30.3. METHODS FOR SIMULATING THE ANALYTICAL LABORATORY

Even if the organizational aspects mentioned in the preceding section are not considered, the analytical laboratory is a complex system. It consists of elements or sub-systems, the analytical procedures and/or instruments. Each of the sub-systems has a certain function, i.e., the ability to produce analytical information. This ability is characterized by the performance characteristics. Usually there is a relationship between the procedures and/or instruments. Often two or more procedures are used to produce the required information. If one method fails, another has to be used. Relationships also exist between two procedures if there is only one technician in charge of both. It is also possible that different procedures in the same laboratory have essentially the same function, but differ in some respects, with the result that in some situations it may be attractive to use one method (for instance, a manual method when there are few samples to analyse) whereas in other situations it may be preferable to use another (automated procedures for a large number of samples). All of these aspects fall under the heading of organization, although they are different from the aspects mentioned in section 30.2.

A laboratory model consisting of the procedures and instruments, probably together with their operators, the functions and the relationships, again is a simplified picture of reality. The important aspects of human behaviour are largely neglected, or at least introduced as a kind of average human behaviour. When keeping this in mind, such models can contribute to the efficiency in the analytical laboratory. It goes without saying that these models should also account for the interaction with the laboratory environment.

Some aspects of such laboratory models, and of the formal optimization techniques that can be applied to these models, have been discussed in this book, most notably in the chapters on operational research methods (Chapters 21-24), but also in Part IV where some aspects of the interaction with the environment were discussed.

In section 21.3 on queueing theory, it was observed that in many instances the theory cannot be applied if the rate of arrival of the samples cannot be described adequately mathematically and/or if the laboratory system is too complicated. Then one should try to construct models suitable for simulation of the (proposed) real laboratory situation. A laboratory model of that kind was described by Schmidt (1976, 1977). This simulation model, SIM-LAB, was written in GPSS-FORTRAN and is especially designed for simulating the clinical laboratory, although it probably can be easily adapted to other routine (control) laboratories. Another model for simulating the clinical laboratory, LABSIMU, also written in GPSS-FORTRAN, was designed by Väänanen et al. (1974).

Without underestimating the complexity of clinical laboratories in general, these laboratories are probably less complex than many other analytical laboratories. As the bulk of the analyses in the clinical laboratory concerns relatively few types of sample, the (quantitative) analytical procedures used are usually well standardized and the analytical problems to be solved are usually well specified (the medical problem may be not). However, even the complexity of such a system is considerable. Obviously the model should be a sufficiently realistic image of reality in order to optimize the laboratory, using for instance laboratory costs or cost per analysis or the average waiting time as a criterion. The model described by Schmidt (1976, 1977) can be used for such optimization studies. It can also be used for determining at any moment the optimal strategy for distributing the samples over the instruments that are available. Some aspects of the model will be described here.

It goes without saying that the model should include options for a variable rate of arrival of the samples. The influence of this rate on the performance of the laboratory has already been observed.

In many clinical laboratories there are several instruments available for the same analysis, with different capacities. Decisions have to be made concerning the instances in which a particular instrument has to be used. Sometimes the same instrument - for instance after changing a module - can be used for different analyses. The laboratory model should allow for changing an instrument from one determination to another and indicating when and how often such a change should be made.

In many instances clinical laboratories are partially equipped with multi-(n-)channel instruments suitable for the simultaneous determination of several (n) components. If a sample is fed into such an instrument, the results of all n determinations become available even if fewer determinations are required. Nevertheless, it may be advantageous to use a multi-channel apparatus rather than performing the analyses separately. Again with the model one should be able to make such decisions.

Finally, but no less important, is the inevitable assignment of priorities to the samples. Allowing for priorities usually has a pronounced influence upon the performance of the laboratory.

From the literature available at present, it appears that predictions about laboratory cost and waiting times can be made for several laboratory organizations (number and type of equipment), but it is not clear whether the results obtained by using simulation models have been confirmed in actual practice. One might expect that for routine laboratories and a high degree of automation the agreement between predictions obtained from the model and practice will be reasonable as the factors accounting for human interference will be less important or at least more or less constant. However, some clinical chemists seem to have doubts about this constancy of the human factors, and probably they are right. When keeping in mind the interactions not occurring in the model and thus rating these models at their true values, a valuable contribution to laboratory optimization can be obtained from simulated laboratory organizations.

To conclude, we shall make some remarks about the analytical laboratory and systems theory. Apparently the black box concept is of little use in studies of the analytical laboratory. The structure or organization of the system has to be studied in order to improve the performance of the system. Mathematical or formal approaches usually can only be applied to isolated parts of the entire system. One should always keep in mind that optimization of part of the laboratory does not always run parallel with an optimization of the entire laboratory. Optimization of the laboratory definitely requires a systems approach. In reality, optimization of the laboratory has to be considered as a sub-optimization because of its interaction with the environment. Such a sub-optimization is not the same as optimizing the larger system of which the laboratory is only a part. For the time being, formal techniques and common sense should go together when studying and trying to improve the performance of laboratories.

REFERENCES

T.J. Allen, Res. Dev. Manage., 1 (1971) 14.
C.F. Cook, Anal. Chem., 48 (1976) 724A.
P.A. Frost and R.D. Whitley, Res. Dev. Manage., 1 (1971) 71.
R. Goulden, Analyst, 99 (1974) 1929.
E.H. Schein, Organisational Psychology, Prentice-Hall, Englewood Cliffs, N.J., 1970.
B. Schmidt, GIT-Fachz., (1976) 11.
B. Schmidt, Z. anal. Chem., 287 (1977) 157.
I. Väänanen, K. Kivirikko, S. Koskenniemi, J. Koskimies and A. Relander, Methods Inf. Med., 13 (1974) 158.

APPENDIX

Table I

The normal distribution

a. The cumulative frequency distribution function

x	F(x)	x	F(x)
0.0	0.5000	2.0	0.9772
0.1	0.5398	2.1	0.9821
0.2	0.5793	2.2	0.9861
0.3	0.6179	2.3	0.9893
0.4	0.6554	2.4	0.9918
0.5	0.6915	2.5	0.9938
0.6	0.7257	2.6	0.9953
0.7	0.7580	2.7	0.9965
0.8	0.7881	2.8	0.9974
0.9	0.8159	2.9	0.9981
1.0	0.8413	3.0	0.9987
1.1	0.8643	3.1	0.9990
1.2	0.8849	3.2	0.9993
1.3	0.9032	3.3	0.9995
1.4	0.9192	3.4	0.99966
1.5	0.9332	3.5	0.99977
1.6	0.9452	3.6	0.99984
1.7	0.9554	3.7	0.99989
1.8	0.9641	3.8	0.99994
1.9	0.9713	3.9	0.99997

F(x) is equal to the area under the normal probability distribution function to the left of x.

As the normal probability distribution function is symmetric around 0 the value of F(x) for a negative x is given by :

$$F(x) = 1 - F(-x)$$

b. The inverse cumulative frequency distribution function

x	$F^{-1}(x)$	x	$F^{-1}(x)$
0.500	0.0000	0.950	1.6449
0.600	0.2533	0.960	1.7507
0.700	0.5244	0.970	1.8808
0.750	0.6745	0.975	1.9600
0.800	0.8416	0.980	2.0537
0.850	1.0364	0.990	2.3263
0.900	1.2816	0.995	2.5758

x is the probability that the value of the normal variable be less than $F^{-1}(x)$. The probability that the normal variable be larger than $F^{-1}(x)$ is 1 - x. When an interval must be found, for which the probability of being outside must be α it is assumed that there is a probability of $\alpha/2$ of being to the left of the interval and $\alpha/2$ of being to the right of the interval. The two extremes of the interval are given by :

$$F^{-1}\left(\frac{\alpha}{2}\right) \quad \text{and} \quad F^{-1}\left(1 - \frac{\alpha}{2}\right)$$

The second of these values can be found in the table above and the first is given by the formula :

$$F^{-1}\left(\frac{\alpha}{2}\right) = - F^{-1}\left(1 - \frac{\alpha}{2}\right)$$

Table II

The chi-square distribution

The inverse cumulative frequency distribution function

k \ χ^2_k	0.01	0.05	0.10	0.25	0.50	0.75	0.90	0.95	0.99
1	0.00016	0.0039	0.0158	0.102	0.455	1.32	2.71	3.84	6.63
2	0.0201	0.103	0.211	0.575	1.39	2.77	4.61	5.99	9.21
3	0.115	0.352	0.584	1.21	2.37	4.11	6.25	7.81	11.3
4	0.297	0.711	1.06	1.92	3.36	5.39	7.78	9.49	13.3
5	0.554	1.15	1.61	2.67	4.35	6.63	9.24	11.1	15.1
6	0.872	1.64	2.20	3.45	5.35	7.84	10.6	12.6	16.8
7	1.24	2.17	2.83	4.25	6.35	9.04	12.0	14.1	18.5
8	1.65	2.73	3.49	5.07	7.34	10.2	13.4	15.5	20.1
9	2.09	3.33	4.17	5.90	8.34	11.4	14.7	16.9	21.7
10	2.56	3.94	4.87	6.74	9.34	12.5	16.0	18.3	23.2
11	3.05	4.57	5.58	7.58	10.3	13.7	17.3	19.7	24.7
12	3.57	5.23	6.30	8.44	11.3	14.8	18.5	21.0	26.2
13	4.11	5.89	7.04	9.30	12.3	16.0	19.8	22.4	27.7
14	4.66	6.57	7.79	10.2	13.3	17.1	21.1	23.7	29.1
15	5.23	7.26	8.55	11.0	14.3	18.2	22.3	25.0	30.6
16	5.81	7.95	9.31	11.9	15.3	19.4	23.5	26.3	32.0
17	6.41	8.67	10.1	12.8	16.3	20.5	24.8	27.6	33.4
18	7.01	9.39	10.9	13.7	17.3	21.6	26.0	28.9	34.8
19	7.63	10.1	11.7	14.6	18.3	22.7	27.2	30.1	36.2
20	8.26	10.9	12.4	15.5	19.3	23.8	28.4	31.4	37.6

21	8.90	11.6	13.2	16.3	20.3	24.9	29.6	32.7	38.9
22	9.54	12.3	14.0	17.2	21.3	26.0	30.8	34.0	40.3
23	10.2	13.1	14.8	18.1	22.3	27.1	32.0	35.2	41.6
24	10.9	13.8	15.7	19.0	23.3	28.2	33.2	36.4	43.0
25	11.5	14.6	16.5	19.9	24.3	29.3	34.4	37.7	44.3

Table III

The t distribution

The inverse cumulative frequency distribution function

k \ t_k	0.60	0.75	0.90	0.95	0.975	0.99	0.995
1	0.33	1.00	3.08	6.31	12.71	31.82	63.66
2	0.29	0.82	1.89	2.92	4.30	6.97	9.93
3	0.28	0.76	1.64	2.35	3.18	4.54	5.84
4	0.27	0.74	1.53	2.13	2.78	3.75	4.60
5	0.27	0.73	1.48	2.02	2.57	3.37	4.03
6	0.27	0.72	1.44	1.94	2.45	3.14	3.71
7	0.26	0.71	1.42	1.90	2.36	3.00	3.50
8	0.26	0.71	1.40	1.86	2.31	2.90	3.36
9	0.26	0.70	1.38	1.83	2.26	2.82	3.25
10	0.26	0.70	1.37	1.81	2.23	2.76	3.17
11	0.26	0.70	1.36	1.80	2.20	2.72	3.11
12	0.26	0.69	1.36	1.78	2.18	2.68	3.06
13	0.26	0.69	1.35	1.77	2.16	2.65	3.01
14	0.26	0.69	1.34	1.76	2.15	2.62	2.98
15	0.26	0.69	1.34	1.75	2.13	2.60	2.95
16	0.26	0.69	1.34	1.75	2.12	2.58	2.92
17	0.26	0.69	1.33	1.74	2.11	2.57	2.90
18	0.26	0.69	1.33	1.73	2.10	2.55	2.88
19	0.26	0.69	1.33	1.73	2.09	2.54	2.86
20	0.26	0.69	1.32	1.72	2.09	2.53	2.85
30	0.26	0.68	1.31	1.70	2.04	2.46	2.75
40	0.25	0.68	1.30	1.68	2.02	2.42	2.70
60	0.25	0.68	1.30	1.67	2.00	2.39	2.66
120	0.25	0.68	1.29	1.66	1.98	2.36	2.62
∞	0.25	0.67	1.28	1.65	1.96	2.33	2.58

Table IV

The F distribution

The inverse cumulative frequency distribution function.

In the following table values of the inverse cumulative frequency distribution function of the F distribution are given for several values of k and m, respectively the number of degrees of freedom of the numerator and denominator. The values correspond to values where the cumulative function is equal to 0.975.

m \ k	1	2	3	4	5	10	15	20	30	40	60	120	∞
1	647.8	799.5	864.2	899.6	921.8	968.6	984.9	993.1	1001.	1006.	1010.	1014.	1018.
2	38.51	39.00	39.17	39.25	39.30	39.40	39.43	39.45	39.46	39.47	39.48	39.49	39.50
3	17.44	16.04	15.44	15.10	14.88	14.42	14.25	14.17	14.08	14.04	13.99	13.95	13.90
4	12.22	10.65	9.98	9.60	9.36	8.84	8.66	8.56	8.46	8.41	8.36	8.31	8.26
5	10.01	8.43	7.76	7.39	7.15	6.62	6.43	6.33	6.23	6.18	6.12	6.07	6.02
10	6.94	5.46	4.83	4.47	4.24	3.72	3.52	3.42	3.31	3.26	3.20	3.14	3.08
15	6.20	4.77	4.15	3.80	3.58	3.06	2.86	2.76	2.64	2.59	2.52	2.46	2.40
20	5.87	4.46	3.86	3.51	3.29	2.77	2.57	2.46	2.35	2.29	2.22	2.16	2.09
30	5.57	4.18	3.59	3.25	3.03	2.51	2.31	2.20	2.07	2.01	1.94	1.87	1.79
40	5.42	4.05	3.46	3.13	2.90	2.39	2.18	2.07	1.94	1.88	1.80	1.72	1.64
60	5.29	3.93	3.34	3.01	2.79	2.27	2.06	1.94	1.82	1.74	1.67	1.58	1.48
120	5.15	3.80	3.23	2.89	2.67	2.16	1.94	1.82	1.69	1.61	1.53	1.43	1.31
∞	5.02	3.69	3.12	2.79	2.57	2.05	1.83	1.71	1.57	1.48	1.39	1.27	1.00

Values for 0.025 can be found in the same table using the formula

$$F_{\frac{\alpha}{2},\ k,m} = \frac{1}{F_{1-\frac{\alpha}{2},\ m,k}}$$

Together these values can be used to find the following interval containing 95% of the total probability :

$$(F_{0.025,\ k,m}\ ;\ F_{0.975,\ k,m})$$

or

$$\left(\frac{1}{F_{0.975,\ m,k}}\ ;\ F_{0.975,\ k,m}\right)$$

Table V

The Kolmogorov-Smirnov distribution

Critical values for the maximum distance D in the Kolmogorov-Smirnov one sample test.

Sample size n	Level of significance α		
	0.10	0.05	0.01
1	.95	.98	.99
2	.78	.84	.93
3	.64	.71	.83
4	.56	.62	.73
5	.51	.57	.67
6	.47	.52	.62
7	.44	.49	.58
8	.41	.46	.54
9	.39	.43	.51
10	.37	.41	.49
15	.30	.34	.40
20	.26	.29	.36
25	.24	.27	.32
30	.22	.24	.29
35	.21	.23	.27
over 35	$\frac{1.22}{\sqrt{n}}$	$\frac{1.36}{\sqrt{n}}$	$\frac{1.63}{\sqrt{n}}$

Adapted from Massey F.J. Jr., J. Amer. Statistical Ass., 46 (1951) 70. With permission from the American Statistical Association.

Table VI

The Wilcoxon distribution

In the following table critical values are given for T in the Wilcoxon two sample test.

Reduced sample size n_0	Level of significance α .05	.02	.01
6	0	-	-
7	2	0	-
8	4	2	0
9	6	3	2
10	8	5	3
11	11	7	5
12	14	10	7
13	17	13	10
14	21	16	13
15	25	20	16
16	30	24	20
17	35	28	23
18	40	33	28
19	46	38	32
20	52	43	38

21	59	49	43
22	66	56	49
23	73	62	55
24	81	69	61
25	89	77	68

Adapted from Table G of the appendix from S. Siegel, Nonparametric Statistics for the Behavioral Sciences, McGraw-Hill Book Company, 1956.

SUBJECT INDEX

Page numbers that are underlined refer to the first page of a chapter or section on the subject.